Molecular Tools for the Detection and Quantification of Toxigenic Cyanobacteria

Molecular Tools for the Detection and Quantification of Toxigenic Cyanobacteria

Edited by

RAINER KURMAYER
Research Institute for Limnology, University of Innsbruck,
Austria

KAARINA SIVONEN
Department of Food and Environmental Sciences, Division of Microbiology and
Biotechnology, University of Helsinki,
Finland

ANNICK WILMOTTE
InBios – Center for Protein Engineering, University of Liège,
Belgium

NICO SALMASO
Research and Innovation Centre, Fondazione Edmund Mach – Istituto Agrario di
S. Michele all'Adige, Italy

EUROPEAN COOPERATION
IN SCIENCE AND TECHNOLOGY

The right of Rainer Kurmayer, Kaarina Sivonen, Annick Wilmotte, and Nico Salmaso to be identified as the authors of this editorial material in this work has been asserted in accordance with law.

Registered Offices

John Wiley & Sons, Inc., 111 River Street, Hoboken, NJ 07030, USA
John Wiley & Sons, Ltd, The Atrium, Southern Gate, Chichester, West Sussex, PO19 8SQ, UK

Editorial Office

The Atrium, Southern Gate, Chichester, West Sussex, PO19 8SQ, UK

For details of our global editorial offices, customer services, and more information about Wiley products visit us at www.wiley.com.

Wiley also publishes its books in a variety of electronic formats and by print-on-demand. Some content that appears in standard print versions of this book may not be available in other formats.

Library of Congress Cataloging-in-Publication Data Applied for

ISBN: 9781119332107

Cover image:
Main image:
Bloom of the filamentous cyanobacterium *Planktothrix rubescens* ("Burgunderblutalge") in Mindelsee (near Constance, Germany) in December 2006 (Photocredit: Guntram Christiansen). The surface water sample contained the hepatotoxin microcystin.
Inserts:
Top left: cyanoHAB occurrence (Photocredit: Elke Dittmann), (further three inserts courtesy of the editors). Second top left: Biomass used as food supplement, third top left: Single cyanobacteria filaments from *Planktothrix rubescens* in the light microscope (400× magnification), bottom left: the essence of molecular tools' application (DNA extraction, (q)PCR amplification, sequencing of PCR products).
Cover design by Wiley

Set in 10/12pt TimesLTStd by SPi Global, Chennai, India
Printed and bound in Malaysia by Vivar Printing Sdn Bhd

10 9 8 7 6 5 4 3 2 1

Hans C.P. Matthijs (1951–2016): A passion for light and life

We are honored that the editors wish to dedicate this book to the memory of our dear colleague Dr. Hans C.P. Matthijs. Hans was an inspirational biochemist, who was fascinated by the life of cyanobacteria. He contributed nearly 40 years of groundbreaking research on their physiology and biochemistry. After many years of primarily fundamental research, his last years were dedicated to research with a distinctive societal relevance.

This book has been initiated and supported by the CYANOCOST action in which Hans has been very active. Besides contributing to the handbooks prepared in this action, Hans organized a training school on the effects of hydrogen peroxide on cyanobacteria in 2014.

Hans studied chemistry at the University of Amsterdam (UvA), the Netherlands, and obtained his PhD degree on the energy metabolism of the cyanobacterium *Plectonema boryanum* at the Vrije Universiteit Amsterdam (VU) in 1983. Afterward, Hans travelled abroad and worked at Brookhaven National Laboratory near New York in the United States, the University of the Mediterranean Aix-Marseille II, and the Washington University of St. Louis back in the United States. In 1991, Hans returned to the University of Amsterdam, where he was appointed Assistant Professor in the research group of Professor Luuc Mur. In 2001, he was promoted to Associate Professor and together with Professor Jef Huisman and Dr. Petra Visser he formed the scientific core of the research group Aquatic Microbiology of the new Institute for Biodiversity and Ecosystem Dynamics. During this time, Hans made significant contributions to fundamental new insights into cyclic electron transport around photosystem I and to structural and regulatory changes in photosynthesis due to stress induced by, for example, nitrogen and iron limitation.

In recent years, Hans explored the possibilities of LED lighting to enhance biomass production and save energy in algal biotechnology and horticulture. Employing the flashing light effect, a study he had performed in 1996, he hypothesized that photosynthesis could be just as efficient with a lot less light input if only one offered cells the right color of light at the right time. Together with a PhD student and a post-doc, the culture rooms were transformed into colorful labs with advanced LED lighting to strengthen his hypothesis.

Another recent topic of Hans' interest was the termination of harmful cyanobacterial blooms with hydrogen peroxide. Owing to a key difference in the Mehler reaction at photosystem I, cyanobacteria are much more sensitive to hydrogen peroxide than eukaryotic algae. Several lakes were treated with great success and two PhD students and a post-doc looked to optimize the procedure and to fully understand the implications of the treatment.

Additionally, Hans was interested in the effects of rising levels of CO_2 on the CCM of cyanobacteria. In April 2016, one of his students defended a PhD thesis on this topic. Although Hans was already seriously ill at that point, he insisted on joining the defense ceremony. Hans presented a very inspirational laudation, one that we will never forget. It was his last public appearance. Hans passed away on April 17, 2016, due to the result of pancreatic cancer. While Hans is no longer with us, his creative insights and ideas for research will live on.

Jef Huisman, Petra Visser, and Merijn Schuurmans
Department of Aquatic Microbiology, University of Amsterdam, The Netherlands

Contents

List of Contributors

Andreas Ballot
Norwegian Institute for Water Research, Oslo, Norway

Stephan Blank
Research Institute for Limnology, University of Innsbruck, Mondsee, Austria

Alexandra Bukowska
Department of Microbial Ecology and Environmental Biotechnology, Faculty of Biology, University of Warsaw, Poland

Camilla Capelli
Research and Innovation Centre, Fondazione Edmund Mach – Istituto Agrario di S. Michele all'Adige, Italy

Qin Chen
Research Institute for Limnology, University of Innsbruck, Mondsee, Austria, and, College of Natural Resources and Environment, Northwest A & F University, Yangling, P. R. China

Guntram Christiansen
Research Institute for Limnology, University of Innsbruck, Mondsee, Austria, and Miti Biosystems GmbH, Max F. Perutz Laboratories, Vienna, Austria

Samuel Cirés
Department of Biology, Autonomous University of Madrid, Spain

Li Deng
Institute of Virology, Helmholtz Zentrum Munich, Munich, Germany, and Research Institute for Limnology, University of Innsbruck, Mondsee, Austria

Elke Dittmann
Institute of Biochemistry and Biology, University of Potsdam, Germany

Rehab El-Shehawy
Institute IMDEA Water, Alcalá de Henares (Madrid), Spain

Elisabeth Entfellner
Research Institute for Limnology, University of Innsbruck, Mondsee, Austria

David P. Fewer
Department of Food and Environmental Sciences, Division of Microbiology and Biotechnology, University of Helsinki, Finland

Ilona Gągała
European Regional Centre for Ecohydrology of the Polish Academy of Sciences, Łódź, Poland

Muriel Gugger
Collection of Cyanobacteria, Institut Pasteur, Paris, France

Sigrid Haande
Norwegian Institute for Water Research, Oslo, Norway

Camilla H.C. Hagman
Norwegian Institute for Water Research, Oslo, Norway

Kaisa Haukka
Department of Food and Environmental Sciences, Division of Microbiology and Biotechnology, University of Helsinki, Finland

Jean-Francois Humbert
Institute of Ecology and Environmental Sciences, UPMC, Paris, France

Iwona Jasser
Department of Microbial Ecology and Environmental Biotechnology, Faculty of Biology, University of Warsaw, Poland

Konstantinos Kormas
Department of Ichthyology and Aquatic Environment, University of Thessaly, Volos, Greece

Ewa Kozłowska
Department of Immunology, Faculty of Biology, University of Warsaw, Poland

Rainer Kurmayer
Research Institute for Limnology, University of Innsbruck, Mondsee, Austria

H. Dail Laughinghouse IV
Fort Lauderdale Research and Education Center, University of Florida/IFAS, Davie, United States of America, and Department of Botany, MRC-166, National Museum of Natural History – Smithsonian Institution, Washington, United States of America

Joanna Mankiewicz-Boczek
European Regional Centre for Ecohydrology of the Polish Academy of Sciences, Łódź, Poland, and Department of Applied Ecology, Faculty of Biology and Environmental Protection, University of Lodz, Poland

Hans C.P. Matthijs
Department of Aquatic Microbiology, Institute for Biodiversity and Ecosystem Dynamics, University of Amsterdam, The Netherlands

Cristiana Moreira
Interdisciplinary Centre of Marine and Environmental Research (CIIMAR/CIMAR), University of Porto, Matosinhos, Portugal

Dagmar Obbels
Laboratory of Protistology and Aquatic Ecology, Department of Biology, Ghent University, Belgium

Igor S. Pessi
InBios – Center for Protein Engineering, University of Liège, Belgium

Antonio Quesada
Department of Biology, Autonomous University of Madrid, Spain

Vitor Ramos
Interdisciplinary Centre of Marine and Environmental Research (CIIMAR/CIMAR), University of Porto, Matosinhos, Portugal, and Faculty of Sciences, University of Porto, Portugal

Anne Rantala-Ylinen
Department of Food and Environmental Sciences, Division of Microbiology and Biotechnology, University of Helsinki, Finland

Rosmarie Rippka
Institut Pasteur, Unité des Cyanobactéries, Centre National de la Recherche Scientifique (CNRS) Unité de Recherche Associé (URA) 2172, Paris, France

Martin Saker
Interdisciplinary Centre of Marine and Environmental Research (CIIMAR/CIMAR), University of Porto, Matosinhos, Portugal, and Alpha Environmental Solutions, Dubai, United Arab Emirates

Nico Salmaso
Research and Innovation Centre, Fondazione Edmund Mach – Istituto Agrario di S. Michele all'Adige, Italy

Henna Savela
Department of Biochemistry/Biotechnology, University of Turku, Turku, Finland

J. Merijn Schuurmans
Department of Aquatic Microbiology, Institute for Biodiversity and Ecosystem Dynamics, University of Amsterdam, The Netherlands

Kaarina Sivonen
Department of Food and Environmental Sciences, Division of Microbiology and Biotechnology, University of Helsinki, Finland

Maxime Sweetlove
Laboratory of Protistology and Aquatic Ecology, Department of Biology, Ghent University, Belgium

Vitor Vasconcelos
Faculty of Sciences, University of Porto, Porto, Portugal, and Interdisciplinary Centre of Marine and Environmental Research (CIIMAR/CIMAR), University of Porto, Matosinhos, Portugal

Elie Verleyen
Laboratory of Protistology and Aquatic Ecology, Department of Biology, Ghent University, Belgium

Wim Vyverman
Laboratory of Protistology and Aquatic Ecology, Department of Biology, Ghent University, Belgium

Annick Wilmotte
InBios – Center for Protein Engineering, University of Liège, Belgium

About the Editors

Dr. Rainer Kurmayer, PhD, Assoc. Prof.

Research Institute for Limnology, University of Innsbruck, Mondsee, Austria

Professional Career

2014–present: Associate Professor, Research Institute for Limnology, University of Innsbruck, University Doctorate, University of Vienna; 2007–2014: Senior Scientist Institute for Limnology, Austrian Academy of Sciences (ÖAW) and University of Innsbruck; 2001–2007, Junior Scientist Institute for Limnology, ÖAW; 1999–2001: post-doctorate position at the Federal Environmental Agency, Berlin, Germany. Training and Mobility of Researchers network "Toxin Production in Cyanobacteria" (TOPIC) within the 4th EU framework program; 1997–1999: PhD fellowship ÖAW "Effects of Cyanobacteria on Zooplankton."

Research

Molecular ecology and evolution of toxin-producing algae (cyanobacteria); consequences of toxins produced by algal blooms; alpine lakes as *in situ* observatories for climate change effects.

Publications

Fifty international peer-reviewed publications, more than 20 popular scientific contributions (newspapers, magazines, TV).

Dr. Kaarina Sivonen, PhD, Prof.

Department of Food and Environmental Sciences, Division of Microbiology and Biotechnology, University of Helsinki, Helsinki, Finland

Professional Career

2010–present: Professor in Microbiology at University of Helsinki; 2000–2010: Academy Professor, Junior (1990–1996) and Senior Scientist (1997–2000) of the Academy of Finland, Research scientist, project leader (1996–1997) at the Helsinki University. Visiting Research Associate at the University of Chicago, Department of Molecular Genetics and Cell Biology, Chicago, Illinois, USA (17.7.1991–31.12.1992). Visiting Scientist at Wright State University, Department of Biological Sciences, Dayton, Ohio, USA (15.11.1990–15.07.1991).

Research

Over 30 years of research on toxic and bioactive compound-producing cyanobacteria in Finnish freshwater and in the Baltic Sea. Maintaining culture collection of 1200 cyanobacteria. Current research covers ecology, physiology, genomics, post-genomics, biosynthesis, method development, structures, and bioactivities of new compounds.

Publications

One hundred and ninety-seven international peer-reviewed publications major as lead and/or corresponding author, number of popular scientific contributions (newspapers, magazines, TV).

Dr. Annick Wilmotte, PhD

InBios – Center for Protein Engineering, University of Liège, Liège, Belgium

Professional Career

1996–present: Research Associate of the FRS-FNRS (Belgian Funds for Scientific Research), University of Liège (eq. Principal Investigator); 1989–1996: Post-doctorate projects at University of Antwerp (BE), Department of Biochemistry (Molecular Biology Group), and the Flemish Institute of Biotechnology (BE) (Genetics and Biotechnology Lab) concerning cyanobacterial and fungal phylogeny, and gene transfers in soil bacteria, respectively; 1983–1989: PhD fellowship, University of Liège on the taxonomy of cyanobacteria; 1982–1983: Research at the University of Groningen (NL), "Plant Systematics" (Prof. Van Den Hoek & Stam). Coordinator of two EC projects on the molecular diversity of Antarctic microbial mats (MICROMAT) and the detection of planktonic cyanobacteria by DNA-chips (MIDI-CHIP) plus national projects on toxic cyanobacterial blooms in Belgium and the molecular diversity and biogeography of cyanobacteria in Polar Regions. Promotor of the BCCM/ULC public culture collection of cyanobacteria.

Research

Biodiversity, taxonomy, evolution, and molecular ecology and biogeography of cyanobacteria. Cultivation and cryopreservation of cyanobacterial strains.

Publications

Seventy-three international peer-reviewed papers (40 as first or last author), regular outreach conferences and expertise missions for the environmental protection of Antarctica.

Dr. Nico Salmaso, PhD

Research and Innovation Centre, Fondazione Edmund Mach – Istituto Agrario di S. Michele all'Adige, S. Michele all'Adige, Italy

Professional Career

2016–present: Head of the Hydrobiology Unit at the Istituto Agrario di S. Michele All'Adige, Fondazione E. Mach (FEM); 2011–2015: Head of the Limnology and River Ecology group at FEM; 2009–2010: Head of the Biocomplexity and Ecosystem Dynamics group at FEM; 2005–2008: Head of the Limnology and Fish Ecology group (FEM); 2007–2008: Deputy Coordinator of the Natural Resources Department (FEM); 1995–2004: Research Assistant at the University of Padua; 31/01/2014: National Scientific Habilitation 05/C1 – II Fascia, Ecology; 06/04/2017: National Scientific Habilitation 05/A1 – I Fascia, Botany. Responsible for several National and European (Central Europe Programme) projects.

Editorial activity

Editor-in-Chief of Advances in Oceanography and Limnology. Associate Editor of Cryptogamie-Algologie.

Research

Ecology of phytoplankton and cyanobacteria. Environmental and biotic factors promoting the development of toxic cyanobacteria. Detection of toxic genotypes. Invasive cyanobacteria.

Publications

Sixty international peer-reviewed publications, more than 75% as lead and/or corresponding author. More than 30 contributions on international non-ISI and national journals and book chapters.

About the Book

The strong interest in participating in the EU-COST Action CYANOCOST throughout Europe reflects: (1) the increasing global occurrence of cyanobacterial blooms; (2) associated adverse health, environmental, and economic impacts throughout the world; (3) the roles of international and national health and environmental agencies in establishing national monitoring and analysis strategies; and (4) the continuing growth of research at the international level on cyanobacterial blooms and cyanotoxins. Therefore, it was decided during the second project meeting in Madrid (November 2–4, 2012) that, in addition to the handbook on analytical methods edited by Jussi Meriluoto, Geoffrey A. Codd, and Lisa Spoof, a handbook on the various molecular detection methods for toxigenic cyanobacteria would be compiled. Rainer Kurmayer, Nico Salmaso, Kaarina Sivonen, and Annick Wilmotte were elected editors.

This book is designed as a handbook describing the molecular monitoring of diversity and toxigenicity of cyanobacteria (blue–green algae) in surface waters including lakes, rivers, drinking water reservoirs, and food supplements. In water bodies, cyanobacteria mass developments = cyanobacteria harmful algal blooms (cHAB) are highly favored by eutrophication and global temperature rise.

Chapters include a series of standard operational procedures (SOPs). The list of SOPs contains the molecular tests that could be routinely used in environmental monitoring. The introductory chapters are concise papers written according to an *a priori* given outline, covering the purpose, methodological details, and advantages and disadvantages of the various tools. Although each introductory chapter refers to the individual SOPs, each SOP should be considered an independent unit with its own (essential) references section.

The introductory chapter gives an overview of the current knowledge on the genetic basis of cyanotoxin synthesis in cyanobacteria (Chapter 1). In particular it provides up-to-date overviews plus the necessary foundation for the subsequent use of molecular tools, analysis, and the interpretation of the results. The other chapters are dedicated to the individual steps from water (food supplement) sampling (Chapter 2), nucleic acid extraction (Chapter 5), and downstream analysis, including PCR- (Chapter 6) and qPCR- (Chapter 7), based methods but also more traditional tools of genotyping (Chapter 9), diagnostic microarrays (Chapter 8), and community characterization by next-generation sequencing techniques (Chapter 10). One chapter is dedicated to the isolation of cyanobacteria strains from water samples (Chapter 3) and another contains guidelines to taxonomic assignment (Chapter 4). The last chapter is dedicated to a review of the application of molecular tools (Chapter 11). The practical chapters (SOPs) contain the necessary protocol details to enable trained operators to perform the described application. All published SOPs have been used in the lab of the respective authors for years and are considered robust. They are based on reference strains which are reliably available from international culture collections.

The book is intended to be used by trained professionals analyzing cyanobacterial diversity and toxigenicity in water samples in the laboratory in both academic and governmental institutions, as well as technical offices and agencies which are in charge of waterbody surveillance and monitoring. Students will learn important methods' standards of essential protocols including steps from sampling to results evaluation. More generally, the book will benefit the reader by (1) increasing the knowledge of the currently available molecular toolbox and (2) enabling them to carry out state-of-the-art analyses on the toxigenicity of cyanobacteria in surface water.

Preface

Cyanobacteria (blue–green algae) are among the oldest organisms on earth, and fossil records of cyanobacteria have been described from stromatolites dating back billions of years. Probably because of their long evolutionary history, cyanobacteria are highly successful and colonize all available habitats. Perhaps unsurprisingly, cyanobacteria also frequently dominate aquatic communities and can form mass accumulations that often appear on the water surface, so-called algal blooms. These algal blooms can cover hundreds of square meters or even square kilometers on the surface of the water and substantially deteriorate water quality and threaten the environment and humans by a number of problems, not least because of the production of poisonous compounds, so-called cyanotoxins.

During the 1970s and 1980s, major progress was made in the identification of the source organisms and the symptoms of toxification. The first chemical structures of toxins were derived during the 1970s and 1980s, for example microcystin-LA and anatoxin-a. The number of fully characterized toxin variants substantially increased during the 1990s as well as the description of cyanobacterial taxa producing the toxin variants. Further major progress was achieved in the detection and quantification of cyanobacterial toxins in the environment as well as in the knowledge concerning their global distribution and occurrence.

In general the cyanobacterial toxins can be classified into three functional groups: hepatotoxins (e.g. microcystin/nodularin, cylindrospermopsin), neurotoxins (e.g. saxitoxin, anatoxin-a), and irritant-dermal toxins (e.g. lipopolysaccharides). Typically, these toxins are inside the cells, but can suddenly be released into the water during cell lysis, for example following algaecide treatment. In 1997, the World Health Organization (WHO) published a guideline value on the hepatotoxic microcystin-LR in drinking water. In the last revision on guidelines for drinking water quality performed by the WHO, in 2004, cyanobacteria were considered a relevant source of organic toxic compounds deteriorating drinking water quality worldwide.

In the last two decades, major progress has been made in elucidating the genetic basis of toxin production in cyanobacteria. In particular the biosynthesis of the abundant cyanobacterial toxins has been discovered: microcystin, nodularin, cylindrospermopsin, saxitoxin, and anatoxin-a. Advances in this field have made it possible to develop detection methods to study producers of these toxins. Since the toxin-producing and non-producing strains cannot be distinguished under the microscope, these molecular tools provide shortcut methods for detection.

Although the knowledge to apply genetic techniques has increased significantly, the use of those techniques for environmental studies, water management, and risk assessment is still in its infancy. We think it would be worthwhile to make an effort to ensure that molecular genetic discoveries are translated into tools that can be used in field and laboratory monitoring. This book, the first of its kind, was written as a handbook for the molecular monitoring of toxigenicity of cyanobacteria in surface waters including lakes, rivers, drinking water reservoirs, and food supplements.

The book is structured into two main parts. Each chapter is linked to a variable number of specific protocols (called standard operational procedures, SOPs). We hope that this book will contribute to a greater acceptability and use of molecular tools in monitoring of potential cyanotoxicity worldwide. We further hope that this book will be useful for graduate students, laboratory technicians, professionals, and supervisors applying molecular tools.

Rainer Kurmayer, Kaarina Sivonen, Annick Wilmotte and Nico Salmaso
Mondsee, August 2016

Acknowledgments

The book is a key output of a major European Cooperation in Science and Technology (COST) Action (ES1105; 2012–2016), www.cyanocost.com. This action, "Cyanobacterial Blooms and Toxins in Water Resources: Occurrence, Impacts and Management" included over 100 active participants from 33 countries and served to a large extent the transfer of know-how between institutions on the topic. The proposal and the coordination of this activity by Triantafyllos Kaloudis, Athens Water Supply and Sewerage Company, Greece and Ludek Blaha, RECETOX, Faculty of Sciences, Czech Republic is gratefully acknowledged.

Beside the many authors, the colleagues who contributed to this handbook are gratefully acknowledged. In particular, as a leader of the relevant Work Package 1 (Occurrence of cyanobacteria and cyanotoxins), Jussi Meriluoto, Åbo Akademi University, Finland is appreciated for his guidance and support toward finalization of this demanding book project.

Many thanks to the team from John Wiley & Sons, Ltd, Chichester, UK for their assistance in contract preparation (Jenny Cossham and Emma Strickland) and project editing (Ashmita Thomas Rajaprathapan and Tim Bettsworth).

COST is a pan-European intergovernmental framework. Its mission is to enable breakthrough scientific and technological developments leading to new concepts and products strengthening Europe's research and innovation capacities. It allows researchers, engineers, and scholars to jointly develop their own ideas and take new initiatives across all fields of science and technology, while promoting multi- and interdisciplinary approaches. COST aims to foster a better integration of less research-intensive countries to the knowledge hubs of the European Research Area. The COST Association, an international not-for-profit association under Belgian law, integrates all management, governing, and administrative functions necessary for the operation of the framework. The COST Association currently has 36 member countries (www.cost.eu).

COST is supported by the EU Framework Programme Horizon 2020

List of other research projects to be acknowledged (Principal Investigators in Alphabetical Order)

Principal Investigator	Project (Title)	Grant Number	Organization
Ilona Gągała, Joanna Mankiewicz-Boczek	Explanation of cause-effect relationships between the occurrence of toxinogenic cyanobacterial blooms, and abiotic and biotic factors with particular emphasis on the role of viruses and bacteria	N N305 096439	National Science Centre
Muriel Gugger	Cyanobacterial toxin production and Photoprotection processes in a changing EnviRonment (CYPHER)	ANR-15-CE34-002	Agence Nationale de la Recherche (ANR)
Rainer Kurmayer	Mobilomics of toxin production in cyanobacteria	P24070	Austrian Science Fund (FWF)
Vitor Ramos	fellowship	SFRH/BD/80153/2011	Fundação para a Ciência e a Tecnologia (FCT
Henna Savela (Timo Lövgren)	nucleoTracker	40013/10	Finnish Funding Agency for Innovation
Vitor Vasconcelos	Innovation and Sustainability in the Management and Exploitation of Marine Resources (INNOVMAR)	NORTE-01-0145-FEDER-000035, Research Line NOVELMAR	Northern Regional Operational Program (NORTE2020), European Regional Development Fund (ERDF)
Elie Verleyen, Annick Wilmotte	Saving Freshwater biodiversity REsearch Data (SAFRED)	BR/154/A6/SAFRED	Belgian Science Policy Office

Principal Investigator	Project (Title)	Grant Number	Organization
Wim Vyverman	Manscape	EV/29	Belgian Science Policy Office
Wim Vyverman	Pondscape	SD/BD/02A and SD/BD/02B	Belgian Science Policy Office
Wim Vyverman, Annick Wilmotte	Climate Change and Antarctic Microbial Biodiversity (CCAMBIO)	SD/BA/03A	Belgian Science Policy Office
Wim Vyverman, Annick Wilmotte	Preservation of microalgae in BCCM collections (PRESPHOTO)	BR/132/A6	Belgian Science Policy Office
Wim Vyverman, Annick Wilmotte	Cyanobacterial blooms: toxicity, diversity, modelling and management (B-BLOOMS2)	SD/TE/01A	Belgian Science Policy Office
Wim Vyverman, Annick Wilmotte	Cyanobacterial blooms: toxicity, diversity, modelling and management (B-BLOOMS1)	EV/34	Belgian Science Policy Office
Annick Wilmotte	High-throughput pyrosequencing of cyanobacterial populations (PYROCYANO)	CR.CH.10-11-1.5139.11	FRS-FNRS
Annick Wilmotte	BIPOLES : Geographical and ecological distribution of cyanobacteria from Antarctica and the Arctic	FRFC-2.4570.09	FRS-FNRS

1

Introduction

Rainer Kurmayer[1]*, Kaarina Sivonen[2], and Nico Salmaso[3]

[1]*Research Institute for Limnology, University of Innsbruck, Mondsee, Austria*
[2]*Department of Food and Environmental Sciences, Division of Microbiology and Biotechnology,
University of Helsinki, Helsinki, Finland*
[3]*Research and Innovation Centre, Fondazione Edmund Mach – Istituto Agrario di S. Michele
all'Adige, S. Michele all'Adige, Italy*

1.1 A Brief Historical Overview

During the last two decades, genetic methods have significantly increased our understanding of the distribution of genes involved in the production of toxins within the phylum of cyanobacteria (e.g. Sivonen and Börner, 2008; Dittmann *et al.*, 2013; Méjean and Ploux, 2013). Early on the synthesis pathways of microcystin in the three genera *Microcystis*, *Planktothrix*, and *Anabaena* (Tillett *et al.*, 2000; Christiansen *et al.*, 2003; Rouhiainen *et al.*, 2004) and of the closely related nodularin have been elucidated (Moffitt and Neilan, 2004). Further, the elucidation of the genes involved in cyanotoxin synthesis increased the understanding of its inheritance and evolution, (e.g. the phylogenetically derived conclusion on the evolutionary age of the microcystin/nodularin synthesis pathway) (Rantala *et al.*, 2004) implying that potentially all cyanobacteria are able to produce microcystins, and, indeed, the number of cyanobacterial genera discovered to produce microcystins is consistently increasing (e.g. Calteau *et al.*, 2014).

*Corresponding author: rainer.kurmayer@uibk.ac.at

Subsequently, the elucidation of the synthesis pathways of other toxins has been achieved, that is first results suggested the involvement of polyketide synthases (PKS) and an amidinotransferase in the synthesis of cylindrospermopsin in *Aphanizomenon* (Shalev-Alon *et al.*, 2002; Kellmann *et al.*, 2006) which then led to the identification of the first putative cylindrospermopsin gene cluster (*cyr*) in *Cylindrospermopsis* (Mihali *et al.*, 2008). Other cylindrospermopsin synthesis gene clusters followed, in particular for *Oscillatoria* (Mazmouz *et al.*, 2010), for *Aphanizomenon* (Stüken and Jakobsen, 2010), *Raphidiopsis curvata*, and *Cylindrospermopsis raciborskii* (Jiang *et al.*, 2014). In general, compared with *mcy* genes (encoding the synthesis of microcystins), there is more shuffling of genes, and eleven genes *cyr*A–K are thought to make the core of the *cyr* gene cluster.

Similarly, candidate genes for saxitoxin biosynthesis have been isolated and the sequence of the complete putative saxitoxin biosynthetic gene cluster (*sxt*) was obtained (Kellmann *et al.*, 2008a,b). This work started with screening of putative saxitoxin biosynthetic enzymes in cyanobacterial isolates, using a degenerate PCR approach, resulting in identification of an O-carbamoyltransferase that was proposed to carbamoylate the hydroxymethyl side chain of saxitoxin precursor. Orthologues of *sxt*1 were exclusively present in paralytic shellfish poisoning (PSP) strains of cyanobacteria and had a high sequence similarity to each other (Kellmann *et al.*, 2008a). The first *sxt* gene cluster was sequenced from *Cylindrospermopsis*, and orthologous gene clusters from *Anabaena*, *Aphanizomenon*, *Raphidiopsis*, and *Lyngbya* followed (Murray *et al.*, 2011). Genetic proof (e.g. by experimental gene inactivation) for the role of this gene cluster in saxitoxin biosynthesis is lacking. However, in the absence of suitable tools of genetic transformation, the functions of the ORF (open reading frame) were bioinformatically inferred, and this prediction was combined with the liquid chromatography-tandem mass spectrometry analysis of the biosynthetic intermediates (Kellmann and Neilan, 2007; Kellmann *et al.*, 2008b).

The first anatoxin-a synthesis gene cluster (*ana*) was sequenced from *Oscillatoria* (Méjean *et al.*, 2009). Subsequently anatoxin-a gene clusters were described from *Anabaena* (Rantala-Ylinen *et al.*, 2011) and *Cylindrospermum* (Calteau *et al.*, 2014). In the following, the genetic basis of microcystin/nodularin, cylindrospermopsin, saxitoxin, and anatoxin synthesis is described in more detail.

1.2 The Genetic Basis of Toxin Production

1.2.1 Microcystin and Nodularin

Microcystins are produced by planktonic freshwater genera *Microcystis*, *Planktothrix*, *Dolichospermum*, *Nostoc*, and *Fischerella* (Dittmann *et al.*, 2013). Early studies, however, also documented microcystin production in a broader range of terrestrial genera, for example in *Hapalosiphon* (Prinsep *et al.*, 1992) and later in *Nostoc* symbionts associated with fungi (Oksanen *et al.*, 2004). In addition numerous freshwater and brackish water genera (e.g. *Arthrospira*, *Oscillatoria*, *Phormidium*, *Pseudanabaena*, *Synechococcus*, *Synechocystis*) have been reported to produce microcystins (Sivonen and Börner, 2008; Fiore *et al.*, 2009; Bernard *et al.*, 2017). In contrast, the closely related nodularin has been characterized from the brackish water species *Nodularia spumigena* and *Nostoc* (Bernard *et al.*, 2017), while in the marine sponge, *Theonella swinhoei*, a nodularin analogue called motuporin has been found (de Silva *et al.*, 1992). The sponge is known to harbor cyanobacterial symbionts. Microcystins are known for their toxicity because of the inhibition of eukaryotic protein phosphatases 1 and 2A resulting in the hyperphosphorylation and breakdown of the structural protein skeleton (Carmichael, 1994). Not at least because of the interference with eukaryotic signaling cascades, microcystins are considered tumor promotors under sublethal exposure conditions (Zhou *et al.*, 2002).

Microcystins are cyclic heptapeptides and share the common structure cyclo (- D-Ala(1) - X(2) - D-MAsp(3) - Z(4) - Adda(5) - D-Glu(6) - Mdha(7)), where X and Z are variable L-amino acids (e.g., microcystin (MC)-LR refers to leucine and arginine in the variable positions), D-MAsp is D-erythro-ß-iso-aspartic acid, Adda is (2S, 3S, 8S, 9S)-3-amino-9-methoxy-2,6,8-trimethyl-10-phenyldeca-4,6-dienoic acid, and Mdha is N-methyl-dehydroalanine (Carmichael *et al.*, 1988). Considerable structural variation has been reported, most frequently in positions 2, 4, and 7 of the molecule, and a large number of structural variants have been characterized (molecular weight 909 – 1115 Da), either from field samples or from isolated strains (e.g. Diehnelt *et al.*, 2006; Spoof and Catherine, 2017). Nodularin (824 Da) and motuporin (812 Da) are both pentapeptides containing N-methyl-dehydrobutyrine (Mdhb) instead of Mdha(7) and lack D-Ala(1) and X(2) when compared with microcystin. Nodularin differs from motuporin due to the substitution of L-Arg(4) by L-Val(4) (Fig. 1.1).

The biosynthesis of microcystin is catalyzed by nonribosomal peptide synthesis (NRPS) via the thio-template mechanism. This biosynthetic pathway has been intensively investigated in different bacteria and fungi, as their end products are often of great pharmaceutical value (Fischbach and Walsh, 2006). At present six gene clusters from five genera (*Microcystis, Planktothrix, Anabaena, Nodularia, Fischerella*) responsible for the biosynthesis of microcystin have been sequenced (Tillett *et al.*, 2000; Christiansen *et al.*, 2003; Rouhiainen *et al.*, 2004; Moffitt and Neilan, 2004; Fewer *et al.*, 2013; Shi *et al.*, 2013) and the involvement in the production of microcystins could be proven by genetic manipulation in *Microcystis* and *Planktothrix* (Dittmann *et al.*, 1997; Christiansen *et al.*, 2003). The whole *mcy* gene cluster comprises a minimum of nine genes (*ca.* 55 kb) consisting of PKS, nonribosomal peptide synthetases (NRPS), and tailoring enzymes. It has a modular structure (Fig. 1.1), each module containing specific functional domains for activation (aminoacyl adenylation (A)-domains), thioesterification (thiolation domains) of the amino acid substrate and for the elongation (condensation (C)-domains) of the growing peptide. McyD, McyE, and McyG are responsible for the production of the amino acid Adda and the activation and condensation of D-glutamate. McyA, McyB, and McyC are NRPS and responsible for the incorporation of the other five amino acids in positions 7, 1, 2, 3, and 4 of the molecule (Tillett *et al.*, 2000). Synthesis is thought to start with activation of phenyllactate through the adenylation domain of McyG (Hicks *et al.*, 2006) followed by extension of the polyketide through McyD and McyE. The polyketide is then condensed with D-glutamate through McyE forming the core of the microcystin peptide (comprising the Adda side chain and the Glutamate). The residual amino acids are then condensed through McyA, B, C proteins and finally the peptide is cyclized through a dedicated type I thioesterase (Tillett *et al.*, 2000). Several tailoring enzymes modify the growing peptide molecule, i.e. McyF (an aspartate racemase; Sielaff *et al.*, 2003), McyI (a dehydrogenase involved in the production of the methyl aspartate unit MeAsp(3); Pearson *et al.*, 2007), McyJ (an O-methyltransferase; Christiansen *et al.*, 2003), McyH (a putative ATP binding cassette (ABC) transporter; Pearson *et al.*, 2004), and McyT (a type II thioesterase; Christiansen *et al.*, 2008).

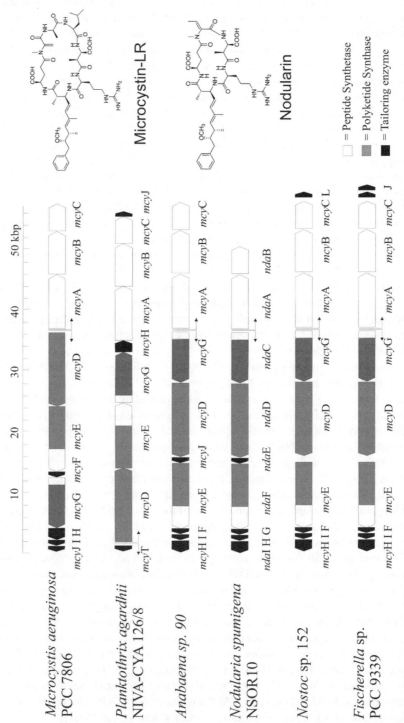

Figure 1.1 Scheme of the genetic basis of microcystin/nodularin synthesis in sequenced cyanobacterial genera. Arrows mark the bi-directional promotor region (from Tillett et al., 2000; Christiansen et al., 2003; Moffitt and Neilan, 2004; Rouhiainen et al., 2004; Fewer et al., 2013; Shih et al., 2013).

Phylogenetic analyses relatively early lead to the conclusion that microcystin synthesis is an evolutionary old feature that has been lost repeatedly during the evolution of cyanobacteria (Rantala *et al.*, 2004). The *nda* genetic cluster involved in the synthesis of nodularin was probably derived from the genes encoding microcystin synthesis via a gene deletion event (Moffitt and Neilan, 2004). The theory that cyanobacteria share a common microcystin-producing ancestor implies that potentially all cyanobacteria are able to produce microcystins and, for example, a functional microcystin synthesis gene cluster has been revealed through genome sequencing in the benthic true-branching cyanobacterium *Fischerella* PCC 9339 (Shi *et al.*, 2013; Calteau *et al.*, 2014).

1.2.2 Cylindrospermopsin

Cylindrospermopsin is a cyclic guanidine alkaloid hepatotoxin (molecular weight 415 Da). It is the pyrimidine ring that was postulated as the molecule component, which is essential for the toxicity of cylindrospermopsin (Banker *et al.*, 2001). Cylindrospermopsin suppresses protein synthesis (Terao *et al.*, 1994) and glutathione synthesis (Runnegar *et al.*, 1995). Apart from cylindrospermopsin toxicity primarily to the liver, kidney, and other organs, carcinogenic activity due to the presence of the uracil and sulfonated guanidino moieties has also been suggested (Murphy and Thomas, 2001). In contrast to the microcystins the structural variability is much lower, that is three variants of the cylindrospermopsin molecule have been described (Sivonen and Börner, 2008).

Schembri *et al.* (2001) were the first that identified genes putatively involved in the biosynthesis of cylindrospermopsin in *Cylindrospermopsis raciborskii* and *Anabaena bergii*. The morphological and genetic distinction between the two CYL producers, *Anabaena bergii* and *Aphanizomenon ovalisporum* was confirmed by Stüken *et al.* (2009). Shalev-Alon *et al.* (2002) identified an amidinotransferase (*aoa*A) in *Aphanizomenon (Chrysosporum) ovalisporum* that is likely to be involved in the formation of guanidinoacetic acid, which is thought to be the starter unit of biosynthesis. Subsequently, the putative cylindrospermopsin biosynthesis gene cluster encoding an amidinotransferase (CyrA), one mixed PKS/NRPS module (CyrB), four PKS modules (CyrC–F), and additional tailoring enzymes have been described, and cylindrospermopsin synthesis has been characterized (Kellmann *et al.*, 2006; Mihali *et al.*, 2008) (Fig. 1.2).

The amidinotransferase (CyrA) is thought to form guanidinoacetate, while the two PKS *cyr*B, *cyr*F are catalyzing polyketide extensions which later form the uracil ring. A spontaneous non-enzymatic formation of three rings is assumed, while biosynthesis is completed with two tailoring reactions (hydroxylation and sulfation) (Mihali *et al.*, 2008). The last step of cylindrospermopsin biosynthesis is catalyzed by a proline dioxygenase protein (CyrI), resulting in the epi-cylindrospermopsin variant (Mazmouz *et al.*, 2011). Remarkable is the transcriptional regulation upstream of *cyr*C (*aoa*C), which is highly similar between *Aphanizomenon (Chrysosporum) ovalisporum*, *Cylindrospermopsis raciborskii*, and *Oscillatoria*, suggesting a common transcriptional regulation of gene expression. CyrK is similar to a family of compound extrusion proteins (i.e. functioning in the transport of organic cationic drugs). Tailoring enzymes, such as CyrJ, encoding a sulfotransferase, are thought to catalyze the sulfation of the C-12 atom, as *cyrJ* was present only in cylindrospermopsin-producing strains, indicating its consistent involvement into cylindrospermopsin synthesis (see also Chapter 6), (Mihali *et al.*, 2008).

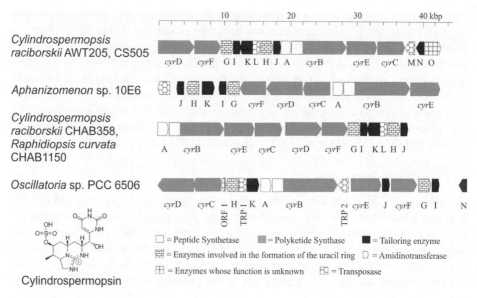

Figure 1.2 *Scheme of the genetic basis of cylindrospermopsin synthesis in various cyanobacterial genera (from Mihali et al., 2008; Mazmouz et al., 2010; Stüken and Jakobsen 2010; Jiang et al., 2014).*

In general, within the *cyr* gene cluster compared with *mcy/nda* genes, there is more shuffling of genes (Mazmouz *et al.*, 2010). Eleven genes *cyr*A-K are thought to make the core of the *cyr* gene cluster. Additional genes can be present, such as ORF1 in *Oscillatoria*, which is an ATP-grasp protein without obvious function (Méjean *et al.*, 2010). CyrL/M are transposases that can be present or absent (in *Oscillatoria*), but *Oscillatoria* has other transposases (TRP1, TRP2), (Méjean *et al.*, 2010). Phylogenetic analysis of the backbone genes *cyr*A-K and 16S rRNA in *Cylindrospermopsis*, *Raphidiopsis*, *Aphanizomenon*, and *Oscillatoria* revealed phylogenetic congruence (Jiang *et al.*, 2014), implying that horizontal gene transfer among the investigated taxa did not occur. Within the individual taxa *Cylindrospermopsis raciborskii* and *Raphidiopsis curvata*, out of 362 strains isolated from Chinese water bodies, six strains with *cyr*A-K were isolated (for all strains the *cyr*N+O genes were absent), implying a rather sporadic distribution of the *cyr* gene cluster (Jiang *et al.*, 2014; Sinha *et al.*, 2014).

1.2.3 Saxitoxin

Saxitoxins (also known as paralytic shellfish poisons) are potent neurotoxins and are produced by cyanobacteria and marine dinoflagellates. Among cyanobacteria, saxitoxins have been found in *Aphanizomenon gracile*, *A. mendotae*, *Aphanizomenon* sp., *Cuspidothrix issatschenkoi*, *Cylindrospermopsis raciborskii*, *Cylindrospermum stagnale*, *Dolichospermum circinale*, *Geitlerinema* spp., *Phormidium uncinatum*, *Raphidiopsis brookii*, and benthic *Lyngbya wollei* (Sivonen and Börner, 2008; Bernard *et al.*, 2017). Saxitoxins (molecular weight 299 Da) are a group of alkaloid tricyclic compounds that are either non-sulfated (saxitoxins and neosaxitoxin), single sulfated (gonyautoxins), or

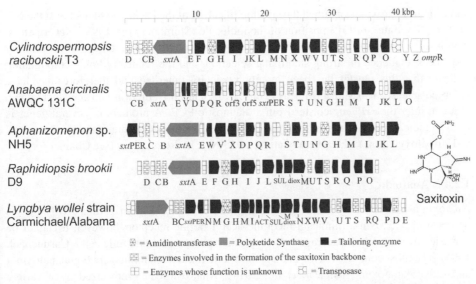

Figure 1.3 *Scheme of the genetic basis of saxitoxin synthesis in various cyanobacterial genera (from Kellmann et al., 2008b; Mihali et al., 2009; Stucken et al., 2010; Murray et al., 2011).*

doubly sulfated (C-toxins), resulting in more than 20 structural analogues (Shimizu, 1993; Mihali *et al.*, 2009). Isotope-labelled precursor feeding experiments suggested that the skeleton of the tricyclic ring system is formed by the Claisen-type condensation of acetate on the α-carbon of arginine with the loss of the carboxyl group of arginine and subsequent amidation and cyclization (Shimizu, 1993).

A first saxitoxin (*sxt*) biosynthesis gene cluster comprising 31 open reading frames has been described from *Cylindrospermopsis raciborskii* (Fig. 1.3), (Kellmann *et al.*, 2008b). Biosynthesis is initiated with SxtA, which has a PKS-like structure and contains four catalytic domains for catalyzing the methylation of acetate, and a Claisen condensation reaction between propionate and arginine. Arginine, acetyl-CoA, SAM, and carbamoylphosphate are the precursors. SxtA provides the substrate for SxtG, which is an amidinotransferase. Then SxtB, a cytidine deaminase-like enzyme, and SxtC (unknown function) are involved in the formation of the first ring. SxtD is a sterol desaturase predicted to introduce a double bond between C1 and C5. SxtS is an oxoglutarate-dependent dioxygenase and consecutive epoxidation and hydroxylation through SxtU, which is an alcohol dehydrogenase, leads to bicyclization, resulting in the tricyclic molecule (Murray *et al.*, 2011). SxtV/W constitute a putative electron transport system, *sxt*V encodes a dehydrogenase, which transfers electrons via SxtW (ferredoxin) to SxtH and SxtT, constituting a terminal oxygenase subunit, resulting in hydroxylation of C12 (saxitoxin molecule). Other tailoring enzymes – SxtI (O-carbamoylation), SxtX (N-1 hydroxylation), SxtN (a sulfo-transferase) – lead to carbamoylated saxitoxin structural variants or N1-hydroxylated variants (such as the neosaxitoxins) or sulfated saxitoxins, respectively, (Murray *et al.*, 2011).

In general the number of tailoring enzymes within the saxitoxin gene cluster is higher when compared with microcystin and cylindrospermopsin synthesis. While within NRPS structural variation is achieved through modification of adenylation domains, structural variation in saxitoxin is obtained through modification of the backbone molecule at the N1, O11, or N-sulfate-, or O13 position of the molecule (Shimizu *et al.*, 1993). Nevertheless, comparative analysis based on gene synteny and phylogenetic analyses of primary sequences, evidence of strong stabilizing selection, and the relatively low recombination indicate that this cluster has been largely vertically inherited and therefore may have emerged at least 2100 MaBP (Murray *et al.*, 2011). Currently, there are no gene loci known that can infer the *sxt* gene cluster from a diagnostic PCR. A protocol of several genes is used to indicate saxitoxin synthesis potential (*sxt*A, PKS, *sxt*G, amidinotransferase, *sxt*H, C-12 hydroxylation, *sxt*I, carbamoylation, *sxt*X, N-1, hydroxylation) (see Chapter 6).

1.2.4 Anatoxin

Anatoxin-a is a low molecular weight secondary amine (Devlin *et al.*, 1977) that is neurotoxic. Anatoxin-a mimics a nicotinic acetylcholine receptor; however, it cannot be degraded by acetylcholinesterase or by any other enzyme in eukaryotic cells (Carmichael, 1994). The most common analogs of anatoxin-a (hereafter, anatoxins) are homoanatoxin-a and dihydroanatoxin-a (Méjean *et al.*, 2014). Anatoxin-a is synthesized by a variety of cyanobacteria, including *Arthrospira, Hydrocoleum, Microcoleus, Oscillatoria, Phormidium, Pseudanabaena, Tychonema, Anabaena, Aphanizomenon, Cuspidothrix, Cylindrospermum,* and *Dolichospermum* (Bernard *et al.*, 2017).

A candidate gene cluster was revealed from *Oscillatoria* by genome sequencing, leading to a first hypothesis of the biosynthetic pathway (Méjean *et al.*, 2009). The entire pathway could be biochemically characterized *in vitro*. In general the peptide synthetases AnaB, C, and D are involved in the formation of the starter unit which is then extended by the PKSs AnaE, F, and G. In a first step, proline is activated by the NRPS AnaC and then bound to AnaD (acyl carrier protein) as Dehydro-Pro (Méjean *et al.*, 2010). AnaB is an oxidase resulting in oxidation of the Dehydro-Pro followed by extension through the three PKS AnaE, F, and G adding a full acetate unit each. AnaG contains a methyltransferase domain probably resulting in methylation (homoanatoxin-a). AnaA encodes for a type II thioesterase that may be involved in chain release.

Comparative analysis of anatoxin synthesis gene clusters from different genera reveals relatively high similarity, suggesting a likely common ancestor of (homo)anatoxin-a synthesis (Fig. 1.4). A transposase is flanking the *ana* gene cluster only in *Oscillatoria* (Calteau *et al.*, 2014). The greater difference is found in gene organization. In fact, for strain Anabaena 37 two clusters are found: *ana*B–G, while *ana*A, I, and J are found about 7 kbp separated and transcribed in the opposite direction (Rantala-Ylinen *et al.*, 2011). Transcription sites were recognized upstream of genes *ana*B–G and ORF1–*ana*A. Finally, *Cylindrospermum* was found to have an extra gene *ana*J encoding a reductase, which is likely to be involved in the reduction of the double band of anatoxin-a, resulting in dihydroanatoxin-a (Calteau *et al.*, 2014).

1.3 Application of Molecular Tools

Although genetic methods are only able to indicate the potential of toxin synthesis and do not provide information about actual toxin concentrations or contents, it is possible that

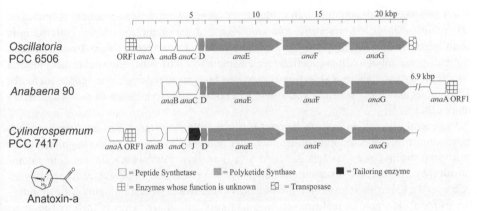

Figure 1.4   *Scheme of the genetic basis of (homo)anatoxin-a synthesis in various cyanobacterial genera (from Méjean et al., 2009; Rantala-Ylinen et al., 2011; Calteau et al., 2014).*

genetic methods will become more widely applicable in risk assessment in the near future (see Chapter 11). If the organisms do not have genes for toxin production, they are not able to produce a specific toxin. If they do have the genes, they can, but only the measurements of actual compounds guarantee that they are, indeed, producing this toxin. That's why it is important to remember that methods based on the presence of genes reveal only the toxigenic potential of specific cyanobacteria or environmental samples. Scientific research concerning the microcystin biosynthesis has revealed that mutations of the genes can occur and cause a genotype to be a non-toxin producer even if it contains the full microcystin synthesis gene cluster (Kurmayer *et al.*, 2004). However, the present understanding is that a minority of mutants in populations carry such mutations (Ostermaier and Kurmayer, 2010), so detecting potentially microcystin-producing cyanobacteria by molecular methods in the environment is worthwhile and will give useful information rather quickly.

During the last years, experience in detecting toxin synthesis genes directly in field samples has increased enormously and robust protocols for the extraction of DNA and subsequent targeted amplification of genes by PCR-based methods are available. By means of PCR, it is possible to amplify gene fragments, part of gene clusters encoding the synthesis of a specific toxin, directly from water samples. In theory, genes that are involved in toxin production can be detected from single cells. Owing to this high sensitivity, it is possible to detect potentially toxic genotypes in single individuals a long time before a toxic cyanobacterial bloom may occur. Water bodies bearing a risk in toxic bloom formation could be identified already early on in the growing season as well as the environmental factors influencing the abundance of toxigenic organisms. Thus, early detection of potentially toxic organisms in the water body can be used to promote appropriate management actions, such as prevention and risk assessment, as well as the adoption of suitable chemical-analytical tools for cyanotoxin identification and quantification (see also Sivonen, 2008 and Meriluoto *et al.*, 2017).

Phytoplankton is routinely counted under the microscope for the surveillance of water bodies used for recreational activities and supply of drinking water. An additional identification of toxic species would aid risk assessment and management of water resources.

For example, single colonies and filaments of planktonic cyanobacteria (*Anabaena (Dolichospermum)*, *Microcystis*, *Planktothrix*) can be identified during microscopic counting according to morphological criteria and then analyzed by PCR-based methods for their potential of toxin production (see Chapters 5 and 6). Since molecular methods can handle many samples in a short time, they can be considered high-throughput methods. Thus molecular tools possibly contribute to making the surveillance of lakes and rivers more efficient.

In general, the genetic approach consists of the following steps: (1) sampling and DNA extraction from collected material (either environmental samples or food supplements or cultivated strains) (see Chapters 2, 3, and 5); (2) analyzing cyanobacteria taxonomic composition by microscopic observations and molecular (PCR and DNA sequencing) analyses (Chapter 4); (3) (q)PCR amplification of genes indicative of toxin production (the monitoring of the PCR signal in real time enables us to quantify those genes in the environment) (Chapters 6 and 7); (4) detection of PCR products, either by agarose gel electrophoresis or qPCR or the ligation detection reaction (LDR) linked to a DNA chip (Chapter 8); and/or (5) profiling (Chapter 9) or deep-sequencing of the PCR products (Chapter 10).

Being able to detect potentially toxin-producing cyanobacteria species now *in situ*, however, is based on isolated strains which have been found to produce the toxin in question and revealing the biosynthesis of these toxins and the genes involved. In fact, strain isolations from nature (e.g. toxic blooms) and their purification enabled (1) to show the toxin production capacity by chemical methods and (2) the development of molecular methods. In the context of this handbook, these strains serve also as reference materials and control measures to make sure that a specific molecular method works (Chapter 3 and Table S6.1 with reference strains in the Appendix). It has been argued by one of the reviewers of this book that isolation and culturing are covered by multiple other sources and somewhat out of place. In the context of referencing PCR results, however, it is believed that good references are essential and typically are obtained from strains. It is the experience of the editors of this book that one part of confusion reported in the scientific literature on the molecular biology and ecophysiology of toxin production in cyanobacteria stems from no or poorly prepared references.

Each chapter of this book represents the general workflow in application of molecular tools for detection of toxigenic cyanobacteria (Fig. 1.5).

In Chapter 2, sampling details and sampling storage are described. Since molecular tools are all very sensitive the first challenge to overcome false positives is to minimize contamination with toxin genes from other sources. For example, using the same plankton net for different water bodies may lead to false positive results if the net is not rinsed carefully between sampling. In general, for PCR amplification of toxin genes from surface water samples, cells are collected (on filter) and DNA is extracted from all the phytoplankton collected. The best storage is at −20°C; however, in addition dry cell biomass can be used for DNA extraction and PCR amplification (see Chapters 5 and 6). In contrast heating of samples (e.g. at 100°C) to determine organic dry weight has been found to decrease the DNA quality and increase the fragmentation of the DNA (Ostermaier *et al.*, 2013).

Chapter 3 describes isolation, purification, and the correct maintenance of strains, also of those obtained from international culture collections. All PCR-based techniques require references, for example in order to find out whether any substance can inhibit the PCR

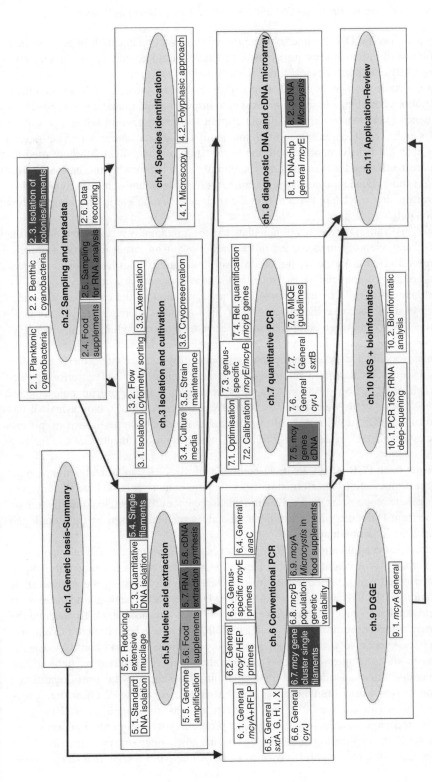

Figure 1.5 Scheme of the handbook's structure, individual chapters, and linked SOPs. SOPs from individual chapters but highlighted by the same background are directly related. (See color plate section for the color representation of this figure.)

reaction in a specific DNA extract obtained from the environment. If it is planned that, in addition to filtration, strains are isolated, this task should be performed directly after sampling under conditions described in Chapter 3.

Chapter 4 describes a general method (polyphasic approach) to be used in the identification of cyanobacterial taxa based on microscopic inspection and molecular analysis (PCR and sequencing) (Komárek, 2016). A list of current species names and synonyms is given in the Appendix. A rapid identification of potentially toxigenic cyanobacteria is critical to evaluate potential risks connected with the exploitation of contaminated waters used for drinking, recreation, and irrigation purposes. Further, allowing for a semi-quantitative estimate of the most abundant cyanobacteria, microscopic observations are equally important when investigating food supplements because toxigenic cyanobacteria may occur in low proportion only.

Chapter 5 details DNA extraction which can be quantitatively extracted using the standard phenol-chloroform extraction procedure following an earlier protocol of Franche and Damerval (1988). As the phenol-chloroform procedure is both laborious and potentially hazardous, commercial DNA isolation kits are now widely available. Those kits, however, may differ in DNA extraction efficiency depending on the cyanobacteria under study, for example because the production of high amounts of mucilage (*Microcystis*) reduces the DNA yield when compared with species with lower amounts of mucilage (i.e. *Planktothrix*; Schober and Kurmayer, 2006). Other extraction protocols use reagents to precipitate exocellular polysaccharide that is produced by many cyanobacteria and may inhibit subsequent PCR amplification (e.g. Tillett and Neilan, 2000). The various procedures to extract DNA from cyanobacteria either from cultural strains or from field samples have been reviewed (McMahon *et al.*, 2007; Srivastava *et al.*, 2007). Not at least it has been shown that PCR amplification of cyanobacterial genes can be performed from fresh cells (Beard *et al.*, 1999; Kurmayer *et al.*, 2002) and colonies/filaments that can be used directly as a template.

In Chapter 6, all PCR assays are described in detail. Besides standard laboratory equipment, the instrumentation to perform molecular analyses comprises basically a PCR cycling machine, gel electrophoresis, and gel documentation devices. For principal information on how to perform PCR, the widely available laboratory manual revised by Sambrook and Russell in 2001 is recommended.

In Chapter 7, the application of qPCR is described, mostly with the aim of quantifying cyanobacterial toxin genes in a given volume of water. In general a standard curve based on predetermined cell/DNA concentrations is established by relating the known DNA concentrations to the threshold cycle of the diluted DNA extract. The threshold cycle is the PCR cycle number at which the fluorescence passes a set threshold level. To generate the fluorescence signal conventional dyes such as SYBR Green have been used (e.g. Vaitomaa *et al.*, 2003). Alternatively, the Taq nuclease assay has been used to quantify toxigenic organisms in water samples (Kurmayer and Kutzenberger, 2003) based on fluorescence signals emitted from an additional molecular probe. Although the semi-logarithmic calibration curves cause limitations with regard to the accuracy in estimating genotype numbers and proportions, the PCR-based techniques in general are still the only quantitative techniques available to estimate the proportion of toxin-producing genes.

In Chapter 8, the application of DNA chips or diagnostic microarrays is described (e.g. Rantala *et al.*, 2008). In general, DNA chips or DNA microarrays contain a large number of short DNA fragments (oligonucleotides) spotted on a very small glass slide (<200 µm in diameter). The principle is PCR amplification of the target gene, followed by a sensitive LDR and hybridization to short oligonucleotides (ZipCodes) spotted on a glass slide. The LDR depends on two probes: one that is discriminative for the genotype and the other that is a general one but that can be recognized by a short synthesized oligonucleotide (the ZipCode). The ZipCode is then recognized via hybridization to complementary DNA spotted on a glass slide. Another microarray without PCR amplification can be used to provide data on the transcription of genes involved in toxin synthesis after mRNA has been extracted and translated to cDNA using reverse transcriptase. The analysis of the rate of the transcription of genes involved in toxin synthesis (i.e. the microcystins) provides insights into their physiological regulation of toxin production. However, owing to the high turnover rates of mRNA in the cells, mRNA sampling is sophisticated and transcript analysis in general is less likely to be useful in the monitoring of water bodies for their potential for cyanotoxin production.

In Chapter 9, one example on PCR product profiling using denaturing gradient gel electrophoresis (DGGE) is provided. DGGE has been used for the monitoring of toxigenic genotype composition for a considerable time (Janse *et al.*, 2003). Although DGGE cannot give quantitative results, this technique could be used to monitor the toxigenic organism composition of a population or a community in the environment. Furthermore, DGGE still has the potential to reveal unknown toxin producers carrying the same conserved gene fragment from environmental samples.

In Chapter 10, ultra-deep sequencing using next-generation sequencing (NGS, metabarcoding) methods is proposed as a tool for monitoring cyanobacteria in the environment. By obtaining at least several thousand sequences from one locus per sample (e.g. 16S rRNA), it would be possible to monitor even less abundant potentially toxigenic species. Methods in sequence analysis and exploration of sequence data are described.

1.4   Laboratory Safety Issues

The book is intended to be used by trained professionals analyzing toxigenicity of water samples in the laboratory in both academic and governmental institutions, as well as technical offices and agencies in charge of water body surveillance. Consequently, standard safety instructions for both sampling and processing of samples in the laboratory apply. In particular, large amounts (grams of wet weight) of cyanobacteria (bloom) cell material collected from water bodies could be hazardous and thus cell material has to be processed using the necessary protection (skin and eye protection). For the processing of dried samples a fume hood and/or inhalation masks should be used.

Some of the solvents used in the provided SOPs (e.g. phenol-chloroform in Chapter 5 and ethidium bromide in Chapters 4 and 6) are toxic and must be used under the fume hood with the necessary skin and eye protection. Toxicity is defined according to the "Globally harmonized system of classification and labelling of chemicals" (GHS). GHS06: Acute toxicity (oral, dermal, inhalation), categories 1, 2, 3, GHS08: health hazard (carcinogenic or mutagenic or reproductive toxicity) (see European Commission Directive (2009) for all inorganic reagents. In this book, for potentially harmful DNA extraction solvents the toxicity (GHS06, GHS08) is indicated.

1.5 References

Banker, R., Carmeli, S., Werman, M., *et al.* (2001) Uracil moiety is required for toxicity of the cyanobacterial hepatotoxin cylindrospermopsin, *Journal of Toxicology and Environmental Health: Part A*, **62**, 281–288.

Beard, S.J., Handley, B.A., Hayes, P.K., and Walsby, A.E. (1999) The diversity of gas vesicle genes in *Planktothrix rubescens* from Lake Zürich, *Microbiology*, **145**, 2757–2768.

Bernard, C., Ballot, A., Thomazeau, S., *et al.* (2017) Appendix 2, Cyanobacteria associated with the production of cyanotoxins, in J. Meriluoto, L. Spoof, and G.A. Codd (eds), *Handbook on Cyanobacterial Monitoring and Cyanotoxin Analysis*, John Wiley & Sons, Ltd, Chichester, UK.

Calteau, A., Fewer, D.P., Latifi, A., *et al.* (2014) Phylum-wide comparative genomics unravel the diversity of secondary metabolism in Cyanobacteria, *BMC Genomics*, **15**.

Carmichael, W.W. (1994) The toxins of cyanobacteria, *Scientific American*, **270**, 78–86.

Carmichael, W.W., Beasly, V., Bunner, D.L., *et al.* (1988) Naming cyclic heptapeptide toxins of cyanobacteria (blue-green algae), *Toxicon*, **26**, 971–973.

Christiansen, G., Fastner, J., Erhard, M., *et al.* (2003) Microcystin biosynthesis in *Planktothrix*: Genes, evolution, and manipulation, *Journal of Bacteriology*, **185** (2), 564–572.

Christiansen, G., Molitor, C., Philmus, B., and Kurmayer, R. (2008) Nontoxic strains of cyanobacteria are the result of major gene deletion events induced by a transposable element, *Molecular Biology and Evolution*, **25** (8), 1695–1704.

de Silva, E.D., Williams, D.E., Andersen, R.J., *et al.* (1992) Motuporin: A potent protein phosphatase inhibitor isolated from the Papua-New-Guinea sponge *Theonella swinhoei* Gray, *Tetrahedron Letters*, **33** (12), 1561–1564.

Devlin, J.P., Edwards, O.E., Gorham, P.R., *et al.* (1977) Anatoxin-a, a toxic alkaloid from *Anabaena flos-aquae* NR-44H, *Canadian Journal of Chemistry*, **55**, 1367–1371.

Diehnelt, C.W., Dugan, N.R., Peterman, S.M., and Budde, W.L. (2006) Identification of microcystin toxins from a strain of *Microcystis aeruginosa* by liquid chromatography introduction into a hybrid linear ion trap-Fourier transform ion cyclotron resonance mass spectrometer, *Analytical Chemistry*, **78** (2), 501–512.

Dittmann, E., Fewer, D.P., and Neilan, B.A. (2013) Cyanobacterial toxins: Biosynthetic routes and evolutionary roots, *FEMS Microbiology Reviews*, **37** (1), 23–43.

Dittmann, E., Neilan, B.A., Erhard, M., *et al.* (1997) Insertional mutagenesis of a peptide synthetase gene that is responsible for hepatotoxin production in the cyanobacterium *Microcystis aeruginosa* PCC 7806, *Molecular Microbiology*, **26**, 779–787.

European Commission (2009) *Regulation (EC) no 1272/2008 of the European Parliament and of the council on classification, labelling and packaging of substances and mixtures, amending and repealing, Directives 67/548/EEC and 1999/45/EC, and amending Regulation (EC) No 1907/2006*, Brussels, Belgium.

Fewer, D.P., Wahlsten, M., Österholm, J., *et al.* (2013) The genetic basis for O-acetylation of the microcystin toxin in cyanobacteria, *Chemistry and Biology*, **20** (7), 861–869.

Fiore, M.F., Genuario, D.B., da Silva, C.S.P., *et al.* (2009) Microcystin production by a freshwater spring cyanobacterium of the genus Fischerella, *Toxicon*, **53** (7–8), 754–761.

Fischbach, M.A. and Walsh, C.T. (2006) Assembly-line enzymology for polyketide and nonribosomal peptide antibiotics: Logic, machinery, and mechanisms, *Chemical Reviews*, **106** (8), 3468–3496.

Franche, C. and Damerval, T. (1988) Test on *nif* probes and DNA hybridizations, *Methods in Enzymology*, **167**, 803–808.

Hicks, L.M., Moffitt, M.C., Beer, L.L., *et al.* (2006) Structural characterization of *in vitro* and *in vivo* intermediates on the loading module of microcystin synthetase, *ACS Chemical Biology*, **1**, 93–102.

Janse, I., Meima, M., Kardinaal, W.E.A., and Zwart, G. (2003) High-resolution differentiation of cyanobacteria by using rRNA-internal transcribed spacer denaturing gradient gel electrophoresis, *Applied and Environmental Microbiology*, **69** (11), 6634–6643.

Jiang, Y.G., Xiao, P., Yu, G.L., *et al.* (2014) Sporadic distribution and distinctive variations of cylindrospermopsin genes in cyanobacterial strains and environmental samples from Chinese freshwater bodies, *Applied and Environmental Microbiology*, **80** (17), 5219–5230.

Kellmann, R. and Neilan, B.A. (2007) Biochemical characterization of paralytic shellfish toxin biosynthesis *in vitro*, *Journal of Phycology*, **43**, 497–508.

Kellmann, R., Mihali, T.K., and Neilan, B.A. (2008a) Identification of a saxitoxin biosynthesis gene with a history of frequent horizontal gene transfers, *Journal of Molecular Evolution*, **67**, 526–538.

Kellmann, R., Mihali, T.K., Jeon, Y.J., *et al.* (2008b) Biosynthetic intermediate analysis and functional homology reveal a saxitoxin gene cluster in cyanobacteria, *Applied and Environmental Microbiology*, **74** (13), 4044–4053.

Kellmann, R., Mills, T., and Neilan, B.A. (2006) Functional modeling and phylogenetic distribution of putative cylindrospermopsin biosynthesis enzymes, *Journal of Molecular Evolution*, **62**, 267–280.

Komárek, J. (2016) A polyphasic approach for the taxonomy of cyanobacteria: Principles and applications, *European Journal of Phycology*, **51** (3), 346–353.

Kurmayer, R. and Kutzenberger, T. (2003) Application of real-time PCR for quantification of microcystin genotypes in a population of the toxic cyanobacterium *Microcystis* sp., *Applied and Environmental Microbiology*, **69**, 6723–6730.

Kurmayer, R., Christiansen, G., Fastner, J., and Börner, T. (2004) Abundance of active and inactive microcystin genotypes in populations of the toxic cyanobacterium *Planktothrix* spp., *Environmental Microbiology*, **6** (8), 831–841.

Kurmayer, R., Dittmann, E., Fastner, J., and Chorus, I. (2002) Diversity of microcystin genes within a population of the toxic cyanobacterium *Microcystis* spp. in Lake Wannsee (Berlin, Germany), *Microbial Ecology*, **43**, 107–118.

Mazmouz, R., Chapuis-Hugon, F., Mann, S., *et al.* (2010) Biosynthesis of cylindrospermopsin and 7-Epicylindrospermopsin in *Oscillatoria* sp. strain PCC 6506: Identification of the *cyr* gene cluster and toxin analysis, *Applied and Environmental Microbiology*, **76** (15), 4943–4949.

Mazmouz, R., Chapuis-Hugon, F., Pichon, V., *et al.* (2011) The last step of the biosynthesis of the cyanotoxins cylindrospermopsin and 7-epi-cylindrospermopsin is catalysed by CyrI, a 2-oxoglutarate-dependent iron oxygenase, *ChemBioChem*, **12** (6), 858–862.

McMahon, K.D., Gu, A.Z., Nerenberg, R., and Angenent, L.T. (2007) Molecular methods in biological systems, *Water Environment Research*, **79** (10), 1109–1151.

Méjean, A. and Ploux, O. (2013) A genomic view of secondary metabolite production in cyanobacteria, in F. Chauvat, and C. Cassier-Chauvat (eds), *Advances in Botanical Research: Genomics of Cyanobacteria*. Academic Press, Elsevier, San Diego, 189–234.

Méjean, A., Mann, S., Maldiney, T., *et al.* (2009) Evidence that biosynthesis of the neurotoxic alkaloids anatoxin-a and homoanatoxin-a in the cyanobacterium *Oscillatoria* PCC 6506 occurs on a modular polyketide synthase initiated by L-Proline, *Journal of the American Chemical Society*, **131** (22), 7512–7513.

Méjean, A., Mann, S., Vassiliadis, G., *et al.* (2010) *In vitro* reconstitution of the first steps of anatoxin-a biosynthesis in *Oscillatoria* PCC 6506: From free L-Proline to acyl carrier protein bound dehydroproline, *Biochemistry*, **49** (1), 103–113.

Méjean, A., Paci, G., Gautier, V., and Ploux, O. (2014) Biosynthesis of anatoxin-a and analogues (anatoxins) in cyanobacteria, *Toxicon*, **91**, 15–22. doi: 10.1016/j.toxicon.2014.07.016

Meriluoto, J., Spoof, L., and Codd, G.A. (2017) *Handbook of Cyanobacterial Monitoring and Cyanotoxin Analysis*, John Wiley and Sons, Ltd, Chichester, UK.

Mihali, T.K., Kellmann, R., and Neilan, B.A. (2009) Characterisation of the paralytic shellfish toxin biosynthesis gene clusters in *Anabaena circinalis* AWQC131C and *Aphanizomenon* sp. NH-5, *BMC Biochemistry*, **10**, 8.

Mihali, T.K., Kellmann, R., Muenchhoff, J., *et al.* (2008) Characterization of the gene cluster responsible for cylindrospermopsin biosynthesis, *Applied and Environmental Microbiology*, **74**, 716–722.

Moffitt, M.C. and Neilan, B.A. (2004) Characterization of the nodularin synthetase gene cluster and proposed theory of the evolution of cyanobacterial hepatotoxins, *Applied and Environmental Microbiology*, **70** (11), 6353–6362.

Murphy, P.J. and Thomas, C.W. (2001) The synthesis and biological activity of the marine metabolite cylindrospermopsin, *Chemical Society Reviews*, **30**, 303–312.

Murray, S.A., Mihali, T.K., and Neilan, B.A. (2011) Extraordinary conservation, gene loss, and positive selection in the evolution of an ancient neurotoxin, *Molecular Biology and Evolution*, **28** (3), 1173–1182.

Oksanen, I., Jokela, J., Fewer, D.P., *et al.* (2004) Discovery of rare and highly toxic microcystins from lichen-associated cyanobacterium *Nostoc* sp strain IO-102-I, *Applied and Environmental Microbiology*, **70** (10), 5756–5763.

Ostermaier, V. and Kurmayer, R. (2010) Application of real-time PCR to estimate toxin production by the cyanobacterium *Planktothrix* sp. *Applied and Environmental Microbiology*, **76** (11), 3495–3502.

Ostermaier, V., Christiansen, G., Schanz, F., and Kurmayer, R. (2013) Genetic variability of microcystin biosynthesis genes in *Planktothrix* as elucidated from samples preserved by heat desiccation during three decades, *PLOS ONE*, **8** (11), e80177.

Pearson, L.A., Barrow, K.D., and Neilan, B.A. (2007) Characterization of the 2-hydroxy acid dehydrogenase, *Mcy*I, encoded within the microcystin biosynthesis gene cluster of *Microcystis aeruginosa* PCC7806, *The Journal of Biological Chemistry*, **282**, 4681–4692.

Pearson, L.A., Hisbergues, M., Börner, T., *et al.* (2004) Inactivation of an ABC transporter gene, *mcyH*, results in loss of microcystin production in the cyanobacterium *Microcystis aeruginosa* PCC7806, *Applied and Environmental Microbiology*, **70**, 6370–6378.

Prinsep, M.R., Caplan, F.R., Moore, R.E., *et al.* (1992) Microcystin-LA from a blue-green alga belonging to the Stigonematales, *Phytochemistry*, **31**, 1247–1248.

Rantala, A., Ermanno, R., Castiglioni, B., *et al.* (2008) Identification of hepatotoxin-producing cyanobacteria by DNA-chip, *Environmental Microbiology*, **10**, 653–664.

Rantala, A., Fewer, D.P., Hisbergues, M., *et al.* (2004) Phylogenetic evidence for the early evolution of microcystin synthesis, *Proceedings of the National Academy of Sciences of the United States of America*, **101**, 568–573.

Rantala-Ylinen, A., Kana, S., Wang, H., *et al.* (2011) Anatoxin-a synthetase gene cluster of the cyanobacterium *Anabaena* sp strain 37 and molecular methods to detect potential producers, *Applied and Environmental Microbiology*, **77** (20), 7271–7278.

Rouhiainen, L., Vakkilainen, T., Siemer, B.L., *et al.* (2004) Genes coding for hepatotoxic heptapeptides (microcystins) in the cyanobacterium *Anabaena* strain 90, *Applied and Environmental Microbiology*, **70** (2), 686–692.

Runnegar, M.T., Kong, S.M., Zhong, Y.Z., and Lu, S.C. (1995) Inhibition of reduced glutathione synthesis by cyanobacterial alkaloid cylindrospermopsin in cultured rat hepatocytes, *Biochemical Pharmacology*, **49**, 219–225.

Sambrook, J. and Russell, D.W. (2001) *Molecular Cloning: A laboratory manual*, 3rd edn, Cold Spring Harbor Laboratory Press, Cold Spring Harbor, NY.

Schembri, M.A., Neilan, B.A., and Saint, C.P. (2001) Identification of genes implicated in toxin production in the cyanobacterium *Cylindrospermopsis raciborskii*, *Environmental Toxicology*, **16**, 413–421.

Schober, E. and Kurmayer, R. (2006) Evaluation of different DNA sampling techniques for the application of the real-time PCR method for the quantification of cyanobacteria in water, *Letters in Applied Microbiology*, **42**, 412–417.

Shalev-Alon, G., Sukenik, A., Livnah, O., *et al.* (2002) A novel gene encoding amidino-transferase in the cylindrospermopsin producing cyanobacterium *Aphanizomenon ovalisporum*, *FEMS Microbiology Letters*, **209**, 87–91.

Shih, P.M., Wu, D.Y., Latifi, A., *et al.* (2013) Improving the coverage of the cyanobacterial phylum using diversity-driven genome sequencing, *Proceedings of the National Academy of Sciences of the United States of America*, **110** (3), 1053–1058.

Shimizu, Y. (1993) Microalgal metabolites, *Chemical Reviews*, **93** (5), 1685–1698.

Sielaff, H., Dittmann, E., Tandeau de Marsac, N., *et al.* (2003) The *mcy*F gene of the microcystin biosynthetic gene cluster from *Microcystis aeruginosa* encodes an aspartate racemase, *Biochemical Journal*, **373**, 909–916.

Sinha, R., Pearson, L.A., Davis, T.W., *et al.* (2014) Comparative genomics of *Cylindrospermopsis raciborskii* strains with differential toxicities, *BMC Genomics*, **15** (1), 1–14.

Sivonen, K. (2008) Emerging high throughput analyses of cyanobacterial toxins and toxic cyanobacteria, in H.K. Hudnell (ed.) *Cyanobacterial Harmful Algal Blooms: State of the science and research needs*, Springer, New York, **619**, 539–557.

Sivonen, K. and Börner, T. (2008) Bioactive compounds produced by cyanobacteria, in A. Herrero and E. Flores (eds), *The Cyanobacteria: Molecular Biology, Genomics and Evolution*. Caister Academic Press, Poole, UK, 159–197.

Spoof, L. and Catherine, A. (2017) Appendix 3: Tables of microcystins and nodularins, in J. Meriluoto, L. Spoof, and G.A. Codd (eds), *Handbook on Cyanobacterial Monitoring and Cyanotoxin Analysis*, John Wiley & Sons, Ltd, Chichester, UK.

Srivastava, A.K., Ara, A., Bhargava, P., *et al.* (2007) A rapid and cost-effective method of genomic DNA isolation from cyanobacterial culture, mat and soil suitable for genomic fingerprinting and community analysis, *Journal of Applied Phycology*, **19** (4), 373–382.

Stucken, K., John, U., Cembella, A., *et al.* (2010) The smallest known genomes of multicellular and toxic cyanobacteria: Comparison, minimal gene sets for linked traits and the evolutionary implications, *PLOS ONE*, **5** (2).

Stüken, A. and Jakobsen, K.S. (2010) The cylindrospermopsin gene cluster of *Aphanizomenon* sp. strain 10E6: Organization and recombination, *Microbiology-SGM*, **156**, 2438–2451.

Stüken, A., Campbell, R.J., Quesada, A., *et al.* (2009) Genetic and morphologic characterization of four putative cylindrospermopsin producing species of the cyanobacterial genera *Anabaena* and *Aphanizomenon*, *Journal of Plankton Research*, **31** (5), 465–480.

Terao, K., Ohmori, S., Igarashi, K., *et al.* (1994) Electron microscopic studies on experimental poisoning in mice induced by cylindrospermopsin isolated from blue-green alga *Umezakia natans*, *Toxicon*, **32**, 833–843.

Tillett, D. and Neilan, B.A. (2000) Xanthogenate nucleic acid isolation from cultured and environmental cyanobacteria, *Journal of Phycology*, **36**, 251–258.

Tillett, D., Dittmann, E., Erhard, M., *et al.* (2000) Structural organization of microcystin biosynthesis in *Microcystis aeruginosa* PCC7806: An integrated peptide-polyketide synthetase system, *Chemistry and Biology*, **7** (10), 753–764.

Vaitomaa, J., Rantala, A., Halinen, K., *et al.* (2003) Quantitative real-time PCR for determination of microcystin synthetase E copy numbers for *Microcystis* and *Anabaena* in lakes, *Applied and Environmental Microbiology*, **69**, 7289–7297.

Zhou, L., Yu, H., and Chen, K. (2002) Relationship between microcystin in drinking water and colorectal cancer, *Biomedical and Environmental Sciences*, **15**, 166–171.

2

Sampling and Metadata

Rainer Kurmayer[1]*, Guntram Christiansen[1,2], Konstantinos Kormas[3], Wim Vyverman[4],
Elie Verleyen[4], Vitor Ramos[5,6], Vitor Vasconcelos[5,6], and Nico Salmaso[7]

[1] Research Institute for Limnology, University of Innsbruck, Mondsee, Austria
[2] Miti Biosystems GmbH, Max F. Perutz Laboratories, Vienna, Austria
[3] Department of Ichthyology and Aquatic Environment, University of Thessaly, Volos, Greece
[4] Laboratory of Protistology and Aquatic Ecology, Department of Biology, Ghent University, Gent, Belgium
[5] Interdisciplinary Centre of Marine and Environmental Research (CIIMAR/CIMAR), University of Porto, Matosinhos, Portugal
[6] Faculty of Sciences, University of Porto, Porto, Portugal
[7] Research and Innovation Centre, Fondazione Edmund Mach – Istituto Agrario di S. Michele all'Adige, S. Michele all'Adige, Italy

2.1 Introduction

This chapter aims to describe the essential steps in collecting and preserving samples to be used for the detection and quantification of toxigenic cyanobacteria. Some of the steps (e.g. stratified vs. integrated-depth sampling, low vacuum-filtering onto GF/C filters) are identical to the processing of samples for cyanotoxin detection and analysis (e.g. Meriluoto and Codd, 2005). However, since, instead of oligopeptides and alkaloids, nucleic acids (either DNA or RNA) are sampled, the specific chemical, physical, and biological properties of

*Corresponding author: rainer.kurmayer@uibk.ac.at

Molecular Tools for the Detection and Quantification of Toxigenic Cyanobacteria, First Edition.
Edited by Rainer Kurmayer, Kaarina Sivonen, Annick Wilmotte and Nico Salmaso.
© 2017 John Wiley & Sons Ltd. Published 2017 by John Wiley & Sons Ltd.

these molecules need to be considered. For example, owing to the short turnover time of mRNA in prokaryotes in general, samples for mRNA must be conserved immediately (e.g. by shock-freezing using liquid nitrogen; −196°C) or fixatives (Park *et al.*, 2013).

Correct sampling and storage depends to some extent on the foreseen downstream procedures and expected outcomes of each survey, which should be known and well defined beforehand. Basically, the molecular tools approach consists of three mandatory steps:

1. DNA (mRNA) extraction from cell material collected on filters (environmental samples or cultures) or pellets, or single colonies/filaments that have been isolated under the microscope. For transcriptional studies mRNA needs to be reversely transcribed (Chapter 5).
2. (q)PCR amplification of genes indicative of toxin production, the monitoring of the PCR signal in real-time enables those genes in relation to an appropriate reference gene to be quantified (Chapters 6, 7, 9).
3. Detection and characterization of PCR products using either conventional agarose gel electrophoresis (Chapter 6) or denaturing gradient gel electrophoresis (Chapter 9) or the ligation detection reaction linked to a DNA chip (Chapter 8).

The sequencing of the amplified PCR product is optional and often useful to confirm the results obtained during steps 2 and 3. Conversely, sequencing is a necessary step in the taxonomic identification of cyanobacteria based on the analysis of 16S rRNA and/or housekeeping genes (Chapter 4). More recently, amplicon ultra-deep sequencing using next-generation sequencing (NGS) techniques has been used for monitoring of phytoplankton (Eiler *et al.*, 2013) and could have potential for monitoring toxigenic genotypes (Chapter 10).

The protocols (see SOPs 2.1–2.5) described here allow for the correct sampling and storage of nucleic acids for applying molecular tools to monitor and study toxigenic cyanobacteria. In addition, the acquisition of relevant environmental metadata is introduced (see SOP 2.6) and its potential to analyze and interpret molecular results is highlighted.

The sampling protocols described here are used for (1) the detection and quantification of genes (genotypes) used for taxonomical determinations, and genes indicative of toxin production (including food supplements), (2) the analysis of the phylogenetic dependence on the distribution of toxin genes among cyanobacteria, and (3) the analysis of the transcriptional regulation of toxin synthesis genes by comparing the transcript amount under various environmental conditions.

2.2 Handling of Samples

While nucleic acids themselves are considered non-hazardous, samples containing toxigenic cyanobacteria are likely to contain cyanotoxins or even mixtures of cyanotoxins. Thus these samples are potentially harmful to animals and humans. Consequently, all precautions for the safe handling and disposal of sampling material must be taken by the responsible personnel, who must have the necessary professional education or training (as is requested for the use of any other potentially hazardous material). Any material should be processed using standard bacteriological precautions and be made nonviable by household bleach or autoclaving prior to discharge in the laboratory or in the environment.

Depending on the respective aim (but independent of the sample strategy), the samples are usually subdivided into aliquots for (1) genotype quantification analysis per volume of filtered water, (2) for toxin gene transcript quantification, and (3) cell microscopic counting and other environmental variables such as chlorophyll or macro nutrients (see SOP 2.1).

2.3 Sample Contamination

As molecular tools described in Chapters 6–10 are highly sensitive, theoretically allowing gene copies from single cells to be detected, sample containers should be new or cleaned (e.g. by overnight incubation in 1 M HCl followed by thorough rinsing with sterile Milli-Q® water (Merck Millipore)). The possibility of sample cross-contamination (e.g. using the same plankton net during a survey of different water bodies) should also be reduced as much as possible. When processing samples in the laboratory, potential cross-contamination between samples should be minimized (i.e. by careful rinsing measuring cylinders, vacuum filtration units, etc. with sterile Milli-Q® water). In some cases it might be useful to use blank samples (sterile Milli-Q® water) and process them exactly the same way as the samples.

2.4 Sampling

2.4.1 Quantitative Depth-Integrated and Discrete Sampling

Water bodies can be sampled quantitatively using depth-integrated sampling over the entire water column (e.g. by means of a Van Dorn sampler or by means of an integrating water sampler). In deep physically stratified lakes, often the epilimnion is sampled (e.g. by sampling every 2 m from the surface down to 20 m) by mixing samples from various depths and taking subsamples for analysis (Fig. 2.1), or using flexible hose-pipe samplers (Lund *et al.*, 1958). Often, when the vertical distribution peaks and biomass shall be defined (e.g. in waters intended for drinking or recreational use), discrete sampling should be performed at several depths in the trophogenic layer (Salmaso *et al.*, 2017).

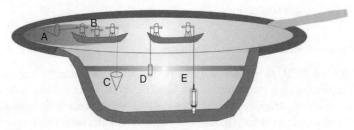

Figure 2.1 *Sampling of toxigenic cyanobacterial communities in lakes and reservoirs from the boat: (A) pulling a plankton net at the surface, (B) grabbing a sample from the surface, (C) pulling a plankton net vertically, (D) using a self-closing water sampling to sample metalimnetic layers, (E) using a gravity corer to sample cyanobacteria settled to the sediment.*

2.4.2 Qualitative Plankton Net Sampling

Since most toxigenic cyanobacteria are growing in relatively large macroscopic colonies, long-coiled trichomes, or bundles, it is possible to concentrate the larger aggregates which might be practicable in less eutrophic lakes. The phytoplankton net (typically ca. 30 μm in mesh size) is either drawn vertically through the water column or moved horizontally in shallow waters. Sometimes water bodies contain a high (in)organic load rendering the filtering of larger volumes on glass fiber filters (e.g. Whatman GF/C) impossible. Using a phytoplankton net, the influence of potential inhibitors for PCR (e.g. by humic acids) can be reduced. It should be kept in mind that plankton net samples cannot be used for quantitative determinations, providing only rough or approximate indications about the relative abundance of the larger cyanobacteria.

2.4.3 Surface (Scum Material) Sampling

For monitoring, grabbing a surface sample from the shore may often be the cheapest way. Since toxigenic cyanobacteria are buoyant and easily drifted by wind action on the water surface, very high cell concentrations often occur along the shoreline. It should be kept in mind that samples from the shore are rarely representative of the whole lake and the presence/absence of scum on the surface close to the shore can change within hours.

2.4.4 Benthic (Terrestrial) Cyanobacteria Sampling

When compared with the floating toxigenic cyanobacteria, sampling of benthic cyanobacteria has to take into account a pronounced horizontal patchiness. Consequently, to minimize this influence, samples from a defined sampling area need to be pooled.

2.4.5 Food Supplement Sampling

Food supplement samples (the various types are indicated in SOP 2.4) should be handled as in food safety programs for microbiological analysis. Ideally, a standard method of sampling from one production lot number should be followed. As a guide, see the ISO/TC34 Standards (e.g. ISO 7002 and ISO 20837). Depending on the form of the sample, an initial pulverization and/or homogenization step can be necessary. Again, care has to be taken to avoid sample (cross-)contaminations.

2.4.6 Isolation of Single Colonies/Filaments

The sensitivity of PCR in general allows toxin genes in colonies and filaments of cyanobacteria isolated under the microscope to be detected. In unicellular colony-forming cyanobacteria such as *Microcystis* spp. the cell division process is accompanied by mucilage production, embedding the cells in a gel-like matrix, and one colony of *Microcystis* sp. is considered a clonal unit. Typically, colonies/filaments are picked randomly under a stereo binocular using forceps or a micropipette and serially transferred to standard solution (e.g. BG11 Medium) for washing several times. The colonies/filaments are then stored in a low volume of sterile Milli-Q® water or PCR buffer (10–20 μL).

2.5 Subsampling Food Supplement Samples

After processing food supplements, subsamples should be picked up from each sample for DNA extraction purposes. Whenever possible (i.e. cells can be distinguished), it is advisable to subsample biomass under an inverted microscope, using a sterile/aseptic technique (if needed, sterile Milli-Q® water can be used as a mounting medium). For this purpose, disposable Petri dishes work better than slides. Through microscopic visualization, it is possible to separate and harvest a part of the sample that is potentially or noticeably contaminated. This may prevent that the genomic DNA to be extracted will correspond only to the dominant organism (i.e. the edible alga), and potential contaminations are indeed overlooked. If feasible, the same above-mentioned technique to isolate single filaments/colonies can be applied.

Whenever dealing with food supplements we have to bear in mind that the "edible algae" will contribute the majority of the biomass to be handled and processed. It is thus advisable to use several replicates (i.e. subsamples) for DNA extraction, since it will reduce the likelihood to artificially hide the presence of possible cyanobacterial contaminants. In SOP 2.4 the main steps needed to be addressed regarding sampling and sample coverage issues are described. Following these steps increases the confidence that the bulk genomic DNA, to be subsequently extracted, embraces the different cyanobacteria that might be present in the sample. Next, the objective is to ensure that the quality of such extracted DNA is not irretrievably affected by food production processes (e.g. drying of algae). For this purpose additional operating procedures, to be followed after SOP 2.4, are given in SOP 5.6 (see also Chapter 11).

2.6 Sampling of Nucleic Acids

In order to obtain sufficient biomass for DNA extraction, a minimum concentration of cells is required. Owing to their buoyancy, cells from bloom-forming cyanobacteria often cannot be concentrated by centrifugation efficiently. Instead, aliquots of well-mixed samples are filtered onto glass fiber filters (GF/C) or membrane filters – usually of 47 mm diameter – until the filters are colored. The vacuum filtration pressure should not exceed 0.4 bar. For quantitative analysis, the volume of water needs to be recorded and the filter is folded inside up, put into a 2 mL Eppendorf reaction tube, and stored at $-20°C$. Filters for DNA analysis can be stored safely for weeks or at $-70°C$ for long-term storage until extraction.

To collect RNA the same low-vacuum filtration technique is used but cells should be filtered within seconds. Immediately afterwards, the filter is folded inside up, put into a cryotube and shock-frozen in liquid nitrogen ($-196°C$). Samples are stored at $-70°C$. The frozen filters are then extracted as soon as possible using protocols described in Chapter 5.

For the first time, Walsby and co-workers have used direct lysis of single filaments in PCR buffer and subsequent PCR amplification of one or several gene loci (reviewed in Hayes *et al.*, 2002). Colonies of *Microcystis* often disintegrate in sterile Milli-Q® water and one aliquot can be used as a template for PCR (see Chapter 6). In contrast, single filaments, such as *Planktothrix* sp., do not disintegrate in sterile Milli-Q® water and further treatment of single filaments suspended in PCR buffer by ultrasound sonication is required (Kurmayer *et al.*, 2004). As compared with filters the overall DNA amounts (µg of DNA) are relatively low (pg of DNA) and the loss of DNA through the degradation process during DNA storage even at –20°C cannot be excluded. The use of PCR sample dilution buffers has been found to enable storage at –20°C for months (Chen *et al.*, 2016; see also Chapter 5).

2.7   General Conclusions

It is important to note that, depending on the sampling strategy, different results are to be expected. While only the depth-integrated or discrete water sampling can reveal true quantitative information which is also representative for a specific water body, shore samples might reveal maximum abundance of a certain toxigenic genotype. On the other hand plankton net samples can filter a tremendous amount of water volume and thus can much better inform on the presence/absence of a certain genotype.

It is emphasized that only quantitative sampling of the trophogenic layers can reveal a representative toxigenic genotype abundance which can be compared seasonally or between water bodies. The information on toxin gene abundance can be informative to monitoring tasks, for example for authorities monitoring the same water body over years. The abundance of a certain genotype could also be related to both biotic and abiotic environmental variables, and, once identified, certain environmental factors (e.g. nutrients) could be used to predict the occurrence of potential toxigenic species in a specific water body.

2.8   References

Chen, Q., Christiansen, G., Deng, L., and Kurmayer, R. (2016) Emergence of nontoxic mutants as revealed by single filament analysis in bloom-forming cyanobacteria of the genus *Planktothrix*, *BMC Microbiology*, **16** (1), 1–12.

Eiler, A., Drakare, S., Bertilsson, S., *et al.* (2013) Unveiling distribution patterns of freshwater phytoplankton by a next generation sequencing based approach, *PLOS ONE*, **8** (1).

Hayes, P.K., Barker, G.L.A., Batley, J., *et al.* (2002) Genetic diversity within populations of cyanobacteria assessed by analysis of single filaments, *Antonie van Leeuwenhoek*, **81** (1–4), 197–202.

ISO (1986) Agricultural food products – Layout for a standard method of sampling from a lot, ISO 7002, International Organization for Standardization, Geneva.

ISO (2006) Microbiology of food and animal feeding stuffs – Polymerase chain reaction (PCR) for the detection of food-borne pathogens – Requirements for sample preparation for qualitative detection, ISO 20837, International Organization for Standardization, Geneva.

Kurmayer, R., Christiansen, G., Fastner, J., and Börner, T. (2004) Abundance of active and inactive microcystin genotypes in populations of the toxic cyanobacterium *Planktothrix* spp., *Environmental Microbiology*, **6** (8), 831–841.

Lund, J.W.G., Kipling, C., and Le Cren, E.D. (1958) The inverted microscope method of estimating algal numbers and the statistical basis of estimations by counting, *Hydrobiologia*, **11**, 143–170, doi: 10.1007/BF00007865.

Meriluoto, J. and Codd, G.A. (eds.) (2005) *TOXIC: Cyanobacterial monitoring and cyanotoxin analysis*, Åbo Akademi University Press, Turku, Finland.

Park, J.J., Lechno-Yossef, S., Wolk, C.P., and Vieille, C. (2013) Cell-specific gene expression in *Anabaena variabilis* grown phototrophically, mixotrophically, and heterotrophically, *BMC Genomics*, **14** (1), 1–21.

Salmaso, N., Bernard, C., Humbert, J.F., *et al.* (2017) Basic guide to detection and monitoring of potentially toxic cyanobacteria, in J. Meriluoto, L. Spoof, and G.A. Codd (eds), *Handbook on Cyanobacterial Monitoring and Cyanotoxin Analysis*, John Wiley & Sons, Ltd, Chichester, UK.

SOP 2.1

Sampling and Filtration (DNA)

Rainer Kurmayer[1]* and Konstantinos Kormas[2]

[1]*Research Institute for Limnology, University of Innsbruck, Mondsee, Austria*
[2]*Department of Ichthyology and Aquatic Environment, University of Thessaly, Volos, Greece*

SOP 2.1.1   Introduction

This SOP describes all the steps to concentrate cyanobacteria cells suspended in the water column of standing and running waters. In general, a known amount of water is filtered by low-vacuum filtration onto filters which are then stored at $-20°C$ for subsequent DNA extraction. The filtration principle can be equally applied to harvesting cells from cultures under experimental conditions in the laboratory.

SOP 2.1.2   Experimental

SOP 2.1.2.1   Materials

- Canisters/carboys from 5 L to 10 L volume depending on the required water volume for filtration.
- Sample bottles (PET), new or cleaned (1 M HCl overnight and subsequently rinsed with sterile Milli-Q® water).
- Glass sample bottles for phytoplankton sampling and storage (from 50 mL to 250 mL).
- Cooling boxes (for sample transport).
- Lugol's solution for fixing cells (dissolve 10 g I_2 (pure iodine) and 20 g KI (potassium iodide) in 200 mL distilled water and 20 mL concentrated glacial acetic acid. Store in ground glass-stoppered, darkened bottle (Wetzel and Likens, 2000, p. 171).

---

*Corresponding author: rainer.kurmayer@uibk.ac.at

- Glass fiber filters for filtration (GF/C type) of 47 mm diameter, forceps to handle and to fold filters.
- 2.0 mL reaction tubes (Eppendorf), or cryotubes, may be more appropriate for liquid N_2.

SOP 2.1.2.2   Equipment

- Plankton net (e.g. 30 μm mesh size).
- Depth-integrated water sampler (e.g. van Dorn sampler, Ruttner sampler, Schindler Patalas sampler); alternatively, automatic integrating water sampler or flexible hose-pipe sampler.
- Quantum sensor for PAR (photosynthetically active radiation) measurement (facultative) and multiprobe to record temperature, conductivity, pH, and oxygen.
- Notebook to record metadata, sampling details, water volume filtered, etc.
- Low-vacuum filtration unit (equipped with a manometer to regulate the vacuum pressure).
- Freezer (–20°C) for storage of DNA samples.

SOP 2.1.3   Procedure

1. Take distinct samples using the water sampler from various depths and mix equal amounts of each depth into a canister (e.g. a 10 L canister is filled with 1 L each depth from zero to nine meter depth).
2. Take facultative vertical net hauls using a plankton net (30 μm mesh size, or any other appropriate size). Try to obtain several 100 mL of the sample showing at least some greenish color. If necessary, repeat the vertical net haul until coloration of the sample is achieved.
3. Keep the samples cool and protected from direct sunlight during transport to the laboratory.
4. In the field, from the mixed integrated sample, take an aliquot for cell microscopic counting. After filling, the glass bottles must be fixed immediately with the Lugol's fixative. Use about 1 mL of Lugol's iodine to preserve 100 mL of phytoplankton sample (or three drops for 20 mL). The resulting sample should be the color of whisky. Samples must be protected from light because that degrades Lugol's solution. Either use brown glass bottles or store in the dark. Further information on storage and other fixatives are reported in Catherine *et al.* (2017).
5. In the laboratory both the quantitative integrated sample and the (facultative) plankton net sample will be subdivided into aliquots: (1) sample for genotype quantification analysis per volume of filtered water; (2) sample for the (facultative) isolation of single live colonies/filaments and genotype quantification on an individual basis (see SOP 2.3) (Fig. 2.2).
6. For DNA analysis volume is filtered onto glass fiber (GF/C) filters until a green color on the filter is obtained (e.g. 50 mL from a highly eutrophic lake to 1000 mL or more from an oligotrophic lake). In parallel a second filter collecting a sample aliquot can be prepared and stored (as a reserve). The vacuum filtration pressure should not exceed 0.4 bar. For quantitative analysis the volume of water needs to be recorded and the filter folded inside up, put into a 2.0 mL reaction tube, and stored at –20°C.

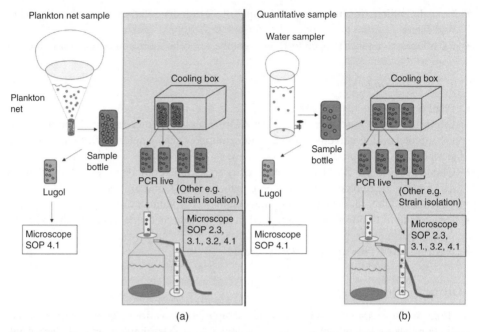

Figure 2.2 *Flowchart showing the processing of water samples for sample filtration from (A) samples collected using a plankton net, (B) depth-integrated samples collected through a water column (see SOP 2.1). Gray shading marks processing of samples in the laboratory.*

7. For live samples morphological (photographic) documentation samples should be processed immediately (see SOP 4.1). Similarly, for isolation of cyanobacteria (Chapter 3) samples should be incubated immediately.

SOP 2.1.4   Notes

Sometimes centrifugation is used for cell concentration. However, owing to gas vesicles, cyanobacteria often do not form a strong pellet even after strong centrifugation pressure. Thus centrifugation of field material is not recommended when the aim is to estimate toxigenic genotype abundance. Gas vesicles in cells can be effectively destroyed using the classic "Hammer, Cork, and Bottle" experiment (Klebahn, 1929). By hitting the cork in a bottle with a hammer a short strong pressure is generated that results in gas vesicle collapse, thus enabling sedimentation during centrifugation.

Glass fiber (GF/C) and membrane filters have been compared in DNA yield after DNA extraction (Schober and Kurmayer, 2006). In contrast to the more rigid nature of membrane filters, the GF/C filters disintegrate and form a distinct pellet after centrifugation subsequent to DNA extraction using phenol-chloroform-isoamyl alcohol. This allows for a distinct separation of the water phase from the phenol-chloroform phase during the DNA extraction procedure (see SOPs 5.1–5.3). In contrast, the phase separation is disturbed by the rigid pieces of membrane filters. Thus using the GF/C filters, separating the water phase (containing the DNA) from the phenolic phase (containing lipophilic molecules) is more convenient.

Note that filters might not fit in some tubes of commercial DNA extraction kits (and thus limit their application, e.g. using mechanical cell disruption by bead beating step). Thus, when using commercial kits the manufacturer's instruction manual should be consulted to explore which filters can be used.

GF/C filters have a porosity of approx. 1.2 µm. They retain most of the cyanobacterial cells, with the exclusion of the smaller individuals (e.g. picocyanobacteria; see also Chapter 4).

To control for DNA extraction efficiency several aliquots can be filtered and extracted in parallel. However, according to the authors' experience, the protocols given in Chapter 5 (see SOP 5.3) yield reproducible DNA amounts.

SOP 2.1.5   References

Catherine, A., Maloufi, S., Congestri, R., *et al.* (2017) Cyanobacterial samples: Preservation, enumeration and biovolume measurements, J. Meriluoto, L. Spoof, and G.A. Codd (eds), *Handbook on Cyanobacterial Monitoring and Cyanotoxin Analysis*, John Wiley & Sons, Ltd, Chichester, UK.

Klebahn, H. (1929) Über die Gasvakuolen der Cyanophyceen, *Verhandlungen der Internationalen Vereinigung für Theoretische und Angewandte Limnologie*, **4**, 408–414.

Schober, E. and Kurmayer, R. (2006) Evaluation of different DNA sampling techniques for the application of the real-time PCR method for the quantification of cyanobacteria in water, *Letters in Applied Microbiology*, **42**, 412–417.

Wetzel, R.G. and Likens, G.E. (2000) *Limnological Analyses*, 3rd edn, Springer-Verlag, New York.

SOP 2.2

Sampling of Benthic Cyanobacteria

Wim Vyverman* and Elie Verleyen

Laboratory of Protistology and Aquatic Ecology, Department of Biology, Ghent University, Gent, Belgium

SOP 2.2.1   Introduction

This SOP describes all the steps to sample benthic cyanobacteria from the bottom sediments of lakes and solid substrates in running waters. Particular points of attention concern

---

*Corresponding email: wim.vyverman@ugent.be

(a) sampling design given the typically high spatial heterogeneity in the distribution of benthic cyanobacteria and (b) the quantification of areal biomass or cell densities. In larger-scale surveys, pooled samples of biofilms containing cyanobacteria are often taken and preserved in Lugol solution or stored at $-20°C$ for subsequent DNA extraction.

SOP 2.2.2   Experimental

SOP 2.2.2.1   Materials

- Sample bottles (PET), new or cleaned (1 M HCl overnight).
- Cooling boxes (for sample transport).
- Lugol's solution for fixing cells (dissolve 10 g I_2 (pure iodine, toxic) and 20 g KI (potassium iodide) in 200 mL distilled water and 20 mL concentrated glacial acetic acid. Store in ground glass-stoppered, darkened bottle (Wetzel and Likens, 2000, p. 171).
- 2.0 mL reaction tubes (Eppendorf) or cryotubes.

SOP 2.2.2.2   Equipment

The type of sampling device will depend on the structure of the sample. A sterilized knife and spatula for mats in streams and the littoral zone of lakes can be used. For samples taken from deeper waters, a glew/gravity corer is used.

- Bathyscope or view box (vessel open to the top, with glass bottom to ensure full visibility of water bottom).
- Light meter (PAR) and multiprobe to record temperature, conductivity, pH, and oxygen.
- Notebook to record metadata, sampling details, etc.
- Camera to document sampling site and sampled locations.
- Freezer ($-20°C$) for storage of DNA samples.

SOP 2.2.3   Procedure

1. In shallow waters, inspect the sampling site with regard to benthic algae growth heterogeneity (e.g. by using a bathyscope).
2. On hard substrates, take samples from several different microsites at one sampling location in order to account for spatial heterogeneity. Depending on sampling objective and available resources, samples can be either kept separate or pooled. Samples from biofilms growing on rocks, concrete, or other firm substrates are most easily collected by scraping mats from a known surface area using a sterilized knife, brush, or spatula into sterile plastic screw-cap bottles.
3. In deep, soft sediments of rivers or lakes, a glew/gravity corer is used; in shallow water sediments a push corer is used. The top 0.5 cm of the sediment is carefully collected, using a sterilized spatula, into the sample bottle. Corer diameter is used to calculate the sampled area.
4. Keep the samples cool and protected from direct sunlight during transport to the laboratory.
5. In the laboratory the pooled samples will be subdivided into the following aliquots: (1) sample for genotype quantification, (2) sample for cell microscopic counting, and

(3) sample for the isolation of single live colonies/filaments and genotype quantification on an individual basis (facultatively see SOP 2.3).

6. Subsampling should be done after first determining the wet weight of the pooled sample (by draining excess water though a GF/F or GF/C filter). Weighted subsamples can then be used to approximately calculate the corresponding surface area to which they correspond.
7. For DNA analysis a known wet weight of biofilm material is put into a 2.0 mL reaction tube and stored at −20°C.
8. For cell microscopic counting, subsamples need to be conserved using Lugol's fixative. Use about 1 mL of Lugol's iodine to preserve 100 mL of phytoplankton sample (or three drops for 20 mL). Samples must be protected from light to prevent photodegradation of the preservative. Either use brown glass bottles or store in the dark.
9. For isolation of single cyanobacteria colonies/filaments, see SOP 2.3

SOP 2.2.4 Notes

For benthic algae water sampling (phytobenthos) in general guidelines used for the execution of the frame water directive of the EU are recommended, on a national level (e.g. Austrian Federal Ministry of Agriculture, Forestry, Environment and Water Management, 2015). Beside warning and safety instructions, these guidelines also cover the appropriate time of sampling and microscopic analysis.

Given the heterogeneous composition of biofilm communities, it is usually advisable to homogenize the pooled samples prior to subsampling. This can be done by gently fragmenting the biofilms using, for example, sterilized tweezers and subsequently mixing the biofilm fragments. Alternatively, separate pooled samples for DNA analysis can be taken following the procedure described above. This will avoid subsampling but might potentially result in discrepancies between microscopic examination of samples and results from DNA analysis.

Another potential error source is the introduction of PCR-inhibiting substances in the course of scratching off cells from stones or other substrates. In general, PCR is inhibited when amplifying DNA extracted from (lake) sediments because of humic acid content (Savichtcheva *et al.*, 2011). Thus purification of DNA extracted from sediments is required.

SOP 2.2.5 References

Austrian Federal Ministry of Agriculture, Forestry, Environment and Water Management (2015) *Guidance on the Monitoring of the Biological Quality Elements: Part A3: Phytobenthos*, Vienna, **94**, http://wisa.bmlfuw.gv.at/fachinformation/ngp/ngp-2015/hintergrund/methodik/bio_lf_2015.html, accessed 6 February 2017.

Savichtcheva, O., Debroas, D., Kurmayer, R., *et al.* (2011) Quantitative PCR enumeration of total/toxic *Planktothrix rubescens* and total cyanobacteria in preserved DNA isolated from lake sediments, *Applied and Environmental Microbiology*, **77** (24, 8744–8753.

Wetzel, R.G. and Likens, G.E. (2000) *Limnological Analyses*, 3rd edn, Springer-Verlag, New York.

SOP 2.3

Isolation of Single Cyanobacteria Colonies/Filaments

*Rainer Kurmayer**

Research Institute for Limnology, University of Innsbruck, Mondsee, Austria

SOP 2.3.1   Introduction

Most toxigenic cyanobacteria grow either as colonies (*Microcystis* sp.) or filaments (*Planktothrix* sp., *Dolichospermum* (*Anabaena*) spp., etc.). This SOP describes how to isolate single filaments or colonies of toxigenic cyanobacteria from the environment using a microscope. Protocols have been used by Kurmayer *et al.* (2002, 2004) and Chen *et al.* (2016).

SOP 2.3.2   Experimental

SOP 2.3.2.1   Materials

- Petri dishes (diameter *ca.* 10 cm).
- BG11 medium (Rippka, 1988).
  - Mechanical pipette (20–200 µL).
- Forceps (fine-pointed tweezers), dissecting needle, or glass thinned micropipettes (made from Pasteur pipettes; SOP 3.1).
- Reaction tubes (0.5 mL).

SOP 2.3.2.2   Equipment

Dissecting microscope (Stereomicroscope) equipped with illumination from below, with a plastic/glass cover that the Petri dishes can sit stably above the light source; continuous magnification (4–40 ×), and enough working space to handle forceps under 20–30 × magnification; ocular micrometer to determine the diameter/length of the isolated colony/filament.

---
*Corresponding author: rainer.kurmayer@uibk.ac.at

Facultatively: Microscope (100–400 × magnification) for more detailed morphological analysis (and imaging analysis).

SOP 2.3.3 Procedure

1. For colony/filament isolation (plankton net/integrated) samples are diluted with culture medium (e.g. BG11) (Rippka, 1988). Individual colonies/filaments are picked out randomly (e.g. from a drop containing approximately five specimens) by means of forceps, dissecting needle, or glass micropipettes under a binocular microscope and transferred into a fresh drop of BG11 medium.
2. Colonies/filaments are washed in BG11 medium to eliminate other colonies/filaments or cyanobacteria (e.g. three times by subsequent transfers between drops of BG11 medium).
3. Using the microscope the morphological characteristics (*sensu* Komárek and Anagnostidis, 1999, 2005; Komárek, 2013) and the colony/filament size (the largest diameter) and the cell size (at 400×) are determined. To avoid disintegration of the colony/filament no coverslip should be used.
4. The colonies/filaments are transferred into a reaction tube containing sterile Milli-Q® water (final sample volume 10 μL). The tubes are stored frozen at –20°C. Thawing and freezing several times improves disintegration of the colonies. Filaments/colonies are mechanically disrupted using a Sonifier® cell disruptor equipped with a microtip (see SOP 5.4).

SOP 2.3.4 Notes

To pick colonies/filaments randomly, samples are diluted with BG11. Then all specimens in one drop of sample are isolated. A good dilution is approximately five specimens in a drop. Some cyanobacteria morphotypes tend to disintegrate under dilution or stick to the surface of the Petri dish. These effects can be minimized by shortening the isolation time.

Forceps, needle, or glass micropipettes are recommended, as pipetting with plastic tips has the disadvantage that filaments/colonies are frequently lost because of sticking at the surface (inside the tip). Longer filaments (several hundreds of micrometers) have a stronger tendency to stick to the walls inside the pipette tip or inside the reaction tube, and without immersion into the PCR buffer will be lost. Using transparent reaction tubes the successful transfer of filaments/colonies can be monitored by inspecting the reaction tube under the microscope from outside.

SOP 2.3.5 References

Chen, Q., Christiansen, G., Deng, L., and Kurmayer, R. (2016) Emergence of nontoxic mutants as revealed by single filament analysis in bloom-forming cyanobacteria of the genus *Planktothrix*, *BMC Microbiology*, **16** (1), 1–12.

Komárek, J. (2013) *Cyanoprokaryota: Part 3: Heterocytous genera*, Springer-Verlag, Berlin.

Komárek, J. and Anagnostidis, K. (1999) *Cyanoprokaryota: 1. Teil Chroococcales*, Gustav Fischer Verlag, Jena, Germany.

Komárek, J. and Anagnostidis, K. (2005) *Cyanoprokaryota: 2. Teil: Oscillatoriales*, Elsevier GmbH, Spektrum Akademischer Verlag, Heidelberg, Germany.

Kurmayer, R., Christiansen, G., Fastner, J., and Börner, T. (2004) Abundance of active and inactive microcystin genotypes in populations of the toxic cyanobacterium *Planktothrix* spp., *Environmental Microbiology*, **6** (8), 831–841.

Kurmayer, R., Dittmann, E., Fastner, J., and Chorus, I. (2002) Diversity of microcystin genes within a population of the toxic cyanobacterium *Microcystis* spp. in Lake Wannsee (Berlin, Germany), *Microbial Ecology*, **43**, 107–118.

Rippka, R. (1988) Isolation and purification of cyanobacteria, *Methods in Enzymology*, **167**, 3–27.

SOP 2.4

Sampling Food Supplements

Vitor Ramos[1,2], Cristiana Moreira[1], and Vitor Vasconcelos[1,2]*

[1]*Interdisciplinary Centre of Marine and Environmental Research (CIIMAR/CIMAR), University of Porto, Matosinhos, Portugal*
[2]*Faculty of Sciences, University of Porto, Porto, Portugal*

SOP 2.4.1   Introduction

When sampling food supplements, the goal is to ensure that we are covering as much as possible the different cyanobacteria that may be present in the sample. Besides the sample coverage issue, sampling and subsampling procedures should also take into consideration the degree of contamination and the type of supplements and their process of fabrication, which in most cases will be unknown. During the manufacturing process of making supplements, (blue–green) algae are dewatered and dehydrated whole (frequently sundried, to decrease operating costs). After that, binders or fillers (e.g. when in the form of pills, tablets, etc.) or other additives can be added (Vichi *et al.*, 2012). These compounds may reduce DNA extraction yields and/or inhibit PCR reactions. In some cases, contamination is clearly perceptible through microscopic inspection, while in others it is not. Thus, the number and recommendable amount of each sample to be subsampled for DNA extraction may vary according to these different aspects.

This SOP is linked to SOP 5.6 and SOP 6.9, and all in combination are primarily intended to help out designing an effective plan for early warning prediction for food supplement

---

*Corresponding author: vmvascon@fc.up.pt

quality and food safety assurance. See also Chapter 11 (Section 11.3.5) for additional rationale for these approaches.

SOP 2.4.2   Experimental

SOP 2.4.2.1   Materials

- Food supplements.
- Forceps, spatula, and mortar and pestle; clean and sterile.
- Disposable Petri dishes, sterile.
- Microcentrifuge tubes (1.5 mL or 2 mL), sterile.
- Sterile Milli-Q® water.
- Freeze-dried biomass from a cyanotoxin producer strain (as a reference for each different potential toxin producer to be monitored).

SOP 2.4.2.2   Equipment

- Inverted microscope.
- Laboratory (semi-analytical) balance.

SOP 2.4.3   Procedure (Fig. 2.3)

1. Whenever possible (i.e. equipment available), samples should be visually inspected and subsamples selected through microscope (see specific comments for food supplements in introductory text of this chapter).
2. In a first step, several portions of the same sample flask (and lot) should be randomly taken. If applicable, pick out different parts from different pills or tablets. Make replicates (3–5) of 5–10 g each.

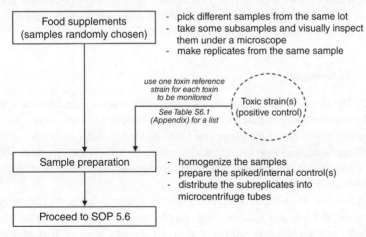

Figure 2.3   *Flow diagram outlining the steps for sampling and sample preparation of food supplements (see SOP 2.4).*

3. One of the replicates shall serve as an internal control for sample preparation and DNA extraction. To do so, add lyophilized biomass from a known cyanotoxin producer strain in a proportion of up to 1% of the total weight. Then, this control sample will undergo the same processing as the other replicates. In PCR testing (see SOP 6.9), the DNA retrieved from the spiked biomass will act as an exogenous reference standard (Coyne *et al.*, 2005).

4. After that, samples (i.e. replicates) should be homogenized after being grinded to a fine powder with a pestle and mortar. Replicates should be treated independently, so be aware of cross-contamination. Collect 2–3 subsamples (subreplicates) to a microcentrifuge tube for DNA extraction. Usually, 1–50 mg of sample is sufficient (the higher the dehydration, e.g. lyophilized material, the lesser amount is needed). Later, after DNA extraction (see SOP 5.6), these subreplicates can be pooled.

5. The remaining biomass can be used for analytical chemistry purposes (this also applies to the control sample).

6. To continue with further procedures for food supplements analysis see SOP 5.6.

SOP 2.4.4 Notes

Notice that the (blue–green) algae biomass is dried (or freeze-dried) and thus morphological features of the organisms (either the edible one or the contaminant(s)) may be slightly different from "normal." Nonetheless, contaminated (i.e. not pure) supplements are likely to be easily recognized under a microscope.

In the case of solid forms (i.e. not in powder), for a broader coverage it is advisable to grasp several different parts of the sample and increase the number of replicates and subreplicates rather than increase the mass weight to be processed in each DNA extraction. The latter will additionally enhance extraction yields.

Samples to be tested should not be limited to blue–green algae food supplements (BGAS) but also to supplements derived from eukaryotic microalgae, such as the widely merchandised *Chlorella* spp. (Görs *et al.*, 2010). Open ponds and raceways commonly used for the cultivation of these microalgae are as susceptible to contamination with cyanobacteria (and by other organisms) as those used for the cultivation of edible cyanobacteria.

SOP 2.4.5 References

Coyne, K.J., Handy, S.M., Demir, E., *et al.* (2005) Improved quantitative real-time PCR assays for enumeration of harmful algal species in field samples using an exogenous DNA reference standard, *Limnology and Oceanography: Methods*, **3**, 381–391.

Görs, M., Schumann, R., Hepperle, D., and Karsten, U. (2010) Quality analysis of commercial *Chlorella* products used as dietary supplement in human nutrition, *Journal of Applied Phycology*, **22** (3), 265–276.

Vichi, S., Lavorini, P., Funari, E., *et al.* (2012) Contamination by *Microcystis* and microcystins of blue–green algae food supplements (BGAS) on the Italian market and possible risk for the exposed population, *Food and Chemical Toxicology*, **50** (12), 4493–4499.

SOP 2.5

Sampling and Filtration (RNA)

Rainer Kurmayer[1]* and Guntram Christiansen[1,2]

[1]*Research Institute for Limnology, University of Innsbruck, Mondsee, Austria*
[2]*Miti Biosystems GmbH, Max F. Perutz Laboratories, Vienna, Austria*

SOP 2.5.1 Introduction

This SOP describes a protocol that can be used for filtration and preservation of cells from both field samples and culture experiments to be used for mRNA extraction (see also SOP 5.7). Protocols have been used by Christiansen *et al.* (2008) and Kurmayer *et al.* (2016).

SOP 2.5.2 Experimental

SOP 2.5.2.1 Materials

- Gloves.
- Glass fiber filters (25 mm diameter), type GF/C (e.g. Whatman).
- Forceps.
- Liquid nitrogen (–196°C, in a special transport container which is not sealed airtight).
- Safe-lock Eppendorf tubes (2.0 mL).

SOP 2.5.2.2 Equipment

- Hand vacuum-pump (equipped with a manometer to regulate the vacuum pressure).
- Bottle-top filter device.
- Freezer (–70°C) for storage of RNA samples.

---

*Corresponding author: rainer.kurmayer@uibk.ac.at

SOP 2.5.3   Procedure

1. Harvest cells quickly by filtration onto GF/C filters by means of vacuum filtration using low vacuum pressure (<0.4 bar). Typically filtrate the volume of 50–100 mL (OD_{800nm} = 0.1, 5 cm cuvette). Record the filtered volume for quantitative analysis.
2. Fold the filter inside up using forceps and transfer the filter into an Eppendorf tube and flash freeze the sample by liquid nitrogen. Store sample −20°C or long term −70°C.
3. For cell microscopic counting aliquots need to be conserved using the Lugol's fixative. Use about 1 mL of Lugol's iodine to preserve 100 mL of cell culture sample (or three drops for 20 mL). The resulting sample should be the color of whisky. Samples must be protected from light because that degrades Lugol's solution. Either use brown glass bottles or store in the dark.

SOP 2.5.4   Notes

Owing to the potential short turnover time of mRNA in general, the potential change in transcript amounts during processing of the samples must be minimized. This influence is particularly relevant during field sampling (e.g. exposing cells from deeper water to direct sunlight would likely influence the transcript amount of light-regulated genes). Using laboratory cultures, it is generally recommended to harvest cultures by filtration directly in the culture room.

The use of centrifugation for cell harvest (at 4°C) might be an alternative but is not recommended and potentially time consuming because of the buoyancy of most toxigenic cyanobacteria. In contrast the entire filtration step until freezing the sample can be performed in 1.5 min. As an alternative to liquid nitrogen a fixative like RNAlater® (Thermo Fisher Scientific) is used.

Using small diameter (25 mm) filters instead of larger diameter (47 mm) filters is recommended when submersing the filter (with the cells) in 1 mL of TRIzol® reagent (see SOP 5.7).

SOP 2.5.5   References

Christiansen, G., Molitor, C., Philmus, B., and Kurmayer, R. (2008) Nontoxic strains of cyanobacteria are the result of major gene deletion events induced by a transposable element, *Molecular Biology and Evolution*, **25** (8), 1695–1704.

Kurmayer, R., Deng, L., and Entfellner, E. (2016) Role of toxic and bioactive secondary metabolites in colonization and bloom formation by filamentous cyanobacteria *Planktothrix*, *Harmful Algae*, **54**, 69–86.

SOP 2.6

Sampling of Abiotic and Biotic Data and Recording Metadata

Elie Verleyen*, Maxime Sweetlove, Dagmar Obbels, and Wim Vyverman

Laboratory of Protistology and Aquatic Ecology, Department of Biology, Ghent University, Ghent, Belgium

SOP 2.6.1 Introduction

Information on environmental conditions is essential to understand the ecological factors affecting toxigenic gene occurrence and toxigenic gene regulation. Here we provide information on procedures for collecting and data management of abiotic and biotic background data. Moreover, the metadata containing information on the protocols used need to be recorded as well. It is beyond the scope to describe standard methods for recording basic limnological factors in detail and we will refer to standard literature instead.

SOP 2.6.2 Experimental

SOP 2.6.2.1 Materials

- Sample bottles (PET) of 500 mL, new or cleaned (1 M HCl overnight and subsequently rinsed with sterile Milli-Q® water).
- Membrane filters (pore size of 0.45 μm).
- Glass fiber (GF/F) filters.
- Lugol's fixative.

SOP 2.6.2.2 Equipment

- Probes for the measurement of pH, dissolved oxygen, temperature, specific conductance.
- Syringe and filter holder.

---

*Corresponding author: Elie.Verleyen@UGent.be

- Coolbox for sample storage.
- PAR-meter and/or Secchi disc.
- Zooplankton net (mesh size 60 μm).

SOP 2.6.3   Type of Metadata and Additional Biotic and Abiotic Data

Metadata contain important information on the biotic and abiotic data themselves and should provide a sound description of the procedures used during sampling and subsequent analyses. The metadata are generally stored in the first sheet of an Excel file (read-me) or form a separate component of a relational database.

SOP 2.6.3.1   General Metadata

For each sampling campaign the following general metadata and information should be collected:

- Short description of the aims of the sampling campaign and the research project.
- Unique number or a standardized code for the sampling campaign which can be used as part of the label (code) of the samples. This should contain a reference to the project, sampling site, station, and if applicable date of sampling.
- Contact details and name of the experimenter that took the samples (or another person responsible for the project).
- Time and date of sampling.
- List with an explanation of the abbreviations used.
- Short description of the rationale behind the different sample labels.
- Notes of observations done during sampling, such as the current weather conditions and those during the preceding days might also be relevant.

SOP 2.6.3.2   Sample Specific Data

Generally, in each sampling station several subsamples are taken for different types of analyses. In addition to the general metadata listed above, the following sample specific data need to be added to the database as well:

- Information on how the different types of samples were transported and stored (e.g. cooled, at $-20°C$, in liquid nitrogen, in fixative such as formaldehyde or buffer enabling DNA analysis).
- A reference to the protocols used for taking the sample and subsequent analyses. The manufacturer of the sampling equipment and analytical instruments is also included, as well as the accuracy and how and when the instruments were calibrated, if applicable.
- The amount of lake water filtered and the type of filter used.
- If applicable, the volume of sample taken.

SOP 2.6.3.3   Additional Biotic and Abiotic Data

Apart from samples for the molecular analysis of the cyanobacterial blooms, samples for the following biotic and abiotic variables can also be taken:

- Vertical depth profile and mixing depth (Wetzel and Likens, 2000, exercise 2), conductivity, pH, and dissolved oxygen are generally recorded simultaneously using the same probe.
- Vertical light profile (measured using a PAR quantum meter), Secchi disc depth.
- Phytoplankton and zooplankton composition. For zooplankton filter a known volume of water over the plankton net and fix the sample; for phytoplankton microscopic counting, cells need to be conserved, using the Lugol's fixative (Wetzel and Likens, 2000).
- Take samples for chlorophyll-pigment analysis. Filter a known volume of water over a GF/C glass fiber filter. The volume depends on the concentration of particles in the water. The filters need to be stored at $-20°C$ immediately to avoid pigment degradation. For the analysis of chlorophyll-a use ISO (1992a).
- Pigment composition with a special focus on cyanobacteria marker pigments (e.g. echinenone, zeaxanthin, myxoxanthophyll) as well as chlorophylls. These samples are typically filtered and the filter is immediately stored at $-20°C$ to prevent pigment degradation (e.g. Van Heukelem and Thomas, 2001).
- Concentration of cyanobacterial toxins (see Meriluoto *et al.*, 2017).
- The concentration of the major ions.
- Take a sample for the analysis of the macronutrient concentrations (TN, TP, SRP, NO_2-N, NO_3-N, NH_4-N). Completely fill a (500 mL) sampling tube. For dissolved nutrients, filter several hundreds of mL water over a 0.45 μm (membrane) filter and store at 4°C. Analyze the filtrate immediately or at the latest the following day. The concentration of the major ions' international standards are available and should be used: phosphorus (ISO, 1998), nitrate + nitrite (ISO, 1992b), ammonia (DIN, 1983).
- Habitat descriptive parameters (catchment size, lake morphometry, water residence time).

SOP 2.6.4   Notes

All physical probes need to be calibrated regularly following the manufacturer's instructions.

In general, for chemical analyses, at least 500 mL should be taken. This ensures a better representativeness of the sample, and a sufficient quantity in case of replicates, etc.

Apart from chlorophyll-a, other photosynthetic pigments can be analyzed following different protocols using high-performance liquid chromatography (e.g. Van Heukelem and Thomas, 2001; Descy, 2017).

Morphological (photographic) documentation of phytoplankton is facultative but can provide important taxonomic information (see SOP 4.1). Estimating the cell number using the microscope is facultative but enables the qPCR-based estimates of toxigenic genotype abundance to be validated (e.g. SOP 7.2–7.4). Samples are enumerated by means of an inverted microscope typically. Cells are counted in transects of a chamber filled with a few milliliters of water (see Lawton *et al.*, 1999 and Catherine *et al.*, 2017).

SOP 2.6.5 References

Catherine, A., Maloufi, S., Congestri, R., *et al.* (2017) Cyanobacterial samples: Preservation, enumeration and biovolume measurements, in J. Meriluoto, L. Spoof, and G.A. Codd (eds), *Handbook on Cyanobacterial Monitoring and Cyanotoxin Analysis*, John Wiley & Sons, Ltd, Chichester, UK.

Descy, J.P. (2017) Estimation of cyanobacteria biomass by marker pigment analysis, in J. Meriluoto, L. Spoof, and G.A. Codd (eds), *Handbook on Cyanobacterial Monitoring and Cyanotoxin Analysis*, John Wiley & Sons, Ltd, Chichester, UK.

DIN (1983) *German standard methods for the examination of water, waste water and sludge; cations (group E); determination of ammonia-nitrogen (E 5)*, Berlin, Verlag.

ISO (1992a) *Water Quality: Measurement of biochemical parameters: Spectrometric determination of the chlorophyll-a concentration*, ISO 10260, International Organization for Standardization, Geneva.

ISO (1992b) *Water Quality: Determination of dissolved fluoride, chloride, nitrite, orthophospate, bromide, nitrate and sulfate ions, using liquid chromatography of ions: Part 1: Method for water with low contamination*, ISO 10304-1, International Organization for Standardization, Geneva.

ISO (1998) *Water Quality: Spectrometric determination of phosphorus using ammonium molybdate*, ISO 6878, International Organization for Standardization, Geneva.

Lawton, L., Marsalek, B., Padisak, J., and Chorus, I. (1999) Determination of cyanobacteria in the laboratory, in Chorus, I and Bartram, J (eds.), *Toxic cyanobacteria in water. A guide to their public health consequences, monitoring and management*, WHO, E and FN Spon, London, 347–368.

Meriluoto, J., Spoof, L., and Codd, G.A. (eds.) (2017) *Handbook of Cyanobacterial Monitoring and Cyanotoxin Analysis*, John Wiley & Sons Ltd, Chichester, UK.

Van Heukelem, L. and Thomas, C.M. (2001) Computer-assisted high-performance liquid chromatography method development with applications to the isolation and analysis of phytoplankton pigments, *Journal of Chromatography A*, **910**, 31–49.

Wetzel, R.G. and Likens, G.E. (2000) *Limnological Analyses*, 3rd edn, Springer-Verlag, New York.

3

Isolation, Purification, and Cultivation of Toxigenic Cyanobacteria

Sigrid Haande[1]*, Iwona Jasser[2], Muriel Gugger[3], Camilla H.C. Hagman[1], Annick Wilmotte[4], and Andreas Ballot[1]

[1]*Norwegian Institute for Water Research, Oslo, Norway*
[2]*Department of Microbial Ecology and Environmental Biotechnology, Faculty of Biology, University of Warsaw, Warsaw, Poland*
[3]*Collection of Cyanobacteria, Institut Pasteur, Paris, France*
[4]*InBios – Center for Protein Engineering, University of Liège, Liège, Belgium*

3.1 Introduction

This chapter summarizes the most commonly used methods for the isolation, purification, and cultivation of toxigenic cyanobacteria. The aim is to give general advice on how to isolate and maintain clonal cyanobacterial cultures in order to use them in genetic studies. The traditional methods for the isolation of cyanobacteria into culture are well established and described and there are several excellent reviews with detailed information on culturing techniques (Rippka *et al.*, 1979, 1981; Waterbury and Stanier, 1981; Rippka, 1988; Waterbury and Willey, 1988; Castenholz and Waterbury, 1989; Andersen, 2005).

Clonal strains of toxigenic cyanobacteria isolated from a natural bloom and maintained in culture have provided the basic material for the development of molecular methods. Knowledge in how to isolate and maintain a clonal culture of any toxigenic cyanobacterium is

*Corresponding author: sigrid.haande@niva.no

Molecular Tools for the Detection and Quantification of Toxigenic Cyanobacteria, First Edition.
Edited by Rainer Kurmayer, Kaarina Sivonen, Annick Wilmotte and Nico Salmaso.

therefore a principal component for the many molecular approaches presented in this handbook. Laboratory strains are typically used in (1) studies where a molecular approach is used to infer the phylogenetic relationships and to perform a proper identification of cyanobacteria; (2) studies of the presence of genes coding for toxin synthesis, revealing the potential of toxin production of a certain clonal strain of cyanobacteria; and (3) design of reliable tools (e.g. primers, probes, assays) for detecting potentially toxin producing cyanobacteria by molecular methods in the environment.

Toxigenic strains of cyanobacteria are maintained in several international culture collections. Worldwide, 95 algal- and cyanobacterial culture collections maintain cyanobacteria (about 9000 strains), based on data for the year 2013 cited by the World Data Center for Microorganisms (WDCM) (www.wfcc.info), and endorsed, among others, by the World Federation for Culture Collections (WFCC).

3.2 Methodical Principles for Cyanobacterial Isolation, Purification, and Cultivation

3.2.1 Sampling, Identification, and Treatments Prior to the Isolation of Cyanobacteria

For the successful isolation of toxigenic cyanobacterial species, it is important to investigate the environmental conditions in the habitat from which the selected species will be isolated. These data allow the description of the natural habitat and facilitate considerations (e.g. salinity, major mineral concentrations, and typical trace metal contents) for the preparation of appropriate culture media and choice of incubation conditions (i.e. light, pH, and temperature). In general, the chances of successful isolation are highest, if these parameters are as close as possible to those encountered by the target organism in its natural environment.

Cyanobacteria from aquatic environments or growing submerged in a water environment are sampled in small sterile translucent containers, covered with the onsite water. When collecting water samples, the sampling devices should be clean and particularly not contaminated with cyanobacteria from other locations. Samples containing cyanobacteria should be stored in isotherm containers to be transported to the laboratory for processing. If travelling for longer periods of time, samples containing cyanobacteria should be stored in translucent containers that can occasionally be exposed to low light. In general, they can be maintained at temperatures of up to 25°C, unless the organisms were collected from cold environments (<18°C), in which case storage in isotherm containers is preferable. Exposure of the samples to extreme temperatures and light is to be avoided. For culturing of cyanobacteria not growing submerged in a water environment, the samples are best stored in small sterile plastic vials containing either a small drop of water or wetted paper to keep the organisms moist. They should not be immersed in water to minimize growth of contaminating microorganisms. General advice on sampling is also given in Chapter 2 and SOPs 2.1 and 2.2.

Prior to isolation, the collected material must be studied under the microscope in order to morphologically identify the present cyanobacterial species. Species identification is described in Chapter 4 and in SOP 4.1.

3.2.2 Traditional Techniques for the Isolation and Purification of Cyanobacteria

Isolation and purification techniques are initially used to obtain a viable culture of a single cyanobacterial species.

The first step of the isolation procedure is to obtain a clonal culture of the target organism, a particular cyanobacterial species, free of other cyanobacteria, micro-alga, fungi, or amoebae eventually present in the sample collected from the natural environment. The second step is to maintain the isolated organism as a strain, alive by continuous subculturing in an appropriate growth medium. The ultimate goal is the establishment of an axenic clonal strain, derived from a single cell, or filament, lacking any detectable contaminants, including other bacteria. The possible presence of lysogenic viruses (cyanophages), however, cannot be excluded, even in an axenic culture. However, a great number of strains are maintained in a non-axenic state, owing to the difficulties involved in obtaining and maintaining

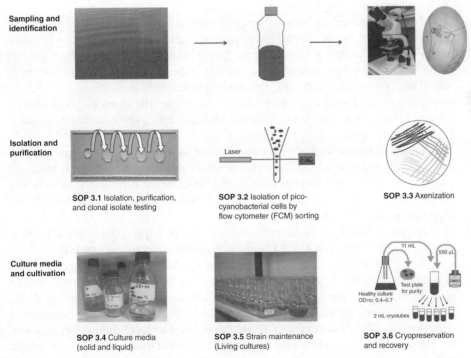

Figure 3.1 *Schematic overview of protocols (see SOPs 3.1–3.6) described in this chapter. (See color plate section for the color representation of this figure.)*

axenic isolates. Strains of most species can easily be isolated and cultured, whereas others are more difficult to grow and to maintain. The different protocols belonging to this chapter include isolation, purification, strain maintenance, media preparation, and cryopreservation (see SOPs 3.1–3.6; Fig. 3.1).

3.2.2.1   Isolation Procedures

3.2.2.1.1   Enrichment Cultures (Isolation by Liquid Enrichment)

In a preliminary step, enrichment cultures can be prepared to increase the number of cyanobacterial cells, colonies, or filaments. Cyanobacterial growth will be stimulated by adding nutrients (i.e. ammonia, nitrate, phosphorous, and major and minor metals) to the natural sample, or by transferring a small aliquot of the sample into a culture medium of choice. However, this procedure should be avoided, unless absolutely necessary, since it increases the risk of supporting the growth of, and thus selecting for, the fastest-growing organisms in the population, thus potentially overgrowing the desired cyanobacterial species.

3.2.2.1.2   Direct Isolation Techniques

1. One of the most applied methods is single cell/filament isolation by using micropipettes, and is performed using an inverted microscope or a stereomicroscope (and, more rarely, a compound microscope) for observing the selected cell or filament. Note, however, that single cell isolation by this method is generally only successful with relatively large unicellular cyanobacteria (>8 μm). This method is described in SOP 3.1.
2. Another common method is isolation with the use of agar. The agar is prepared with the selected medium in a concentration between 0.8 and 2.0%. The isolation technique is similar to that for the isolation of bacteria. This method is described in SOP 3.1.
3. The dilution technique is effective for abundant organisms in a natural sample. According to the approximate cell or filament concentration, the sample will be diluted until a small volume (e.g. one droplet) contains theoretically one cell. This volume is then transferred into a culture vessel with the selected medium and incubated at the preferred temperature and light conditions.
4. Flow cytometry (FCM) is increasingly used for isolation of picoplanktonic (cell diameter 0.2–2.0 μm) (Sieburth *et al.*, 1978), and ultraplanktonic cyanobacteria (cell diameter < 5.0 μm) (Campbell *et al.*, 1994). That is because cells smaller than 5 μm are difficult to isolate by traditional microbiological methods, as they often do not form colonies on agar, or by micropipettes. FCM allows for identification and separation of different cells in heterogeneous mixtures suspended in fluids according to their basic optical characteristics after excitation with a laser beam. For the identification of cyanobacteria, blue lasers are used, while a red-emitting diode is used to identify phycocyanin-rich cells. Two main types of flow sorting mechanisms are used: fluidic and droplet (Sieracki *et al.*, 2005). All kinds of cytometers offer three sorting modes: (1) sorting optimized for purity, when some cells are lost during the process; (2) enrich sorting, in which cells are sorted at the expense of purity, but the number of sorted cells is high; and (3) single-cell

mode, used when high count accuracy is required as, for example, for clonal cultures sorting (Davis, 2007). After sorting collected cells are grown in culture vessels with the selected medium. Both sorting methods proved to work for isolation of picocyanobacteria and larger cell sized cyanobacteria from environmental samples (Reckermann, 2000; Crosbie *et al.*, 2003; Cellamare *et al.*, 2010) and can provide axenic, single-cell isolation, but in case of fluidic sorting a subsequent dilution step is needed (Jasser *et al.*, 2010). The purity of the acquired cultures should be checked by the DGGE analysis or cloning. Flow sorting is described in SOP 3.2.

3.2.2.2   Purification Methods

The establishment of axenic cyanobacterial cultures needs additional steps besides the techniques mentioned above. A great variety of methods have been developed and tested for this purpose. This includes making use of gliding movement and phototaxis of motile cyanobacteria, treatments of cyanobacterial cultures with antifungal compounds and antibiotics, heat, ultraviolet light, or gamma irradiation (not advisable because of genetic mutations), and mechanical separations, such as streaking procedures on agar plates, micromanipulation, filtration, and equilibrium centrifugation (Vaara *et al.*, 1979; Sarchizian and Ardelean, 2010).

Although the ideal culture is derived from a single cell or filament without bacterial contamination, it has been observed that often cyanobacterial cultures perform better when grown with bacteria than under axenic conditions. Axenization is described in detail in SOP 3.3.

3.2.3   Culture Media Preparation

There is no medium that is universally suitable for isolation and cultivation of all types of cyanobacteria. To culture cyanobacteria from different environments, various culture media are required and have been developed.

Media for cyanobacterial cultures are generally composed of the component's macronutrients and trace elements. For some marine cyanobacteria, vitamin B12 is added. All three components are prepared as stock solutions in quantities of 100 mL to 1 L. Media can be made liquid or solidified by 0.8–2% agar. SOP 3.4 describes the general steps in preparing culture media and presents some common media and trace solution media used for culturing cyanobacteria. The media mainly differ in their content of combined nitrogen, phosphorus, type of chelating agents, iron, and trace metals. It is generally advised to use media with a relatively low trace metal content since higher concentrations of these compounds may be toxic to cyanobacteria. One example is copper, as even relatively low concentrations can inhibit growth of some cyanobacterial strains. Most cyanobacteria prefer neutral to alkaline environments (pH 7–10) and hence most cyanobacterial media are alkaline. The water used for the preparation of cyanobacterial media must be deionized and all organic compounds must be removed by double distillation or by passage through a Millipore (Merck Millipore) purifier (Milli-Q® water).

BG11, Z8, and WC medium are commonly used for the culturing of freshwater cyanobacteria and ASN-III is widely used for culturing marine cyanobacteria. These media are presented in SOP 3.4. For the culturing of N_2-fixing cyanobacteria, the N-source will be removed from the media. Preferably the sources of combined N-sources ($NaNO_3$ or KNO_3) should be replaced by its corresponding chlorides, although the latter is not always done. For the autoclaving of solid media, the agar is sterilized in water separately from the minerals to avoid the formation of toxic compounds. One of the vessels must be sufficiently big to mix the agar suspension with the mineral solutions.

3.2.4 Cultivation Conditions

3.2.4.1 Maintenance of Living Cultures

The culturing of cyanobacteria needs a perpetual maintenance under controlled environmental conditions. An inoculum from a stationary phase culture is transferred into fresh pre-sterilized liquid or solid medium under aseptic conditions. For liquid cyanobacterial cultures, an aliquot of the older culture is transferred to a culture vessel with fresh pre-sterilized medium, in a ratio of 1/20 (v/v). Cyanobacterial colonies from solid cultures are transferred using a sterile inoculation loop to a fresh solid medium into a tube or on a Petri dish. Light and temperature are important factors for the cultivation of cyanobacteria. Although cyanobacterial species have different light requirements, appropriate light conditions for culturing cyanobacteria are around 10–30 µmol photons m^2 s^{-1} with a light–dark cycle of 12–14 h of light and 10–12 h of darkness. Temperatures for cultivation should be selected on the basis of the temperature range of the natural environment where the sample derives from. For example, thermophilic strains can be maintained at 37°C, while the ones originating from deep alpine lakes will be kept at 18°C and the polar ones at 12°C. However, temperatures between 15 and 20°C are suitable for most cyanobacterial cultures (Waterbury, 1992; Lorenz *et al.*, 2005). Large-scale cultures should only be inoculated with heavily grown cultures at a dilution of approximately 1/20 (v/v). This implies to grow, for example, a 50 mL culture, and to transfer this one once it is well dense into a 1 L culture. SOP 3.5 describes the maintenance of living cultures both in liquid and solid media.

3.2.4.2 Cryopreservation

A special way to cultivate cyanobacteria or other living organisms is cryopreservation. It is the storage of a living organism at an ultra-low temperature where the biological activity is substantially slowed down. In cryopreservation, important parameters include the choice and concentration of the cryoprotectant, freezing rate, physiological status of the culture, and the thawing procedure (Rastoll *et al.*, 2013). Cyanobacteria (and eukaryotic organisms) are cryopreserved as large populations. The survival rate or the percent viability of a frozen cyanobacterial population is the part of the culture that remains viable and resumes normal physical activities after thawing. It has been shown that cyanobacteria can be frozen for long periods at relatively low temperatures (e.g. –20°C to –70°C) when proper precautions are taken during freezing and thawing (Day and Brand, 2005). However, it is believed that a population of living organisms which survives the freezing and thawing process can be stored for a long time when kept below a critical temperature of approximately –130°C. This low temperature is the so-called glass transition temperature where water does not change status and ice crystallization is precluded. Most cyanobacteria are susceptible to cell damages caused during freezing and thawing. This includes chilling injury, mechanical stress during extracellular or intracellular ice formation and osmotic stress (Day and Brand, 2005), which can be minimized when using special cryopreservation protocols. This includes a regulated rate of temperature change and the addition of a suitable cryopreservation agent prior to freezing. A permeating cryopreservation agent is a water-soluble compound, which displaces the intracellular water and reduces intracellular ice formation and osmotic stress in the cells (Day and Brand, 2005). Commonly used cryopreservation agents are methanol (MeOH), dimethyl sulfoxide (DMSO), and glycerol (Day and Brand, 2005; Wood *et al.*, 2008; Rastoll *et al.*, 2013). Cryopreservation is described in SOP 3.6.

3.3 General Conclusions

A large number of methods and protocols on isolation, purification, and culturing techniques for cyanobacteria are available. In this chapter, our goal is to present the most common methods for the successful isolation and maintenance of cyanobacterial cultures and to give practical advice on critical steps in the procedures.

3.4 References

Andersen, R.A. (ed.) (2005) *Algal Culturing Techniques*, Elsevier Academic Press, New York.

Campbell, L., Shapiro, L.P., and Haugen, E. (1994) Immunochemical characterization of eukaryotic ultraplankton from Atlantic and Pacific Ocean, *Journal of Plankton Research*, **16**, 35–51.

Castenholz, R.W. and Waterbury, J.B. (1989) Group I: Cyanobacteria, in *Bergey's Manual of Systematic Bacteriology*, N.R. Krieg and J.G. Holt (eds), Williams and Wilkins, Baltimore, **3**, 1710–1728.

Cellamare, M., Rolland, A., and Jacquet, S. (2010) Flow cytometry sorting of freshwater phytoplankton, *Journal of Applied Phycology*, **22**, 87–100.

Crosbie, N., Pöckl, M., and Weisse, T. (2003) Rapid establishment of clonal isolates of freshwater autotrophic picoplankton by single-cell and single-colony sorting, *Journal of Microbiological Methods*, **55**, 361–370.

Davis, D. (2007) Cell sorting by flow cytometry, in *Flow Cytometry: Principles and applications*, M.G. Macey (ed.), Humana Press Inc, Totowa, NJ.

Day, J.G. and Brand, J.J. (2005) Cryopreservation methods for maintaining microalgal cultures, in *Algal Culturing Techniques*, R.A. Andersen (ed.), Elsevier Academic Press, New York, 165–188.

Jasser, I., Karnkowska-Ishikawa, A., Kozłowska, E., *et al.* (2010) Isolation of picocyanobacteria from Great Mazurian Lake System: Comparison of two methods, *Polish Journal of Microbiology*, **59**, 21–31.

Lorenz, M., Friedl, T., and Day, J.G. (2005) Perpetual maintenance of actively metabolizing microalgal cultures, in *Algal Culturing Techniques*, R.A. Andersen (ed.), Elsevier Academic Press, New York, 145–156.

Rastoll, M.J., Ouahid, Y., Martín-Gordillo, F., *et al.* (2013) The development of a cryopreservation method suitable for a large cyanobacteria collection, *Journal of Applied Phycology*, **25**, 1483–1493.

Reckermann, M. (2000) Flow sorting in aquatic ecology, *Scientia Marina*, **64**, 235–246.

Rippka, R. (1988) Isolation and purification of cyanobacteria, *Methods in Enzymology*, **167**, 3–27.

Rippka, R., Deruelles, J., Waterbury, J.B., *et al.* (1979) Generic assignments, strain histories and properties of pure cultures of cyanobacteria, *Journal of General Microbiology*, **111**, 1–61.

Rippka, R., Waterbury, J.B., and Stanier, R.Y. (1981) Isolation and purification of cyanobacteria: Some general principles, in *The Prokaryotes*, Starr, MP, Stolp, H, Trüper, HG, Balows, A and Schlegel, HG (eds.), Springer-Verlag, Berlin, **1**, 212–220.

Sarchizian, I. and Ardelean, I.I. (2010) Improved lysozyme method to obtain cyanobacteria in axenic cultures, *Romanian Journal of Biology – Plant Biology*, **55**, 143–150.

Sieburth, J., Smetacek, V., and Lenz, J. (1978) Pelagic ecosystem structure: Heterotrophic compartments of the plankton and their relationship to plankton size fractions, *Limnology and Oceanography*, **23**, 1256–1263.

Sieracki, M., Poulton, N., and Crosbie, N. (2005) Automated isolation techniques for microalgae, in *Algal Culturing Techniques*, R.A. Andersen (ed.), Elsevier Academic Press, New York, 101–116.

Vaara, T., Vaara, M., and Niemela, S. (1979) Two improved methods for obtaining axenic cultures of cyanobacteria, *Applied and Environmental Microbiology*, **38**, 1011–1014.

Waterbury, J.B. (1992) The cyanobacteria: Isolation, purification and identification, in *The Prokaryotes*, A. Balows, H.G. Trüper, M. Dworkin (eds), 2nd edn, Springer-Verlag, New York, **2**, 2058–2078.

Waterbury, J.B. and Stanier, R.Y. (1981) Isolation and growth of cyanobacteria from marine and hypersaline environments, in *The Prokaryotes*, M.P. Starr, H. Stolp, H.G. Trüper, *et al.* (eds.), Springer-Verlag, Berlin, **1**, 221–223.

Waterbury, J.B. and Willey, J.M. (1988) Isolation and growth of marine planktonic cyanobacteria, *Methods in Enzymology*, **167**, 100–105.

Wood, S.A., Rhodes, L.L., Adams, S.L., *et al.* (2008) Maintenance of cyanotoxin production by cryopreserved cyanobacteria in the New Zealand culture collection, *New Zealand Journal of Marine and Freshwater Research*, **42**, 277–283.

SOP 3.1

Isolation, Purification, and Clonal Isolate Testing

Sigrid Haande*, Camilla H.C. Hagman, and Andreas Ballot

Norwegian Institute for Water Research, Oslo, Norway

SOP 3.1.1   Introduction

A multitude of isolation and purification techniques can be used in order to prepare clonal cyanobacterial cultures (e.g. Waterbury and Stanier, 1981; Rippka *et al.*, 1981; Rippka, 1988; Andersen and Kawachi, 2005). This SOP describes the standard method of single filament/cell isolation and purification by using (a) micropipettes and (b) solid media plating.

SOP 3.1.2   Experimental

SOP 3.1.2.1   Materials

SOP 3.1.2.1.1   Isolation Using Micropipettes

- Natural sample of cyanobacteria.
- Sterile long glass Pasteur pipettes or micro hematocrit capillary glass tubes for making micropipettes.
- Rubber bulb or silicon tubing with a mouthpiece (such as the end of 1 mL plastic syringe).
- Forceps.
- Sterile Petri dish/sterile glass slide.
- Small culture tubes, or a microtiter plate.
- 70% (v/v) ethanol.

*Corresponding author: sigrid.haande@niva.no

- Sterile water.
- Culture medium of choice.

SOP 3.1.2.1.2   Isolation Using Solid Media with Agar or Agarose

- Natural sample of cyanobacteria.
- Sterile inoculation loop.
- Sterile long Pasteur pipettes or micro hematocrit capillary glass tubes for making micropipettes.
- Rubber bulb or silicon tubing with a mouthpiece (such as the end of 1 mL plastic syringe).
- Sterile Petri dish/sterile glass slide.
- Agar or agarose plate made with culture medium of choice.
- 70% (v/v) ethanol.
- Sterile water.
- Culture medium of choice.

SOP 3.1.2.2   Equipment

SOP 3.1.2.2.1   Isolation Using Micropipettes or Isolation Using Solid Media

- Stereomicroscope, microscope, or inverted microscope.
- A sterile lab bench area.
- Bunsen burner or flame.

SOP 3.1.3   Procedure

For the direct isolation techniques described below, it is important that the field samples are examined under the microscope on arrival at the laboratory in order to evaluate their content of cyanobacterial species and other prokaryotic and eukaryotic organisms (Fig. 3.2).

SOP 3.1.3.1   Isolation Using Micropipettes

1. Micropipettes are made from sterile long Pasteur pipettes or micro hematocrit capillary glass tubes. The glass is heated in a Bunsen flame, while holding their ends between one's right and left hand (forceps can also be used). As soon as the part exposed to the hottest region in the flame starts softening (indicated by a slight bending of the glass), they are removed from the heat source. Immediate pulling with both hands will extend the softened region to yield a thinner capillary with wider parts on either end. The latter is then broken at one of the tapering ends with fine forceps (passed in 70% ethanol and flamed for sterilization) to obtain the desired dimensions (thickness and length). The optimal thickness will have to be determined experimentally, and may vary with the size of the organism to be isolated. The recommended length for use of micropipettes made from Pasteur pipettes is about 1–2 cm. If made too long, the capillary of the micropipette will be too flexible (indicated by waving movements in the field of observation), which will hinder the precision of targeting the desired organism.

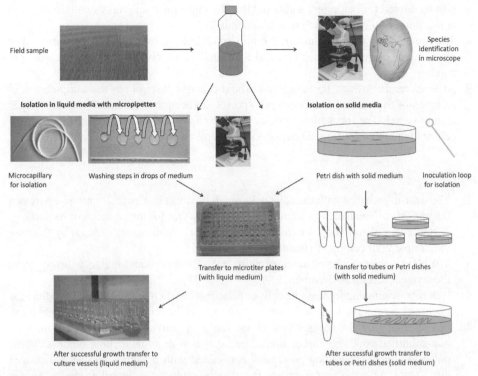

Field sample

Species identification in microscope

Isolation in liquid media with micropipettes

Isolation on solid media

Microcapillary for isolation

Washing steps in drops of medium

Petri dish with solid medium

Inoculation loop for isolation

Transfer to microtiter plates (with liquid medium)

Transfer to tubes or Petri dishes (with solid medium)

After successful growth transfer to culture vessels (liquid medium)

After successful growth transfer to tubes or Petri dishes (solid medium)

Figure 3.2 *Flowchart showing the methods of isolation presented in SOP 3.1. (See color plate section for the color representation of this figure.)*

2. The natural/field sample of cyanobacteria may be very concentrated (as often is the case for bloom-forming cyanobacteria or mat-forming cyanobacteria), and it is best to dilute it with sterile water or culture medium (if extensive dilution is necessary) to find only a low number of the desired organism (preferably $< 10^3$ mL^{-1}) in the field of observation.
3. Place several droplets of sterile medium of choice on either an empty sterile Petri dish or the surface of an agar plate (if using a stereomicroscope) or on a sterile glass slide (if using a microscope).
4. Attach a rubber bulb or silicon tubing with a mouthpiece (such as the end of a 1 mL plastic syringe) to the micropipette and a cyanobacterial cell/colony/filament will be isolated by the capillary action from the sample.
5. The selected target organism is expelled from the micropipette into a droplet of sterile medium by gentle pressure using the rubber bulb, or by gently blowing into the tube extension.
6. The selected cell or filament is then picked again with a new micropipette from the first droplet and transferred to a second droplet. This procedure is repeated until one is sure that only the desired cyanobacterial cell or filament is contained in the final droplet of medium.
7. Given that with each step of isolation unwanted smaller cyanobacteria, as well as bacteria, that may not be detectable under the stereomicroscope (or inverted microscope)

will be diluted further, a few additional transfers into fresh droplets of culture medium at this point of the procedure are recommended.

8. The organism is finally transferred into a small culture tube, or well of a microtiter plate, containing culture medium, and incubated at the chosen temperature and light conditions.

9. After successful growth, the isolate has to be checked by microscopy to assure successful isolation of the desired cyanobacterial species, and can then be transferred to a larger culture vessel. The incubation time necessary to achieve growth starting from a single cell or filament varies from a few weeks to several months, depending on the species.

SOP 3.1.3.2   Isolation Using Agar

1. The agar is prepared with the selected growth medium in a concentration of between 0.8 and 2.0%. The isolation technique is similar to that for the isolation of bacteria.

2. A sterile inoculation loop is used to spread a small sample across the agar by thinning the sample until cells or filaments are separated.

3. The plate is then incubated at temperature and light conditions similar to the one from the original natural environment.

4a. Colonies originating from single cells or filaments can then be transferred from the agar plate with a bacterial loop or micropipette into an isolation vessel with sterile medium.

4b. For species that do not grow well on the agar surface but embedded in agar, non-solidified cool agar can be used. Special agar with a low melting point is cooled down to almost the gelling point and then mixed with a natural sample and poured into a plate. After the agar solidifies, the plate is incubated as described above. Agar is known to contain impurities which act as agents against cyanobacteria. For sensitive cyanobacteria, those agents have to be removed by washing procedures.

For isolating planktonic cyanobacteria, agarose will be used instead of agar, and the concentration will be lowered to 0.8–1%.

For isolating benthic cyanobacteria and several mat-forming cyanobacteria, solid media plating will be preferred. Exposition of the inoculated plates to low light (e.g. < 60 μmol m^{-2} s^{-1}) will be important to consider.

SOP 3.1.4   Notes

All operations should be carried out with sterile instruments, water, and media.

If using a Bunsen burner/flame to sterilize equipment (e.g. bacterial loops) or creating a capillary Pasteur pipette, make sure the equipment is properly cooled before using it. If necessary, cool it in sterile medium or water before touching the cells. Isolating cyanobacteria by micropipette has the risk that obtained cultures are not clonal or contaminated by bacteria and other less apparent cyanobacteria. Thus, it is advisable to purify strains isolated by micropipette by streaking on agar.

SOP 3.1.5   References

Andersen, R.A. and Kawachi, M. (2005) Traditional microalgae isolation techniques, in *Algal Culturing Techniques*, R.A. Andersen (ed.), Elsevier Academic Press, New York, 83–100.

Rippka, R., Waterbury, J.B., and Stanier, R.Y. (1981) Isolation and purification of cyanobacteria: Some general principles, in *The Prokaryotes*, M.P. Starr, H. Stolp, H.G. Trüper, *et al.* (eds.), Springer-Verlag, Berlin, **1**, 212–220.

Rippka, R. (1988) Isolation and purification of cyanobacteria, *Methods in Enzymology*, **167**, 3–27.

Waterbury, J.B. and Stanier, R.Y. (1981) Isolation and growth of cyanobacteria from marine and hypersaline environments, in *The Prokaryotes*, M.P. Starr, H. Stolp, H.G. Trüper, *et al.* (eds), Springer-Verlag, Berlin, **1**, 221–223.

SOP 3.2

Isolation of Picocyanobacterial Cells by Flow Cytometer (FCM) Sorting

Ewa Kozłowska[1] *and Iwona Jasser*[2]*

[1]*Department of Immunology, Faculty of Biology, University of Warsaw, Warsaw, Poland*
[2]*Department of Microbial Ecology and Environmental Biotechnology, Faculty of Biology, University of Warsaw, Warsaw, Poland*

SOP 3.2.1 Introduction

Flow cytometer (FCM) sorting provides a good alternative to classic isolation methods, especially in case of picoplankton and ultraplankton (Reckermann, 2000). The differentiation among cells occurs according to their basic optical characteristics such as light scatter and multicolor fluorescence emitted by cells excited with a laser beam (Campbell, 2001). Flow cytometers if equipped with sorting tools are used to isolate various types of cells from mixed populations from natural samples. Two types of flow sorting mechanisms (fluidic and droplet) can be used for such isolation. Below we present exemplary procedures used for fluidic (FACSCalibur; BD Biosciences, USA) and droplet (FACSAria; BD Biosciences, USA) sorting.

*Corresponding author: jasser.iwona@biol.uw.edu.pl

SOP 3.2.2   Experimental

SOP 3.2.2.1   Materials

- Sample of water with cells, which are to be sorted.
- Reference strain of PE picoplankton, standard for compensation of FL2 channel Reference strain of PC picoplankton standard for compensation of FL3 or FL4 channel.
- 12 × 75 mm BD Falcon tubes (for FACSCalibur).
- 50 mL conical BD Falcon tubes (for FACSCalibur) or 15 mL conical BD Falcon tubes or 96 well plate for FACSAria.
- Nitex mesh 53 micrometer pore size or 35 micrometer Nitex-capped tubes (BD Falcon) – for larger cyanobacteria.
- 10 micrometer or even 3 micrometer pore-size polycarbonate membranes for picocyanobacteria.
- Multi-well culture plates for culturing the cells, e.g. 96 wells.
- Culture medium, e.g. WC, BG11.
- Sheath fluid – 0.1% PBS.
- Distilled water.

SOP 3.2.2.2   Equipment

- FACSCalibur Becton Dickinson (Fig. 3.3) or FACSAria Becton Dickinson. For identification of cyanobacteria populations usually blue lasers (488 nm excitation wavelengths).
- Additionally red-emitting diode (635 nm excitation) may be used to identify phycocyanin-rich cells.
- Centrifuge, e.g. Eppendorf 5810R with rotor for 50 mL tubes.
- Autoclave.
- Sterile filtration units (e.g. Sartorius) with filters to remove larger cells.
- Culture chamber with adjustable light intensities and temperature.

SOP 3.2.3   Procedure

SOP 3.2.3.1   Preparation of Cytometer and Sorting Unit

1. For optimal performance, switch on the FACSCalibur or FACSAria and warm up the lasers at least 15 min before sorting, then switch on the computer.
2. Prime the sort line to ensure that the sort lines are free of clogs and air bubbles. Install a tube.

SOP 3.2.3.2   Setup of cytometer settings (Shapiro, 1995)

1. Prepare the protocol for data acquisition. All adjustments to the FACSCalibur can be made in CELLQuest and on FACSAria in Diva software.
2. Prepare compensation using reference strains. In case of picocyanobacteria sorting *Synechococcus rubescens* (SAG 3.81) for PE picocyanobacteria (FL2) and PCC 6307 or PCC 7918 for PC picocyanobacteria (FL3) can be used.
3. Run your sample and save data to prepare gating for sorting (Fig. 3.4).

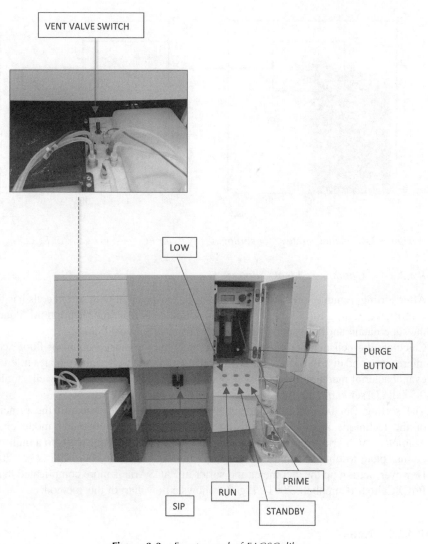

Figure 3.3 *Front panel of FACSCalibur.*

SOP 3.2.3.3 Sorting

1. In case of FACSCalibur, adjust the density of your samples for desired cells sorting at a rate of up to 300 per second with the maximum event rate not exceeding 2000 events per second at low flow (12 μL min^{-1}). Flow of 2000 events per second at low flow needs a sample concentration of 10^{7}cells mL^{-1} (FACSCalibur, 1996). This step is not necessary in case of FACSAria.
2. Prepare the sorting parameters:
   a. Sort gates: G1 or G2
   b. Allow continuous sorting or set up the desire number of cell to be sorted
   c. Start sorting.

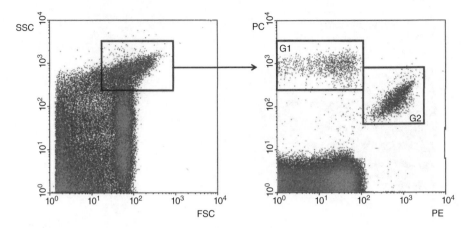

Figure 3.4   *Gating strategy for sorting. G1 gate for PC cells; G2 gate for PE cells.*

SOP 3.2.3.4   Obtaining Clonal Cultures

1. After sorting, remove sorting fluid. To do so, the suspension with sorted cells has to be centrifuged 3500 g for 20 min, to separate sorted cells from the sheath fluid. Pour out the supernatant and re-suspend the cells in chosen culture medium.
2. Calculate the cell number either on the basis of the cytometer count or fluorescence microscope and distribute the cells with a pipette to multi-wells culture plates filled with cyanobacterial medium in such a way that the obtained cell number per well should be 0.5 cell (Jasser *et al.*, 2010).
3. The sorting can be performed using any sorting cytometer because the principles of the technique are the same. Using droplet sorting with one-cell mode on, for example, FACSAria will give an advantage of sorting target single cells to a multi-well culture plate to obtain clonal cultures in a one-step procedure (Crosbie *et al.*, 2003). However, setting up the cytometer and sorter in FACSAria is more complicated than in FACSCalibur (Campbell, 2001). Fig. 3.5 shows a flowchart of this method.

SOP 3.2.4   Notes

Forward scatter (FSC) is characterized mainly by cell size, while side scatter (SSC) depends on cell ultrastructure and refracting index.

To avoid clogging up the tubing system, samples should be pre-filtered through Nitex mesh (53 micrometer Niltex when using a 76 micrometer jet-in-air nozzle in droplet sorting or 35 micrometer Nitex-capped tubes (Falcon, Campbell)) or in sterile filtration units by gravity filtration through 10 micrometer pore size membranes.

In the case of FACSCalibur, priming of sorting unit could be done with distilled water, for more efficient dissolving of possible salt crystal residues in the tubing system and better performance.

To obtain viable cells the sheath fluid has to be well suited for cell growth, the whole sheath and tubing system has to be washed and free from any toxic substances (used, for example, to sterilize the system).

Flow sorting

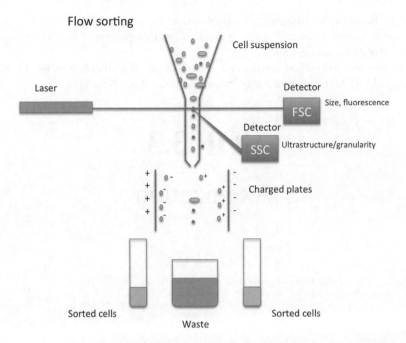

Figure 3.5 *Flowchart showing the method for flow cytometer sorting presented in SOP 3.2.*

To achieve the highest purity in fluidic sorting (FACSCalibur) the "exclusion mode" of the FCM sorting function should be used. This means that the sort occurs only when a target cell is identified in the sort envelope and no non-target cells are in the envelope. In the case of droplet sorting (FACSAria) the "purity" or "one-cell" mode should be followed, when clonal cultures are targeted after sorting.

From our experience the fluidic sorting is more economical as well as easier and less time- and labor-consuming during the cleaning procedure compared to droplet sorting, especially in the case of sorting environmental samples.

The WC medium proved to be the best for isolation of picocyanobacteria, although for longer cultivation a BG11 medium is more suitable (Jasser *et al.*, 2010).

SOP 3.2.5 References

Campbell, L. (2001) Flow cytometric analysis of autotrophic picoplankton, in *Methods in Microbiology: Marine microbial ecology*, J.H. Paul (ed.), Academic Press, New York, **30**, 317–343.

Crosbie, N., Pöckl, M., and Weisse, T. (2003) Rapid establishment of clonal isolates of freshwater autotrophic picoplankton by single-cell and single-colony sorting, *Journal of Microbiological Methods*, **55**, 361–370.

FACSCalibur (1996) *FACSCalibur System User's Guide*, Becton Dickinson, **164**, http://rd.mc.ntu.edu.tw/bomrd/cytometry/upfile/files/201481151817.pdf, accessed 6 February 2017.

Jasser, I., Karnkowska-Ishikawa, A., Kozłowska, E., *et al.* (2010) Isolation of pico-cyanobacteria from Great Mazurian Lake System: Comparison of two methods, *Polish Journal of Microbiology*, **59**, 21–31.

Reckermann, M. (2000) Flow sorting in aquatic ecology, *Scientia Marina*, **64**, 235–46.

Shapiro, H.M. (1995) *Practical Flow Cytometry*, Wiley-Liss, New York.

SOP 3.3

Axenization

*Muriel Gugger**

Collection of Cyanobacteria, Institut Pasteur, Paris, France

SOP 3.3.1   Introduction

Most axenic cultures of cyanobacteria have been isolated on solid media by a standard plating procedure useful for immotile and unicellular cyanobacteria like it can be done for other bacterial isolation. Axenization of cyanobacteria is undertaken on a natural sample. The aim is to obtain a culture of the targeted cyanobacterial isolate exempted of any other bacteria.

SOP 3.3.2   Experimental

SOP 3.3.2.1   Materials

- Either a fresh natural sample should be taken or axenization could also be performed on an already clonal organism, but it should be a young culture. If not, the number of cyanobacterial cells in the culture would be greatly outnumbered by the other bacteria.
- Solid media in agar or agarose plates. A large variety of media have been designed for the cultivation of cyanobacteria (see Rippka, 1988, for review). A way of preparing the solid media from the agar and the mineral solution is well presented in Allen (1968).
- A test plate is a solid growth medium supplemented with glucose (0.2%) and casamino acids (0.02%). The test plates once inoculated should be incubated in the dark for 2–3 days at the room temperature prior to microscopic examination using phase contrast objectives and oil immersion.

*Corresponding author: muriel.gugger@pasteur.fr

SOP 3.3.2.2   Equipment

- A sterile lab bench area where you can work with a flame or Bunsen burner.
- For microscopic observations, a microscope and a stereomicroscope or an inverted microscope.
- For spreading and streaking the colonies, wire loops or spatulas that can be sterilized in a flame will be used. Some examples of wire loops and spatulas are shown in Fig. 3.6.

For picking the colonies, micropipettes are made from sterile Pasteur pipettes (see SOP 3.1).

SOP 3.3.3   Procedure

The sample or culture containing the clonal cyanobacteria and the cohabiting bacteria should be first subjected to a careful microscopic observation, (1) to check the good state of the cyanobacterium to render axenic and (2) to estimate the degree of contamination. A culture in bad condition and with a consortium of bacteria has little chance to become a culture containing healthy cyanobacterial cells without other bacteria via this method. If the culture contains more than one cyanobacterium (e.g. different cyanobacterial filaments or a filament with unicellular cyanobacterial cells), the best thing to do is first to isolate the wanted cyanobacterium from the other cyanobacteria through the isolation procedure (see SOP 3.1).

1. A drop of the impure cyanobacterial culture is deposited onto an agar plate and streaked through the plate (Fig. 3.7). For spreading and streaking the colonies, a wire loop or a spatula sterilized with flame is cooled down and tested on the side of the agar plate before touching the bacteria.
2. The spread and streak is done as illustrated in Fig. 3.7. First, the drop is spread over Area 1 using the loop or spatula only touching the surface of the plate (do not dig into the agar). Second, the streaks are made on the rest of the surface plate, between each the loop or spatula is sterilized with flame. The first streak is made over Area 2 by passing the tip of the spatula through spreading Area 1, holding the tip of the spatula between each streak. A second streak is made over Area 3 by passing the tip of the

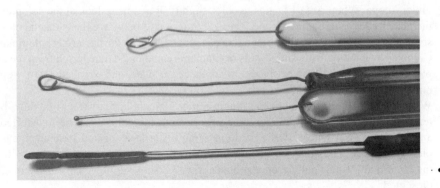

Figure 3.6   *Wire loops and spatulas that can be sterilized in a flame.*

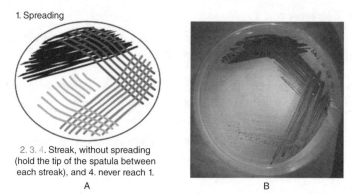

1. Spreading

2. 3. 4. Streak, without spreading (hold the tip of the spatula between each streak), and 4. never reach 1.

A   B

Figure 3.7 *Spread and streak on an agar plate a drop of a young impure cyanobacterial culture. (A) Principle of spread and streak of an impure colony on an agar plate within four areas. Area 1: For spreading on 1/3 of the plate; Area 2: To streak from Area 1; Area 3: To streak through Area 2; Area 4: To streak through Area 3 without reaching Area 1. (B) Example of a well-grown axenization plate, from the most concentrated Area 1 to the most diluted one (Area 4). Other examples for spread and streak on an agar plate are presented in Rippka (1988).*

spatula through Area 2. A third streak is made over Area 4, first by passing the tip of the spatula through Area 3 and, second, by passing the tip of the spatula only on the agar surface.

3. Then, the lid of the plate is closed with a band of parafilm, and the plate is placed upside down in the growth chamber, close to the original culture.

4. Individual colonies will appear in Areas 2, 3, or 4, depending on the initial concentration or cell number in Area 1. In our example, Area 4 is where the isolated cyanobacterial colonies have more chance to be separated from the isolated bacterial colonies. As the bacteria grow faster than the cyanobacteria, a regular observation of the plate under a stereomicroscope or an inverted microscope is necessary. The colonies will be visible in the microscope in Area 4 but barely distinguished by direct observation of the plate, while Area 1 will begin to show some cyanobacterial growth.

5. Look at the isolating plate regularly, notably using an inverted microscope or a stereomicroscope. When Area 1 begins to color slightly, it is time to look at the colonies on the plate under a stereomicroscope. Areas 2, 3, or 4 will be where the isolated colonies should be seen, and where the potential one is separated from the bacterial contaminants. Colonies originating from single cells or filaments can then be transferred from the agar plate with a bacterial loop or spatula to another agar plate, and streaked again. Repeat until no more bacterial colonies can be seen on the plate.

6. When the colony originating from a single cell or filament appears free from other bacterial colonies, as seen by microscopic observation, the colony can be transferred to an isolation vessel with a sterile medium using a sterile elongated Pasteur pipette to pick the colony. The purity of the new culture will be tested at next transfer by placing an aliquot of new culture on test plates. Fig. 3.8 shows a flowchart for the method presented.

Environmental
sample

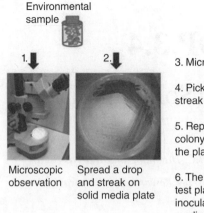

3. Microscopic observation of the plate

4. Pick a single colony to spread and streak on a new solid media plate

5. Repeat 3. and 4. until you get a single colony and apparently nothing else on the plate

Microscopic observation

Spread a drop and streak on solid media plate

6. The purity of the culture is tested on a test plate and one single colony is used to inoculate a new culture into a liquid medium

Figure 3.8 *Flowchart showing the method for axenization presented in SOP 3.3.*

SOP 3.3.4 Notes

This protocol is not suitable for motile filaments.

This method will need several transfers if the cyanobacterial culture is highly contaminated. It is better to work with a fresh sample, a young culture, or a fresh transfer.

Other axenization methods have been attempted notably for planktonic cyanobacteria containing numerous gas vacuoles like using agarose of low melting temperature for *Microcystis* (Shirai *et al.*, 1989) and using antibiotics for *Spirulina* (Thacker *et al.*, 1994). Another example to produce axenic culture for soil-borne and endophytic cyanobacteria is described in Bowyer and Skerman (1968). More about the purification techniques and the steps to scaling up the recently axenized culture is available in Rippka *et al.* (1981).

The planktonic strains prefer agarose to agar.

SOP 3.3.5 References

Allen, M.M. (1968) Simple conditions for growth of unicellular blue-green algae on plates, *Journal of Phycology*, **4**, 1–4.

Bowyer, J.W. and Skerman, V.B.D. (1968) Production of axenic cultures of soil-borne and endophytic blue-green-algae, *Journal of General Microbiology*, **54**, 299–306.

Rippka, R. (1988) Isolation and purification of cyanobacteria, *Methods in Enzymology*, **167**, 3–27.

Rippka, R., Waterbury, J.B. and Stanier, R.Y. (1981) Isolation and purification of cyanobacteria: Some general principles, in *The Prokaryotes*, M.P. Starr, H. Stolp, H.G. Trüper, *et al.* (eds), Springer-Verlag, Berlin, **1**, 212–220.

Shirai, M., Matumaru, K., Ohotake, A., *et al.* (1989) Development of a solid medium for growth and isolation of axenic *Microcystis* strains (cyanobacteria), *Applied and Environmental Microbiology*, **55**, 2569–2571.

Thacker, S.P., Kothari, R.M., and Ramamurthy, V. (1994) Obtaining axenic cultures of filamentous cyanobacterium *Spirulina*, *BioTechniques*, **16**, 216–217.

SOP 3.4

Culture Media (Solid and Liquid)

Sigrid Haande*, Camilla H.C. Hagman, and Andreas Ballot

Norwegian Institute for Water Research, Oslo, Norway

SOP 3.4.1   Introduction

In order to culture cyanobacteria from different environments (i.e. fresh, marine, saline water) various culture media are required and have been developed. This SOP describes the general procedure for preparing a culture media and presents recipes for some selected cyanobacterial growth media that are most commonly used in major cyanobacterial culture collections.

SOP 3.4.2   Experimental

SOP 3.4.2.1   Materials

- Heat-resistant glassware like Pyrex (Corning Co. Ltd.) or Duran (Schott Co. LTD), 1 L, 0.5 L bottles.
- Solid media plates or slant of solid media in a tube (agar or agarose).
- Distilled water/Deionized water (Milli-Q® water).
- Chemicals (specified in Tables 3.1 and 3.2 for the selected culture media).
- Agar, which is composed of agarose and agaropectin, is used to prepare solidified culture media. For this purpose 0.8–2% agar or agarose is used.
- Plastic Petri dishes or glass tubes.
- Plastic boxes.

SOP 3.4.2.2   Equipment

- Autoclave.
- Once or double distilling apparatus with a Pyrex or quartz glass condenser/Millipore (Merck Millipore) purifier.
- Laminar flow hood.

---

*Corresponding author:sigrid.haande@niva.no

Table 3.1   *Recipes for commonly used cyanobacterial media*

Ingredient (g liter^{-1}/mM)	Medium designation			
	BG11[1]	Modified Z8[2]	ASN-III[3]	WC[4]
Deionized water (mL)	to 1000	to 1000	to 1000	to 1000
NaCl (g L^{-1}/mM)	—	—	25/428	—
MgSO$_4$ · 7H$_2$O (g L^{-1}/mM)	0.075/0.30	0.025/0.10	3.5/14.2	0.037/0.15
MgCl$_2$ · 6H$_2$O (g L^{-1}/mM)	—	—	2/9.8	—
KCl (g L^{-1}/mM)	—	—	0.5/6.7	—
CaCl$_2$ · 2H$_2$O (g L^{-1}/mM)	0.036/0.25	—	0.5/3.4	0.037/0.25
NaNO$_3$ (g L^{-1}/mM)	1.5/17.67	0.47/5.49	0.75/8.8	0.085/1
Ca(NO$_3$)$_2$ · 4H$_2$O (g L^{-1}/mM)	—	0.06/0.25	—	—
NH$_4$Cl (g L^{-1}/mM)	—	—	—	—
K$_2$HPO$_4$ (g L^{-1}/mM)	0.04/0.18	0.03/0.18	0.02/0.09	0.0087/0.05
Na$_2$CO$_3$ (g L^{-1}/mM)	0.02/0.19	0.021/0.20	0.02/0.19	
NaHCO$_3$ (g L^{-1}/mM)	—	—	—	0.013/0.15
Na$_2$SiO$_3$ · 9H$_2$O (g L^{-1}/mM)	—	—	—	0.028/0.1
Na$_2$ EDTA · 2H$_2$O (g L^{-1}/mM)	—	3.9/0.010	—	
Ferric ammonium citrate (g L^{-1}/mM)	0.006/0.030	—	0.003/0.015	—
Citric acid (g L^{-1}/mM)	0.006/0.029	—	0.003/0.014	—
EDTA K$_2$ Mg · 2H$_2$O (g L^{-1}/mM)	0.001/0.0024	—	0.0005/0.0012	—
Fe EDTA (mL/mM)	—	10[5]	—	—
Micronutrients (mL)	1.0	1.0	1.0	1.0
TES Buffer (g L^{-1})	—	—	—	0.115
Trace metal solution used (see Table 3.2)	A5+Co	Gaffron	A5+Co	WC
Vitamin mix (see Table 3.2)	—	—	—	—
Final pH after autoclaving	7.4	6.5–7.7	7.5	7.8

[1] Rippka *et al.* (1979);
[2] Kotai (1972);
[3] Waterbury and Stanier (1981);
[4] Guillard and Lorenzen (1972);
[5] Solution A, 2.8 g FeCl$_3$ in 100 mL 0.1 N HCl; solution B, 3.9 mg EDTA-disodium in 100 mL 0.1 N NaOH. Add 10 mL solution A and 9.5 mL solution B plus water to 1 L.

SOP 3.4.3   Procedure

Media are usually prepared from stock solutions of macronutrients, trace elements, and vitamins, and brought to their final volume with deionized water. Media may be used as liquid or solidified by 0.8–2% agar.

Table 3.1 presents some of the most commonly used cyanobacterial growth media and Table 3.2 presents the trace metal solutions used in these growth media. We describe the general procedure for preparing culture media below. However, the original references provided as footnotes to Tables 3.1 and 3.2 should be consulted for technical details concerning the exact preparation of these media (e.g. preparation of stock solution, the order in which to dissolve the individual chemicals, pH).

Table 3.2 *Recipes for trace metal solutions*

Ingredient (g liter^{-1}/mM)	Trace metal solution designation		
	A5 + Co[1]	Gaffron[2]	WC[3]
Deionized water (mL)	1000	1000	1000
H_3BO_3 (g L^{-1})	2.86/46	3.1	1
$FeCl_3 \cdot 6H_2O$ (g L^{-1})	—	—	3.15
$MnCl_2 \cdot 4H_2O$ (g L^{-1})	1.81/9	—	0.18
$MnSO_4 \cdot 4H_2O$ (g L^{-1})	—	2.23	—
$ZnSO_4 \cdot 7H_2O$ (g L^{-1})	0.222/0.77	0.287	0.022
$Na_2MoO_4 \cdot 2H_2O$ (g L^{-1})	0.39/1.6	—	0.006
$(NH_4)_6Mo_7O_{24} \cdot 4H_2O$ (g L^{-1})	—	0.088	—
$CuSO_4 \cdot 5H_2O$ (g L^{-1})	0.079/0.3	—	0.01
$Co(NO_3)_2 \cdot 6H_2O$ (g L^{-1})	0.0494/0.17	0.146	—
$CoCl_2 \cdot 6H_2O$ (g L^{-1})	—	—	0.01
$VOSO_4 \cdot 6H_2O$ (g L^{-1})	—	0.054	—
$Al_2(SO_4)_3K_2SO_4 \cdot 2H_2O$ (g L^{-1})	—	0.474	—
$NiSO_4(NH_4)_2SO_4 \cdot 6H_2O$ (g L^{-1})	—	0.198	—
$Cd(NO_3)_2 \cdot 4H_2O$ (g L^{-1})	—	0.154	—
$Cr(NO_3)_3 \cdot 7H_2O$ (g L^{-1})	—	0.037	—
$Na_2WO_4 \cdot 4H_2O$ (g L^{-1})	—	0.033	—
KBr (g L^{-1})	—	0.119	—
KI (g L^{-1})	—	0.083	—
Na_2 EDTA $\cdot 2H_2O$ (g L^{-1})	—	—	4.36/11,7
Thiamine (g L^{-1})	—	—	0.1
Biotin (g L^{-1})	—	—	0.0005
B_{12} (g L^{-1})	—	—	0.0005
Amount added per liter of medium (mL)	1.0	0.1	1.0

[1] Rippka *et al.* (1979);
[2] Hughes *et al.* (1958);
[3] Guillard and Lorenzen (1972).

SOP 3.4.3.1   General Steps for the Preparation of Liquid Media

1. Prepare stock solutions of macronutrients and trace elements (consult references in foot-notes of Tables 3.1 and 3.2).
2. Prepare deionized water. Organic compounds in the water are removed by either double distillation or passage through a Millipore (Merck Millipore) water purifier.
3. Mix correct amounts of the different stock solutions and add water to their final volume following the exact instructions of the specified culture media.
4. Sterilize by autoclaving at 121°C for 20 min and cool down.
5. Store at room temperature, and let it stabilize for 1–2 days before use.

SOP 3.4.3.2   General Steps for the Preparation of Solid Media

1. Prepare double strength agar in deionized water.
2. Prepare double strength medium.

3. Sterilize mineral medium and agar separately by autoclaving at 121°C for 20 min.
4. Mix equal volumes of media and agar or agarose after cooling to 50°C.
5. Agar and/or agarose are dissolved in the cyanobacterial media until a final concentration of 0.8–2% (ex. 0.8–2 g agar to 100 mL media).
6. Fill either glass tubes with 10 mL inclined like a slope (named slant) or plastic Petri dishes (amount dependent on the size of the Petri dish) and cool in a dry room or under a laminar flow hood.
7. The freshly poured plates can be stacked on top of each other in order to prevent the formation of condensation under their cover.
8. Plates can be stored at room temperature in plastic boxes.

Other available media recipes for the culturing of cyanobacteria are presented in Andersen (2005) or on the cyanosite website: http://www-cyanosite.bio.purdue.edu/media/table/media.html.

1. Aerate 800 mL deionized water with CO_2 for approximately 10 minutes to avoid precipitations.

2. Add 10 mL of Z8 I solution.

3. Add 10 mL of Z8 II solution.

4. Add 10 mL of Z8 III solution.

5. Add 1 mL of trace metals solution.

6. Adjust volume to 1000 mL with deionized water

7. Sterilize by autoclaving (121°C for 20 min).

8. The Z8 medium will have a pH range of 6.5–7.7.

Figure 3.9 *Flowchart showing the method for preparation of culture medium Z8.*

SOP 3.4.4   Notes

The chemicals used for the media preparation should be of the highest quality. Several chemicals are contaminated with trace metals or other contaminants. This could inhibit the growth of sensitive species.

The thick Petri dishes are better than the thin ones. First, they can have more volume of solid medium and ensure that the inoculated strain can grow properly and stay healthy in the defined incubation period (before the next transfer). Second, a thick layer of solid medium will not dry.

For some media, certain stock solutions, such as phosphorus, may need to be autoclaved separately and added after cooling to avoid precipitation. The media recipe should inform of this when necessary.

To avoid precipitation during autoclaving of the Z8 media, about half the amount of water is bubbled with CO_2 (15–20 min) prior to the mixing of the stock solution.

Fig. 3.9 shows a flowchart for the preparation of culture medium Z8 as an example of how to prepare a culture medium. The preparation of each media is slightly different and the original references should always be consulted.

SOP 3.4.5   References

Andersen, R.A. (ed.) (2005) *Algal Culturing Techniques*, Elsevier Academic Press, New York.

Guillard, R.R.L. and Lorenzen, C.J. (1972) Yellow-green algae with chlorophyllide c, *Journal of Phycology*, **8**, 10–14.

Hughes, E.O., Gorham, P.R., and Zehnder, A. (1958) Toxicity of a unialgal culture of *Microcystis* aeruginosa, *Canadian Journal of Microbiology*, **4**, 225–236.

Kotai, J. (1972) *Instructions for Preparation of Modified Nutrient Solution Z8 for Algae*, Norwegian Institute for Water Research, Oslo, Publication B-11/69.

Rippka, R., Deruelles, J., Waterbury, J.B., *et al.* (1979) Generic assignments, strain histories and properties of pure cultures of cyanobacteria, *Journal of General Microbiology*, **111**, 1–61.

Waterbury, J.B. and Stanier, R.Y. (1981) Isolation and growth of cyanobacteria from marine and hypersaline environments, in *The Prokaryotes*, M.P. Starr, H. Stolp, H.G. Trüper, *et al.* (eds), Springer-Verlag, Berlin, **1**, 221–223.

SOP 3.5

Strain Maintenance (Living Cultures)

Sigrid Haande*, Camilla H.C. Hagman, and Andreas Ballot

Norwegian Institute for Water Research, Oslo, Norway

SOP 3.5.1   Introduction

Most commonly, cyanobacterial cultures are maintained by perpetual transfer and under controlled environmental conditions (e.g. Rippka *et al.*, 1981; Waterbury and Stanier, 1981; Rippka, 1988; Andersen and Kawachi, 2005). Routine serial subculturing is performed by using aseptic microbiological techniques where an inoculum from a culture in a stationary growth phase is transferred into fresh sterilized liquid or a solid medium. Successful maintenance will assure active and healthy cultures which are representative for its population. This SOP describes transfer techniques for cultures kept in liquid and solid media and general maintenance conditions.

SOP 3.5.2   Experimental

SOP 3.5.2.1   Materials

- Glassware such as Erlenmeyer flasks (Pyrex) with silicon or cotton plugs.
- Test tubes with screw caps and rack.
- Agar plates.
- Liquid or solid media.
- Microscope slides and coverslips.
- Pasteur pipettes (plugged with non-absorbent cotton wool in the wide end).
- Wire loops.
- Marker pen.

SOP 3.5.2.2   Equipment

- Autoclave.
- Transfer hood/laminar flow cabinet.

*Corresponding author:sigrid.haande@niva.no

- Bunsen burner.
- Stereomicroscope, microscope, or inverted microscope.

SOP 3.5.3 Procedure

SOP 3.5.3.1 Transfer Techniques

Before the transfer is started, it is important that all labels are carefully checked and all new flasks or agar plates are labelled accordingly in order to avoid mistakes. After labelling, the cultures which are to be transferred should be organized in parallel with the corresponding uninoculated flasks or agar plates.

Standard aseptic microbiological methods must be used and all transfers should be performed in a transfer hood/laminar flow cabinet.

All glassware, pipettes/loops, and culture media which are needed for the transfer must be sterilized before starting the procedure.

All flasks/solid media plates/culture media flasks should be prepared in advance, and opened for as short a time as possible in order to limit contamination.

See Fig. 3.10 for some of the equipment needed for the transfer of liquid cultures.

SOP 3.5.3.1.1 Transfer of Liquid Cultures

1. All glassware must be sterilized by autoclaving at 121°C for 20 min and then allowed to cool down.

Figure 3.10 *Some equipment for the transfer of liquid cultures.*

2. Prepare hood; use UV light overnight, let fan run for about 10 min before use. If used earlier same day, wash with 70% ethanol before the fan is started and run for at least 10 min.
3. Use sterilized culture media of choice.
4. The cultures must be examined carefully under the microscope to (a) check the state and growth of the culture and (b) to check for possible contaminations. The culture growth shall be in stationary phase when transferred.
5. All flasks must be labelled.
6. Turn on the burner. (A burner under the hood is prohibited in some labs, and in this case the transfer can be done on a bench in 20 cm proximity of a Bunsen flame or under a sterile hood.)
7. Take one culture into the hood at a time and transfer it to a new Erlenmeyer flask. The new flask contains 50 mL liquid medium prepared in advance. When it is opened in close proximity to a flame, sterilize the opening of the flask before and after filling.
8. Sterilize the opening of the flask with the old culture each time it is opened. An aliquot of the culture will be transferred to the new flask; 1–2 droplets of culture are sufficient to establish new cultures.
9. First, open the old culture flask and take out a small volume (some drops or mL) using a pipette while avoiding touching the opening of the flask. Close the old culture flask. Open the new flask and pour the small volume contained in the pipette without touching the opening of the flask. Put the lid back on the new flask.
10. Take the old and new culture out of the working area close to the flame or under the hood, and proceed with the next culture.
11. If any drop falls into the working area, clean it with 70% ethanol.
12. At the end, clean the working area with 70% ethanol.

SOP 3.5.3.1.2   *Transfer of Solid Media Cultures*

1. Prepare hood (see point 2, Section 3.5.3.1.1).
2. Use prepared sterile solid medium plates with sterilized agar or agarose and the medium of choice.
3. Examine the cultures (see point 4, Section 3.5.3.1.1).
4. All plates must be labeled.
5. Turn on the burner (see point 6, Section 3.5.3.1.1).
6. The wire loop is sterilized in the flame; hold the wire at an angle such that the wire glows red, and cool down in air or by leaning it on a plate.
7. The lid of the plate with the new sterilized medium is lifted at a slight angle.
8. The lid of the plate of the established culture is lifted and a portion of the culture is removed with the loop; put lid back on.
9. The inoculum is often 1–10% (v/v) of the original culture.
10. The material from the established culture is streaked onto the solid medium with the loop. Make sure the material is distributed evenly (no clumping); put lid back on.

Figure 3.11 shows a flowchart for the methods presented in this SOP.

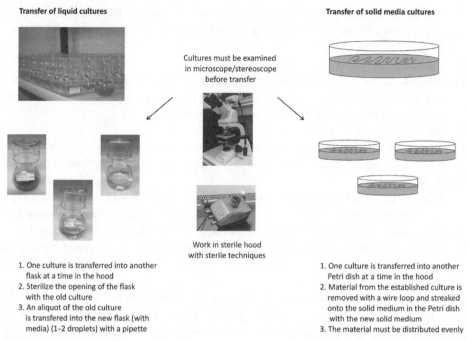

Figure 3.11 *Flowchart showing the methods for maintenance of living cultures presented in SOP 3.5. (See color plate section for the color representation of this figure.)*

SOP 3.5.3.2   Maintenance of Cultures

1. Stock cultures are held in appropriate culture rooms/culture cabinets, and the two important conditions for incubation are:
   a. Appropriate light conditions of around 10–30 µmol photons m² s⁻¹ with light–dark cycle of 12–14 h of light and 10–12 h of darkness.
   b. Appropriate temperature conditions, which are based on the temperature range of the natural environment where the sample derives from. Temperatures between 15 and 20°C are suitable for most cyanobacterial cultures. Thermophilic strains can be maintained at 37°C, while the ones originating from deep alpine lakes will be kept at 18°C and the polar ones at 12°C.
2. The cultures need perpetual maintenance and the intervals between the transfer of each culture are variable and depend on growth and sensitiveness. Some strains must be transferred every two weeks, while other strains may grow for months before transfer is needed.

SOP 3.5.4   Notes

If you are not allowed to use the flame under the hood, better to choose to work in close proximity to the flame when working on solid media than to use the hood, as you will need to sterilize the loop in the flame.

Cultures of cyanobacteria growing on substrate (e.g. on the culturing flask) may need to be scraped from the flask walls/bottom with a (sterile) pipette or inoculating loop in order to get hold of any cell material.

The amount of inoculum may need to be adjusted for every culture, according to density and speed of growth.

SOP 3.5.5   References

Andersen, R.A. and Kawachi, M. (2005) Traditional microalgae isolation techniques, in *Algal Culturing Techniques*, R.A. Andersen (ed.), Elsevier Academic Press, New York, 83–100.

Rippka, R., Waterbury, J.B., and Stanier, R.Y. (1981) Isolation and purification of cyanobacteria: Some general principles, in *The Prokaryotes*, M.P. Starr, H. Stolp, H.G. Trüper, *et al.* (eds.), Springer-Verlag, Berlin, **1**, 212–220.

Rippka, R. (1988) Isolation and purification of cyanobacteria, *Methods in Enzymology*, **167**, 3–27.

Waterbury, J.B. and Stanier, R.Y. (1981) Isolation and growth of cyanobacteria from marine and hypersaline environments, in *The Prokaryotes*, M.P. Starr, H. Stolp, H.G. Trüper, *et al.* (eds.), Springer-Verlag, Berlin, **1**, 221–223.

SOP 3.6

Cryopreservation and Recovery

Muriel Gugger*

Collection of Cyanobacteria, Institut Pasteur, Paris, France

SOP 3.6.1   Introduction

Several methods have been tested for the cryopreservation of the cyanobacteria (Hubalek, 2003; Day, 2007; Amaral *et al.*, 2013) and some with a great recovery success. At the Pasteur Culture Collection of Cyanobacteria (PCC), the 750 axenic strains are cryopreserved and recovered according to the following recommendations and protocol.

*Corresponding author:muriel.gugger@pasteur.fr

SOP 3.6.2   Experimental

SOP 3.6.2.1   Materials

SOP 3.6.2.1.1   For Cryopreserving

- A well-grown and healthy culture.
- Dimethyl sulfoxide (DMSO); this organosulfur compound of formula $(CH_3)_2SO$ is toxic for the cyanobacteria and to humans: be cautious with its use. It can be purchased from Sigma-Aldrich Co. as a sterile solution.
- A test plate is a solid growth medium supplemented with glucose (0.2%) and casamino acids (0.02%). The test plates, once inoculated, should be incubated in the dark for 2–3 days at the room temperature prior to microscopic examination using phase contrast objectives and oil immersion.
- Liquid nitrogen.

SOP 3.6.2.1.2   For Recovery

- Liquid nitrogen.
- Water bath at 37°C.
- 40–50 mL of fresh medium for the strains to recover.

SOP 3.6.2.2   Equipment

SOP 3.6.2.2.1   For Cryopreserving

- A bench hood.
- Pipets to transfer 11 mL and 1.5 mL with a Pipetman®.
- Sterile tubes of 18 mL with a cap.
- A vortex.
- A centrifuge.
- A marker that resists the cooling in the ice.
- Cryotubes or cryovials that can contain 2 mL.
- A handy reservoir to first freeze the tubes in liquid nitrogen with gloves and eyes or face protection against the splashes of liquid N_2, and a clamp to grab the cryotubes.
- A freezer at –150°C or liquid nitrogen tanks.
- Box for the cryotubes in the freezer, or cryocanes, to hold the cryotubes in the liquid nitrogen tanks.

SOP 3.6.2.2.2   For Recovery

- A dark box made of a box inside a black garbage bag.
- A culture room or a bench at room temperature and natural light in the laboratory.

SOP 3.6.3   Procedure

SOP 3.6.3.1   The Cryopreservation

For most of the axenic cyanobacterial strains, the cryoconservation can be performed by storage in liquid nitrogen using 5% (v/v) of DMSO as cryoprotectant.

1. Under the bench hood cleaned with 70% ethanol prior to and after use, prepare all the necessary materials, with the vortex and reservoir of liquid nitrogen kept handy (Fig. 3.12 and Fig. 3.13).
2. If the strain is axenic, test the purity of the culture on a test plate.
3. In the 18 mL sterile tubes, mix 11 mL of the culture to cryopreserve and 550 μL of DMSO to obtain a final concentration of 5% DMSO v/v.
4. Vortex gently to homogenize the culture with the DMSO.
5. Dispatch 1.5 mL of the homogenate into 2 mL cryotubes, close the cap firmly and plunge the cryotubes directly into the liquid nitrogen.
6. The cryotubes can be stored in the −150°C freezer or in a liquid nitrogen reservoir once mounted on the cryocanes.

The cryopreservation of planktonic strains – such as *Arthrospira*, *Microcystis*, and *Planktothrix* – are more sensitive and need additional steps (2a–2c). Before freezing them, it is necessary to collapse the gas vesicles by pressure, for example by centrifuging them.

2a. After testing the purity of the culture on a test plate, transfer under the hood 25 mL of the planktonic culture in a sterile tube.
2b. Centrifuge the 25 mL at 10000 rpm (8700 g) for 15 min.

Figure 3.12 *All the equipment needed to prepare the tube for the cryopreservation of a healthy cyanobacterial culture under the hood.*

Figure 3.13 *Near the hood: vortex, a handy reservoir of liquid nitrogen, gloves and face protection for the manipulator, the clamps, the cryotube rack, and the cryocanes.*

2c. Remove 14 mL of the supernatant, and mix the remaining 11 mL gently to resuspend the pellet.

Then proceed as described above in Step 3.

SOP 3.6.3.2   The Recovery

Recovery is done by quick thawing at 37°C, followed by an immediate transfer of the cell suspension into a fresh medium, respecting a dilution that yields in a carry-over of DMSO of not more than 0.5% (v/v) under the bench hood.

1. Bring the cryotubes in ice once taken out of the −150°C freezer or the cryotank.
2. Quickly thaw the cryotubes into a water bath at 37°C.
3. Transfer the cell suspension in the cryotube into 40 mL fresh medium to result into a final DMSO concentration below 0.5%. Above 0.5% of DMSO, the product is toxic for the cyanobacteria. Indeed, three successive transfers will be necessary to recover a healthy culture that can be used for research or to redo a new cryopreservation batch.
4. Place the freshly inoculated culture into darkness for 2–3 days, by placing the culture into the black box (surrounded by the garbage bag, which will efficiently protect it from the light, avoid aluminum foil, which reflects any light entering in the supposed black box space). This allows the strain time to adapt to its nutrient-rich surroundings, without activating the photosystem apparatus.
5. Then, open the bag a little at a time (over a number of days) to adapt the culture to the light. The recovery is performed in the room at 22°C or at room temperature.
6. Give the culture time to grow. Note that it is clear green at the beginning and week after week will change to a dense green culture, but will still contain a quantity of DMSO.

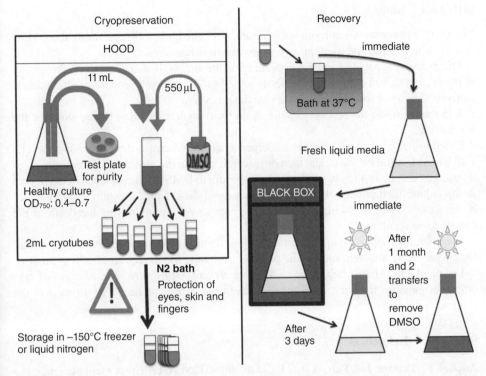

Figure 3.14 *Flowchart showing the methods for cryopreservation and recovery presented in SOP 3.6. (See color plate section for the color representation of this figure.)*

7. To recover the nice and healthy culture with a negligible quantity of DMSO, at least three successive transfers will be necessary. Fig. 3.14 shows a flowchart for the method presented in this SOP.

SOP 3.6.3.3 *Method Efficiency*

For research projects, several cryopreserved PCC strains were recovered between 2008 and 2012 (Table 3.3).

Table 3.3 *Recovery rate of cryopreserved strains between 2008–2012*

Year	Number of strains	Number of strains recovered	Rate (%)
2008	123	118	96
2009	63	62	98
2010	22	21	95
2011	46	45	98
2012	35	34	97

SOP 3.6.4   Notes

The DMSO is conserved at room temperature. If opened, the tube containing the DMSO will be stored at 4°C and used quickly within the next few days.

The recovery is highly variable, depending on the strains. It also depends on the quality of the cryopreserved culture, which needs to be healthy and well grown. A culture that is either too young or too old will not survive the recovery.

It is useful to test the recovery of the strain from cryopreservation before stopping the live culture.

Cyanobacteria are very sensitive to detergents, which sometimes dry on the Erlenmeyer's wall when cleaning the glass, and then diffuse quickly into the medium. The recovered cryopreserved culture is already weakened by exposure to DMSO. The presence of detergents in the culture medium can therefore lead to a rapid die-off of the culture: within 12–24 h, the freshly inoculated culture turns from green to yellow (indicating the death of the cyanobacterial culture).

Some strains support cryopreservation very well and are able to recover within a month. Others are really affected, such as the planktonic strains, which recover after four months.

For planktonic strains belonging to *Microcystis* and *Arthrospira*, five attempts of cryotubes originating from the same cryopreserved batch may be necessary to recover the original culture alive.

SOP 3.6.5   References

Amaral, R., Pereira, J.C, Pais, A.A.C.C., and Santos, L.M.A. (2013) Is axenicity crucial to cryopreserve microalgae?, *Cryobiology*, **67**, 312–320.

Day, J.G. (2007) Cryopreservation of Microalgae and Cyanobacteria, in *Methods in Molecular Biology, 368: Cryopreservation and Freeze-Drying Protocols*, J.G. Day and G.N. Stacey (eds), Humana Press Inc, Totowa, NJ, 141–151.

Hubalek, Z. (2003) Protectants used in the cryopreservation of microorganisms, *Cryobiology*, **46**, 205–229.

4

Taxonomic Identification of Cyanobacteria by a Polyphasic Approach

Annick Wilmotte[1]*, H. Dail Laughinghouse IV[2,3], Camilla Capelli[4], Rosmarie Rippka[5‡], and Nico Salmaso[4]*

[1] *InBios – Center for Protein Engineering, University of Liège, Liège, Belgium*
[2] *Fort Lauderdale Research and Education Center, University of Florida/IFAS, Davie, Florida, United States of America*
[3] *Department of Botany, MRC-166, National Museum of Natural History – Smithsonian Institution, Washington DC, United States of America*
[4] *Research and Innovation Centre, Fondazione Edmund Mach – Istituto Agrario di S. Michele all'Adige, S. Michele all'Adige, Italy*
[5] *Institut Pasteur, Unité des Cyanobactéries, Centre National de la Recherche Scientifique (CNRS), Unité de Recherche Associé (URA) 2172, Paris, France*

4.1 Introduction

Cyanobacteria are oxygenic photosynthetic bacteria, and have a large morphological and ecological diversity. This heterogeneous group is also known as Cyanophyceae, blue–green algae (Staley *et al.*, 1989), or Cyanoprokaryotes (Komárek and Anagnostidis, 1999). Analyses undertaken on clonal cultures and samples collected from water blooms

‡Retired.
*Corresponding authors: awilmotte@ulg.ac.be; nico.salmaso@fmach.it

Molecular Tools for the Detection and Quantification of Toxigenic Cyanobacteria, First Edition.
Edited by Rainer Kurmayer, Kaarina Sivonen, Annick Wilmotte and Nico Salmaso.

and other environments have led to the identification of more than 100 cyanobacterial taxa able, or suspected, to produce a broad range of toxic secondary metabolites (Bernard *et al.*, 2017). Representatives of such genera and species are shown in Fig. 4.1. However, the number of toxic cyanobacterial taxa is still underestimated and is expected to increase as a result of screening a larger number of environmental populations and the discovery of new toxins (Meriluoto and Codd, 2005; Shams *et al.*, 2015). As a principle and precaution in water management, all cyanobacteria should be considered potentially toxic (Codd *et al.*, 1999; Laughinghouse *et al.*, 2012; Salmaso *et al.*, 2017). Consequently, a rapid identification of cyanobacteria is essential to estimate potential risks for animal and human health associated with their presence in waters used for drinking, recreation, and irrigation. Toxic and nontoxic representatives have been observed among different species

(A)   (B)   (C)

(D)   (E)   (F)

Figure 4.1   *(A) Bloom of* Microcystis aeruginosa *in a lake of the Institute of Botany of São Paulo (photo: A. Tucci). (B) Yellow coloration from a bloom of* Cylindrospermopsis raciborskii *(photo: H.D. Laughinghouse IV). (C) Bloom of* Planktothrix rubescens; *Lake Ledro (photo: A. Boscaini). (D) Bloom of* Planktothrix rubescens, *showing a particular and enlarged section of Lake Ledro (photo: A. Boscaini). (E) Bloom of* Dolichospermum lemmermannii; *Lake Garda (photo: N. Salmaso). (F) Bundles of* Aphanizomenon flos-aquae *looking like larch needles in Bergnappweiher, a small pond used for bathing in Bavaria (photo: K. Teubner). (G) Emus crossing a mixed bloom of cyanobacteria and euglenophytes at the Zoo of Rio Grande do Sul (photo: H.D. Laughinghouse IV and V.R. Werner). (H)* Microcystis aeruginosa, *Lake Garda, net sample, magnification 100× (photo: N. Salmaso). (I)* Cylindrospermopsis raciborskii, *scale bar 10 μm (photo: V.R. Werner). (J)* Planktothrix rubescens, *Lake Como, scale bar 10 μm (photo: A. Boscaini). (K)* Dolichospermum lemmermannii *with attached vorticellids, Lake Garda, scale bar 40 μm (photo: S. Shams). (L) Cluster of akinetes of* Dolichospermum lemmermannii, *Lake Garda, scale bar 40 μm (photo: N. Salmaso). (M) Aggregate of* Aphanizomenon flos-aquae *filaments, Lake Winnipeg; flakes are between 3 and up to 10–20 mm (photo: H.J. Kling). (N)* Dolichospermum crassum *(planktic), scale bar 50 μm (photo: H.D. Laughinghouse IV). (O)* Tychonema bourrellyi, *Lake Garda, scale bar 20 μm (photo: N. Salmaso). Source: Courtesy of Tucci, Boscaini, Teubner, Werner, Shams, Kling. (See color plate section for the color representation of this figure.)*

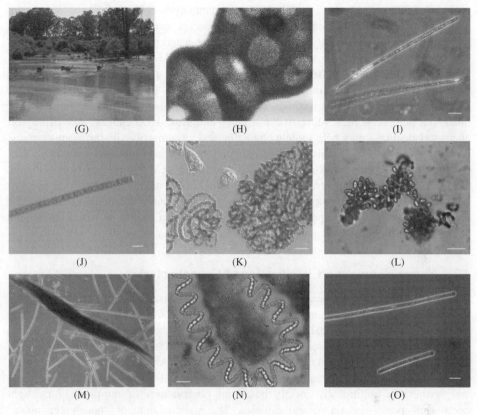

Figure 4.1 *(Continued)*

and populations, but also within the same species. Further, there is a high variability in the quantity of cyanotoxin per cell and of the relative ratios, if several toxin variants are present (Janse *et al.*, 2005; Briand *et al.*, 2008; Kosol *et al.*, 2009; Wood *et al.*, 2012; Salmaso *et al.*, 2014, 2016). Therefore, generic (and, if possible, specific) identification of putative toxic morphotypes by light microscopy is only sufficient for a quick and tentative assessment of their danger and impact on the environment. A complete and reliable risk evaluation relies on the use of molecular and analytical methods for a more precise identification to assess genetic and chemotypic differences at and below the species level. Methods relevant for the characterization of "genotypes" and "chemotypes" among potentially toxic cyanobacteria include DNA amplification by PCR and sequencing of the resulting amplicons (this chapter and SOP 4.2), the determination of genes encoding cyanotoxin biosynthesis (Chapters 6–7), and the analysis of cyanotoxins (Meriluoto and Codd, 2005; Meriluoto *et al.*, 2017).

In this chapter, we discuss the criteria and methods that should be adopted for the taxonomic identification of cyanobacteria. This includes a brief introduction on the two Codes of Nomenclature governing in parallel the nomenclature of these organisms. We will then present the major steps that are important for the taxonomy of cyanobacteria. These include: (1) determination of morphology by light microscopy; (2) genetic characterization by analysis of individual gene sequences, and/or multilocus sequence typing (MLST) or its variant, multilocus sequence analysis (MLSA); (3) the assignment of the organism to a taxonomic entity (genus, species, eco- and/or genotype within a species) by phylogenetic analysis of single or concatenated multiple genes with inclusion of cyanobacterial nucleotide sequences available in public repositories for comparison. In some cases, taxon identification can also be achieved on the basis of allelic profiles determined by MLST, providing the respective organisms are represented in the MLST databases. Ideally, all these methods should be accompanied by the determination of other relevant properties (ultrastructural, physiological, biochemical, and ecological characteristics) that may help to define/redefine and circumscribe the taxon under study. More extensive and time-consuming investigations toward a polyphasic taxonomy (Vandamme *et al.*, 1996), also known as "polyphasic approach" (Abed and Garcia-Pichel, 2001) or "polyphasic characterization" (Abed *et al.*, 2002) are not absolutely essential or feasible for a fast toxicity risk assessment. However, polyphasic taxonomy is essential for a reliable identification of species producing toxins, which should be based on isolated strains (Bernard *et al.*, 2017), and for determining their biogeography.

4.2 Nomenclature and Classification of Cyanobacteria

For historical reasons, the systematics of cyanobacteria (which includes identification, nomenclature, classification, and phylogeny; Vandamme *et al.*, 1996) is treated differently in botanical and bacteriological literature. This is primarily due to different approaches, and opinions, with respect to the definition of a species (Gold-Morgan and González-González, 2005; Johansen and Casamatta, 2005; Rosselló-Móra and Amann, 2015). In the botanical approach, dating from Linnaeus (1753), identification at the binominal level (genus and species) is traditionally mainly based on observations of morphology by light microscopy performed on natural samples. The nomenclature of "blue–green algae" or "Cyanophyceae" follows the rules of the International Code of Nomenclature for algae, fungi, and plants (ICN) (Melbourne Code) (formerly known as the International Code of Botanical Nomenclature – ICBN or "Botanical Code") (McNeill *et al.*, 2012). With the introduction of electron microscopy for the study of ultra-structural features, the bacteria-like nature of these organisms was clearly documented. Consequently, a proposal was made (Stanier *et al.*, 1978) to bring the nomenclature of blue–green algae as "cyanobacteria" under the International Code of Nomenclature of Prokaryotes (ICNP) (formerly the International Code of Nomenclature of Bacteria – ICNB or "Bacteriological Code"). According to the provisions of this Code (Lapage *et al.*, 1992), living type strains free from any other organisms (i.e. "axenic") are mandatory for the valid description of new species, and subcultures derived from the type strains have to be deposited in at least two publicly accessible culture collections in two different countries (De Vos and Trüper, 2000).

For a number of different reasons (see Oren, 2004; Oren and Garrity, 2014), the full integration of cyanobacterial taxa names under the "Bacteriological Code" has, with a few exceptions, still not been achieved. Taxonomic studies on axenic cultures by bacteriologists with this ultimate goal in mind (e.g. Stanier *et al.*, 1971; Waterbury and Stanier, 1978; Rippka *et al.*, 1979) resulted in revised, though provisional, identification and classification schemes, later updated in Bergey's Manual of Systematic Bacteriology (Boone and Castenholz, 2001; Staley *et al.*, 1989). These "bacteriological classifications" were not adopted by botanists, but the knowledge gained on axenic cultures was partly taken into account in subsequent taxonomic revisions, and in the major botanical taxonomic keys currently in use (Komárek and Anagnostidis, 1999, 2005; Komárek, 2013). A detailed comparison between the bacteriological and botanical classification is beyond the scope of this manual. It should, however, be stated that the number of genera and species known from the botanical literature (see Nabout *et al.*, 2013) is much higher than the named taxa included in the "bacteriological classifications." This is not surprising, since the latter mainly covered genera represented by the very limited number (about 200) of axenic strains available at the time, and specific epithets were in most cases omitted, owing to the lack of sufficient molecular data supporting species distinctions.

For other bacterial groups, DNA/DNA re-association methods, also known as DNA/DNA hybridization, are used as the basis for new species' descriptions. This technique is still required under the "Bacteriological Code" and relative binding ratios of 70% (or more), and a melting temperature increment of less than 5°C, have been set as the limits for strains related at the species level (Wayne *et al.*, 1987). The usefulness of this method for cyanobacteria is exemplified by the study of Otsuka *et al.* (2001), who demonstrated that morphotypes identified as five different species of the genus *Microcystis* according to the botanical nomenclature, and including toxic and nontoxic representatives, are in fact members of a single nomenspecies. However, DNA/DNA hybridizations are only meaningful if performed on axenic strains, and this method is difficult and time-consuming. Thus, there are little data available for cyanobacteria. Earlier work was reviewed by Wilmotte (1994) and more recent data can be found in Suda *et al.* (2002) and Gaget *et al.* (2015). Furthermore, these data cannot be stored and compared via public databases. For this reason, this molecular method will not be included in the descriptions below.

As a simpler means for assessing genetic relatedness at the species level, Taton *et al.* (2003) subdivided Antarctic cyanobacterial strains (and environmental clones) into "phylotypes" (Taton *et al.*, 2003) or "operational taxonomic units" (OTUs) (Taton *et al.*, 2006). Both "phylotype" and "OTU" are interchangeable and were defined as a group of strains (or environmental clones) sharing more than 97.5% 16S rRNA gene sequence identity in pairwise comparison. This threshold value was chosen, since bacterial strains exhibiting less than approximately 97.5% sequence identity for the full lengths of this gene generally give less than 70% relative binding in DNA/DNA re-association experiments (Stackebrandt and Goebel, 1994), and thus most likely represent different species (see above). A higher cut-off value (98.7– 99% 16S rRNA gene sequence identity) for delimitating a bacterial species has more recently been suggested by Stackebrandt and Ebers (2006) but is not yet widely applied. On the other hand, support for the latter more stringent species boundary, at least among some cyanobacterial taxa, has been obtained in a recent study on several species of the genus *Planktothrix* (Gaget *et al.*, 2015).

Finally, we emphasize that, similar to other microorganisms, cyanobacterial systematics, regardless of the Codes, is in constant evolution, and will change over time as new knowledge is gained. This is quite clear from the numerous recent publications on new, or emended, genera and species supported by molecular studies, or from the even higher number of unnamed taxonomic entities revealed by such approaches. Therefore, care should be taken to thoroughly consult the most recent literature, prior to proposing the description of new taxa.

4.3 Microscopy

4.3.1 Light Microscopy

The identification of cyanobacteria has traditionally been based on detailed analyses of morphological characteristics observed by light microscopy (Fig. 4.1), and the resulting classifications have been revised numerous times over the last two centuries (for classical literature and more recent references, see Castenholz and Waterbury in Staley *et al.*, 1989; Komárek and Anagnostidis, 1999, 2005; Komárek, 2013). Light microscopy is an inexpensive method that requires only basic equipment found in many laboratories. It provides a lot of information on morphology and cellular differentiation of the morphotypes present in a sample and, if counted, on their abundance. However, careful examination is time-consuming and appropriate identification requires knowledge of cyanobacterial taxonomy. Many taxa have overlapping morphological features, and cannot be identified with certainty, even by experts (Whitton and Potts, 2012). In addition, phenotypic characteristics can have more or less plasticity depending on the organism, and may vary with environmental conditions (e.g. *Cylindrospermopsis raciborskii*; Saker *et al.*, 1999; Soares *et al.*, 2013). The increasing number of taxonomic revisions causes additional difficulties for the naming of cyanobacteria. Therefore, we strongly advise taking pictures of the identified taxa and to preserve some material for later analyses or comparisons (see SOP 4.1).

Both upright and inverted microscopes can be used to identify cyanobacteria following the methods described in SOP 4.1. The examination of the samples and cyanobacterial specimens should always be carried out on freshly collected live material, in order to avoid any alterations of the morphological characters and natural pigmentation. Such material will also allow a direct comparison of the examined samples with those described in publications, manuals and illustrated photo guides, as well as web resources (Salmaso *et al.*, 2017) (Table 4.1).

The simultaneous identification and enumeration of cyanobacteria is usually carried out using the Utermöhl method (Utermöhl, 1931), which is based on the sedimentation of measured aliquots of preserved samples (with Lugol's solution or formaldehyde) into settling chambers of a determined volume. After a sufficient sedimentation period, eukaryotic phytoplankton and cyanobacteria are identified and counted with an inverted microscope. Biovolume can be calculated using standard formulas based on cell dimensions and shapes (Hillebrand *et al.*, 1999; Sun and Liu, 2003). Alternatively, the counting can be performed with an upright microscope using a Sedgwick-Rafter chamber (see SOP 4.1), haematocytometers, or a Petroff-Hausser bacterial counting slide, depending on the size range of the organisms (Guillard and Sieracki, 2005). In contrast, picocyanobacteria are often too light and too small (<1 µm) to settle well in counting chambers/slides. Their identification and enumeration need to be performed by autofluorescence microscopy (see Section 4.3.2) or by flow cytometry (Marie *et al.*, 2015; see also SOP 3.2). General procedures for the quantification of phytoplankton and cyanobacteria, including preservation and sample storage, are recommended in a European standard (EN-15204, 2006). Specific procedures for the counting and estimation of biovolumes of cyanobacterial cells, colonies, and filaments are reported in Catherine *et al.* (2017) and Salmaso *et al.* (2017). Gas-vacuolated taxa require special pretreatments to avoid their misidentification and quantitative underestimation (Salmaso *et al.*, 2017).

Table 4.1 Selection of Web databases and sites relevant for the taxonomic classification and nomenclature of toxigenic cyanobacteria

Target	Database/web site	Link	References
Taxonomic classification	AlgaeBase[2] CyanoDB[1]	www.algaebase.org www.cyanodb.cz	Guiry and Guiry (2017) Komárek and Hauer (2013)
	Index Nominum Algarum[2]	http://ucjeps.berkeley.edu/CPD	Silva (2016)
General molecular databases INSDC (International Nucleotide Sequence Database Collaboration)	European Molecular Biology Laboratory (EMBL)[5]	http://www.ebi.ac.uk	European Bioinformatics Institute (2014)
	GenBank at NCBI (National Center for Biotechnology Information)[5]	www.ncbi.nlm.nih.gov/genbank	Benson et al. (2013)
	DNA Data Bank of Japan (DDBJ)[5]	www.ddbj.nig.ac.jp	Mashima et al. (2016)
Ribosomal rRNA databases	GreenGenes[4]	greengenes.lbl.gov greengenes.secondgenome.com	DeSantis et al. (2006)
	SILVA (from Latin silva, forest)[5]	http://www.arb-silva.de	Quast et al. (2013), Yilmaz et al. (2014)
	The Ribosomal Database Project (RDP)[4]	http://rdp.cme.msu.edu	Cole et al. (2014)
	EzBioCloud EzTaxon/EzTaxon-e[3]	http://www.ezbiocloud.net/eztaxon	Kim et al. (2012)
Ribosomal rRNA databases and other selected target genes (e.g. rpoB, gyrB)	leBIBI[QBPP] (Quick Bioinformatic Phylogeny of Prokaryotes)[3]	https://umr5558-bibiserv.univ-lyon1.fr/lebibi/lebibi.cgi	Flandrois et al. (2015)
Genomes	Genomes OnLine Database (GOLD)[5] EzBioCloud EzGenome[3]	https://gold.jgi.doe.gov/index http://www.ezbiocloud.net/ezgenome	Reddy et al. (2015)
Molecular probes	probeBase (rRNA-targeted oligonucleotide probes)[5]	http://probebase.csb.univie.ac.at	Greuter et al. (2016)
Electronic Atlases	fytoplankton.cz[2] Cyanosite[1]	www.fytoplankton.cz/fytoatlas.php www-cyanosite.bio.purdue.edu	

[1] Only cyanobacteria; [2] prokaryotic and eukaryotic algae; [3] bacteria and archaea; [4] bacteria, archaea and fungi; [5] bacteria, archaea and eukaryotes.

The examination of water samples by microscopy is often the first step for assessing the presence of putative toxin producers. Nevertheless, excluding a few reports (Via-Ordorika *et al.*, 2004; Kokociński *et al.*, 2011), there is no clear relation between morphotype and toxin production, and thus toxic strains (or populations) cannot be distinguished under the microscope. The identification of well-known cyanotoxin producers in samples indicates that there is a potential risk to animal and human health, and that further analyses need to be considered. However, it is not a proof of toxicity.

There are many studies in fundamental and environmental microbiology, as well as research in toxicology, in which toxic, or potentially toxic, cyanobacterial species have been identified based on their morphology and colony formation as observed in the natural samples. Given the limited information from bacteriological identification and classification schemes (see above), botanical taxonomic treatises are generally used (and should be used) for identification and nomenclature (Komárek and Anagnostidis, 1999, 2005; Komárek, 2013). However, assigning taxonomic names, both at the genus and species level, to a cyanobacterial morphotype observed in the field may be quite subjective, and dependent on the identifier's education, experience, and the availability of the original descriptions of the taxa. For example, Lee *et al.* (2014) show that only 18 out of 39 of the cyanobacterial strains isolated from Australian waters were identified to species level using the light microscope by at least one of two trained taxonomists, and only three species out of 17 were in complete agreement between the two examiners. This illustrates the need to analyze one or more molecular markers for a more reliable identification of cyanobacterial taxa.

4.3.2 Autofluorescence Microscopy

The majority of cyanobacteria harvest light for photosynthesis mainly by pigmented water-soluble proteins, the phycobiliproteins, assembled into large supramolecular complexes called phycobilisomes (Bryant, 1982; Grossman *et al.*, 1993). Exceptions are "prochlorophytan" cyanobacteria (*Prochloron*, *Prochlorothrix*, and *Prochlorococcus*) and *Acaryochloris* that lack phycobilisomes (though minor amounts of phycobiliproteins may be produced) and have chlorophyll a/b (chl a_2/chl b_2) or chlorophyll a/b-chl d containing light harvesting complexes (see reviews by Ting *et al.*, 2002 and Kühl *et al.*, 2007, respectively). The blue phycobiliproteins – i.e. phycocyanin (PC) and allophycocyanin (APC) – are present in all phycobilisome-forming cyanobacteria, whereas the reddish phycoerythrins (PE) and phycoerythrocyanin (PEC) are only synthesized, in addition to the latter pigments, in some cyanobacterial representatives (Bryant, 1982; Ong and Glazer, 1991). Light energy captured at wavelengths specific to each phycobiliprotein is sequentially transferred from PE (or PEC), if present, to PC and APC, and finally to a special pair of chlorophyll a in the reaction center of photosystem II (Glazer, 1989). Phycobiliproteins and chlorophylls have characteristic *in vivo* absorption and fluorescence emission maxima in the visible to near-infrared range of light (400–750 nm) (Thorne *et al.*, 1977; Bryant, 1982; Ong and Glazer, 1991; Chen *et al.*, 2012). Consequently, individual cells (or filaments) of cyanobacteria can be visualized by autofluorescence microscopy, and in parts be distinguished based on their pigment signatures, by choosing appropriate spectral wavelengths for fluorescence excitation and emission. This method has most widely been applied for the study of picocyanobacteria, which cannot be readily distinguished in cell size from other bacteria and eukaryotic picoplankton by standard microscopy. The water-soluble phycobiliproteins are easily lost from damaged cells. Thus, it is essential that the observations be made on freshly collected material, or on samples preserved in fixatives that assure cell integrity (Tsuji *et al.*, 1986; Waterbury *et al.*, 1986). For quanti-tative enumerations of picocyanobacteria, plankton samples are generally concentrated by filtration onto polycarbonate membranes and detection is performed by near-violet to blue (395–490 nm) or green (510–560 nm) light excitation (MacIsaac and Stockner, 1993). The filter sets that have been used for delimiting the bandwidth of excitation and emission wavelengths vary among different authors (e.g. Tsuji *et al.*, 1986; Waterbury *et al.*, 1986; Crosbie *et al.*, 2003; Jasser *et al.*, 2010; Ohki *et al.*, 2012), and also depend on the manufacturer of the microscope. If the photosynthetic pigments in the cells are excited with blue light, PE will emit a yellow to orange light, whereas PC and chlorophylls emit red light. The excitation by green light causes emission of an orange–red light in PE-rich picocyanobacteria, and red fluorescence from PC in those lacking PE (often considered as "PC-rich"), but it is not good for the detection of chlorophylls (Waterbury *et al.*, 1986) (Fig. 4.2). Therefore, the presence of photosynthetic picoeukaryotes, or prochlorophy-tan picocyanobacteria (*Prochlorococcus*) and *Acaryochloris* (lacking phycobilisomes, see above), needs to be examined with both light excitations (Chisholm *et al.*, 1988; Felföldi *et al.*, 2009). This will also distinguish cyanobacteria from non-photosynthetic bacteria and non-photosynthetic eukaryotes, which do not fluoresce. Distinction from anoxygenic photosynthetic bacteria requires special blue filter sets and a camera for detecting and recording fluorescence emission in the infrared region of the light spectrum, characteristic of the latter phototrophs (Ohki *et al.*, 2012). Common picocyanobacteria in freshwater environments (but also found in marine and oceanic habitats) belong to the genera *Synechococcus* and *Cyanobium*. Their correct identification requires the application of molecular methods on isolated strains. Contrary to previous works, which assumed lack of toxicity, a few recent investigations demonstrated that picocyanobacteria are actually able to produce different toxic secondary metabolites, including hepatotoxins, neurotoxins, and dermatotoxins, as well as bad-tasting and odorous compounds (Jasser and Callieri, 2017).

4.4 Molecular Markers: Single Loci

Laboratories that have access to a molecular biology facility (e.g. PCR and sequencing equipment; Chapter 6 and SOP 4.2) can apply molecular techniques for the identification of cyanobacteria. Amplification of genomic DNA by PCR is the first step for most of the molecular methods currently employed for the identification of taxa (see SOP 4.2). The PCR amplicons can then be sequenced, and the resulting sequences compared with those deposited in databases (see Section 4.8). The sequences will also serve for tree construction by phylogenetic analyses.

Among the single molecular markers that have been used to study the phylogeny and identify cyanobacteria since the 1980s, the 16S rRNA gene is the most widely used. It recognizes the most closely related strains (or sequences of uncultivated cyanobacteria) by the degree of sequence identity, and following phylogenetic analyses allows the position-ing of any new sequence into the appropriate branch of the cyanobacterial evolutionary radiation. The usefulness of this molecular marker is supported by the genome study of

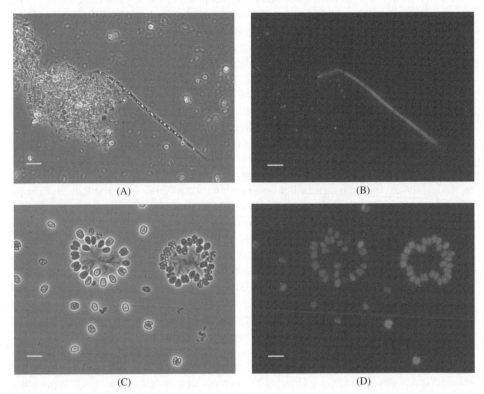

Figure 4.2   *Identification of cyanobacteria in a bloom sample from a freshwater lake (Etagnac, Poitou-Charentes, France, 10/08/2000) by phase contrast (A, C) and autofluorescence (B, D) microscopy. A green filter (Zeiss 15, excitation band path 546/12 nm, beam splitter 580 nm, long path emission 590 nm) was used for fluorescence detection. Bar markers: 10 µm. In (B) note the red fluorescence of "picocyanobacteria" of various different cell dimensions, in addition to the filamentous cyanobacterium of unknown taxonomic position. The presence of the unicellular cyanobacteria cannot be differentiated from other bacteria within and around the mineral precipitate and cellular debris in the corresponding phase contrast image (A). In the absence of the autofluorescence image (B), even the identification of the filamentous cyanobacterium containing large light refractile inclusions (most likely polyalkanoate) would have been difficult, given that morphologically similar non-cyanobacterial taxa exist. The phase contrast image (C) shows a colony of Woronichinia, and some individual cells of this cyanobacterial genus. Note the doublets of lengthwise dividing cells and light refractile gas vesicle clusters ("aerotopes") typical of this planktonic taxon. Gas vesicle collapse resulting from the pressure exerted by the cover slip gives some of the cells a darker appearance. Note also the mucilaginous appendages interconnecting the Woronichinia cells within the colony, and associated thin rod-shaped cells. From the corresponding autofluorescent image (D), it is clear that in this field of view Woronichinia is the only representative of the cyanobacterial lineage. Differences in the intensity of red fluorescence may be attributable to partial loss of phycocyanin, and/or the fact that not all Woronichinia cells are located in the same plane of observation (photos: R. Rippka). (See color plate section for the color representation of this figure.)*

Shih *et al.* (2013), which shows that the major subclades obtained with a large number (31) of concatenated conserved proteins agreed well with phylogenetic trees based on 16S rRNA gene sequences. The total number of cyanobacterial sequences of this gene (from isolates and uncultured representatives) deposited in the Ribosomal Database Project (RDP; see Section 4.8) in September 2016 was nearly 50,000 (*ca.* 8300 corresponding to good quality sequences of cyanobacterial isolates). This gene is about 1500 base pairs (bp) long, and both partial and complete sequences can be used. This marker includes conserved and variable domains, which permits the design of primers for the whole phylum (Table 4.2), as well as for specific genera (e.g. *Microcystis*, *Planktothrix*, *Nostoc*), or distinct phylogenetic lineages (e.g. particular clades of "*Synechococcus*") (Rudi *et al.*, 1997; Ragon *et al.*, 2014). Although in such cases the presence of an amplicon obtained with specific primers is indicative of a particular taxon (or group), the ultimate proof is only given by its sequence. The disadvantages of the 16S rRNA gene are that (1) it is very conserved, and thus may not clearly distinguish within and between closely related taxa, and (2) it is often present in several copies in a same genome. Such intragenomic copies differ sometimes in sequence, leading to the identification of multiple ribotypes (Engene and Gerwick, 2011; Sun *et al.*, 2013), and thus overestimation of molecular diversity. However, Engene and Gerwick (2011) conclude that the degree of divergence between multiple rRNA gene copies is generally low in cyanobacterial genomes and that the intragenomic differences will, in most cases, have only a low impact on the taxonomic and phylogenetic inferences based on this molecular marker. Compared to other housekeeping genes/loci (see below), there are very few cases of horizontal gene transfer (HGT) (Tian *et al.*, 2015) documented for the 16S rRNA genes, making them particularly suitable and reliable for the computation of phylogenetic trees, which assumes that genes are only vertically inherited. If bioinformatic analyses reveal that a new 16S rRNA gene sequence has a doubtful nucleotide signature (and secondary structure), or is positioned in an unexpected phylogenetic clade (or cluster), possible HGT should be analyzed by examining the relationships obtained with other unlinked molecular markers. To distinguish within and between closely related taxa, other molecular markers and techniques have proven to be useful. They often target genetic loci that are supposed to be single-copy (Palenik, 1994; Case *et al.*, 2007). Those most widely used for cyanobacteria, including potentially toxic taxa, are presented below (Table 4.2).

The internal transcribed spacer (ITS) between the 16S and 23S rRNA genes can be amplified in one PCR reaction together with the 16S rRNA gene. This genomic region is more variable than the latter molecular marker, both in sequence and length, ranging from about 330–1010 bp (Iteman *et al.*, 2000; Boyer *et al.*, 2001; Baurain *et al.*, 2002; Rocap *et al.*, 2002; Malone *et al.*, 2015). The secondary structure of the ITS was determined by Iteman *et al.* (2000) for several different cyanobacterial taxa and shown to consist of conserved domains (D1–D5), tRNA gene(s) and antiterminator boxes (box B and box A). These domains are helpful for aligning the ITS sequences, which is particularly important for more distantly related organisms. Moreover, the secondary structure of this region has been proposed as a diagnostic character (Johansen *et al.*, 2011; Malone *et al.*, 2015). The intragenomic variability is higher for the ITS than for the 16S rRNA gene, but this depends on the taxon (Piccin-Santos *et al.*, 2014). If direct sequencing of the amplicon yields multiple peaks on the resulting chromatogram, cloning before sequencing is needed (Gugger *et al.*, 2005). For amplification of the complete ITS, the forward and reverse

Table 4.2 *Selection of genetic markers and primers used for the identification of cyanobacteria and phylogenetic studies. The list is not exhaustive and includes only some of the most frequently targeted genes and loci. The primer specificity quoted for cyanobacteria in general, or particular groups/taxa is based on the currently available references. Some primers target a locus in all bacteria but may be used for PCR amplification, if the second primer is specific for the cyanobacterial phylum, or specific lineages therein. The primer sequences, PCR conditions, and procedure details are reported in the cited references, which should be consulted*

Genetic marker (product)	Primers	Specificity	PCR/ Sequencing	Amplicon (approx. bp)	References
16S rRNA (16S ribosomal RNA)	CYA106F, CYA359F, CYA781R(a), CYA781R(b)	cyanobacteria	PCR and sequencing	450/700	Nübel et al. (1997)
	pA	bacteria	PCR[1] and sequencing	depends on the reverse primer	Edwards et al. (1989)
	23S30R	cyanobacteria	PCR[1] and sequencing	depends on the forward primer	Taton et al. (2003)
	16S979F, 16S544R (16S545R), 16S1092R	bacteria	sequencing[1]		Hrouzek et al. (2005), Rajaniemi et al. (2005), Rajaniemi-Wacklin et al. (2005)
	pC, pE, pD•, pF•, pH•	bacteria	sequencing		Edwards et al. (1989), Gkelis et al. (2005)

16S-23S ITS (16S-23S ribosomal RNA internal transcribed spacer region)	322F, 340R	cyanobacteria	PCR and sequencing	complete sequence of the ITS plus 200 bp of the 16S rRNA	Iteman et al. (2000), Fathalli et al. (2011), Piccin-Santos et al. (2014)
	23S30R	cyanobacteria	PCR and sequencing	complete sequence of 16S rRNA and ITS, if the forward primer is pA	Taton et al. (2003)
cpcBA-IGS (PC-IGS) (partial β and α subunits of the phycocyanin operon, and intergenic spacer)	PCβf, PCαr	cyanobacteria	PCR and sequencing	variable (Anabaena, 685)	Neilan et al. (1995), Ballot et al. (2008)
rpoB (RNA polymerase β subunit)	rpoBF, rpoBR	cyanobacteria	PCR and sequencing	520-635	Rajaniemi et al. (2005)
	rpoBanaF, rpoBanaR	Anabaena, Aphanizomenon	PCR and sequencing	520-635	Rajaniemi et al. (2005)
	RPObF1, RPObR1	Planktothrix, Oscillatoria	PCR and sequencing	600	Gaget et al. (2011)

(Continued)

Table 4.2 *(Continued)*

Genetic marker (product)	Primers	Specificity	PCR/ Sequencing	Amplicon (approx. bp)	References
rpoC1 (RNA polymerase β′ subunit)	RF, RR	cyanobacteria	PCR	731	Rantala et al. (2004)
	RF, RR, RintF, RintR	cyanobacteria	sequencing	731	Rantala et al. (2004)
	RPOC145F, RPOC683R, RPOC1006R	cyanobacteria, except *Microcystis*	PCR and sequencing	555/880	Valério et al. (2009)
	RPOCM61F, RPOCM624R	*Microcystis*	PCR and sequencing	580	Valério et al. (2009)
rbcLX (RuBisCO large subunit and assembly chaperone)	CW, CX	cyanobacteria	PCR and sequencing	782–1003	Rudi et al. (1998)
	CW, CX, DN	*Nostoc*	sequencing	variable	Rudi et al. (1998)
	CW, DF	*Planktothrix*	PCR	variable	Rudi et al. (1998)
	CW, DF, DL	*Planktothrix*	sequencing	variable	Rudi et al. (1998)
nifH (dinitrogenase reductase)	nifHf, nifHr	Nitrogen-fixing genera	PCR and sequencing	400	Gugger et al. (2005), Thomazeau et al. (2010)
hetR (heterocyte differentiation)	hetR1, hetR2	Heterocytous genera	PCR and sequencing	500	Orcutt et al. (2002) Thomazeau et al. (2010)

[1] See SOP 4.2.

primers generally target nucleotide motifs situated within, but at variable positions of, the 16S rRNA gene, and at the 5' end of the 23S rRNA gene. The amplicon length is thus variable, depending on both the ITS length of the organism under study (see above) and the primers used for amplification.

The intergenic sequence (IGS) between the terminal end of the *cpcB* gene and the proximal end of the *cpcA* gene, called *cpcBA*-IGS or PC-IGS (Dyble *et al.*, 2002; Tan *et al.*, 2010), is part of the operon encoding the synthesis of PC, an antenna pigment that is only present in cyanobacteria, red algae, glaucophytes, and cryptophytes (Overkamp *et al.*, 2014). Therefore, this operon can be used to design primers for specific amplification of the cyanobacterial sequences from non-axenic but unicyanobacterial cultures. Possible recombination events, or lateral transfers involving this locus, have been shown for some taxa; however, other studies indicate that this locus is a suitable marker for other taxa (Piccin-Santos *et al.*, 2014). For example, primers designed for specific amplification of *Cylindrospermopsis* isolates amplified a sequence of about 685 bp (Dyble *et al.*, 2002).

The *rbc*LX locus (Rudi *et al.*, 1998; Rajaniemi *et al.*, 2005) is part of the operon encoding the small (*rbc*S) and large subunits (*rbc*L) of the D-ribulose 1,5-bisphosphate carboxylase-oxygenase. A region of about 800–1000 bp is amplified with primers annealing to the *rbc*L and *rbc*S genes and includes two intergenic spacers and a chaperone gene, *rbc*X.

The *rpo*C1 (Palenik, 1994) gene encodes the gamma subunit of the cyanobacterial (or chloroplast) RNA polymerase. The primers designed for the detection of an *rpo*C1 gene fragment by Palenik (1994) gave rise to amplicons of 612 bp for marine strains of *Synechococcus*.

The *rpo*B (Case *et al.*, 2007) gene encodes the beta subunit of the RNA polymerase. Different primer pairs resulted in partial *rpo*B sequences of 600 bp (Gaget *et al.*, 2015) or 520–635 bp (Rajaniemi *et al.*, 2005). An example of agarose gel electrophoresis of PCR products (*rpo*B) is shown in Fig. 4.3.

Other markers targeting genes encoding proteins of important function, such as *nif*H and *het*R, are specific to heterocytous and non-heterocytous N_2-fixing taxa (Steunou *et al.*, 2006; Tomitani *et al.*, 2006; Thomazeau *et al.*, 2010).

Additional genetic markers useful for the classification of cyanobacterial populations, including potentially toxic members, have been reported (e.g. Tanabe *et al.*, 2007; Garcia-Pichel, 2008; Mazard *et al.*, 2012; Moreira *et al.*, 2013). Like most of the genetic markers cited above, they proved to be useful for the discrimination of genotypes, at or below the species level (e.g. Lin *et al.*, 2010; Tan *et al.*, 2010; Toledo and Palenik, 1997). However, their usefulness will depend on the cyanobacterial taxa and the taxonomic resolution desired (e.g. Dall'Agnol *et al.*, 2012).

A potential drawback in the use of housekeeping genes and ITS loci is that they may be subject to HGT, which redistributes genes among different bacterial genomes (Yerrapragada *et al.*, 2009; Bolhuis *et al.*, 2010). On the basis of 51 cyanobacterial genomes and 324 single-copy protein families, Dagan *et al.* (2013) estimate that 66% of cyanobacterial protein families have been affected by at least one HGT event, causing problems in the reconstruction of major evolutionary transitions.

An additional group of molecular markers is based on the gene clusters encoding the non-ribosomal peptide (NRPS) and polyketide synthases (PKS) involved in the production of cyanotoxins. As explained in Chapter 6, PCR primers have been designed

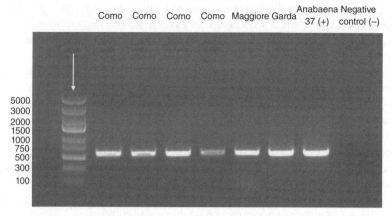

Figure 4.3 *PCR products of rpoB analyzed by electrophoresis on agarose gel, and staining with ethidium bromide. The amplicons were obtained with primers rpoBanaF and rpoBanaR (Table 4.2). The bands refer to strains isolated from samples collected in lakes Como (October 2014), Maggiore (November 2014) and Garda (November 2014). "+" positive (Anabaena strain 37 UHCC) and "–" negative controls. The first line reports the bands obtained from the DNA ladder; the size of the ladder (left scale) is in base pairs; the white arrow indicates the direction of movement of the ladder DNA. The rpoB amplicons are all located at a level corresponding to the region between the molecular marker bands of 500 and 750 bp. The sequencing of the PCR products and phylogenetic analyses allowed associating the rpoB amplicons to* Dolichospermum lemmermannii *(Salmaso et al. 2015b). Source: Salmaso (2015b). Reproduced with permission of Elsevier.*

for amplification of specific taxa or groups. However, the presence of these genes indicates only that the cyanobacteria have the potential to produce the toxins, but there may be mutations or regulatory mechanisms that hinder the synthesis (Christiansen *et al.*, 2008; Ostermaier *et al.*, 2012).

In general, it is good practice to verify the taxonomic identification of a species and results of 16S rRNA gene sequence analyses (or other genes/loci) by examining more than one isolate, preferably collected from different sampling points and at different seasons, prior to drawing conclusions with respect to phylogeny and biogeography.

4.5   Molecular Markers: Multiple Loci

4.5.1   Multilocus Sequence Typing (MLST) and Multilocus Sequence Analysis (MLSA)

The concurrent analysis of multiple genes is performed using different approaches. Compared to the analysis of single genetic markers, these techniques have been applied less for the characterization of cyanobacterial taxa, though their use is increasing. The MLST method is an unambiguous procedure for characterizing isolates of bacterial species, and requires the determination of nucleotide sequences of approximately 400–500 bp for a range of housekeeping genes common to the organisms being analyzed. For each gene, sequences ("alleles") that differ from each other are designated by a random integer number, whereas those that are identical to one or another of them will receive the same respective allele number. Then, the combinations of alleles ("allelic profiles" or "sequence types") for the ensemble of gene loci in the dataset are compared and tabulated, and those that are distinct will be denominated by a sequence type (ST) number (see supplementary Table S1 in Tanabe *et al.*, 2007, as an example). Strains sharing the same allelic profile can then be recognized/identified by the respective ST number. In their original work on MLST, Maiden *et al.* (1998) analyzed the pairwise differences between allelic profiles by computing dendrograms using a cluster analysis. In MLSA, the sequences of the housekeeping genes determined for each strain are concatenated and used in phylogenetic analyses. MLSA can also be coupled with MLST analysis (MLST/MLSA), where only the sequences corresponding to the different allelic profiles (unique STs) are concatenated prior to phylogenetic analyses (since it is already known which strains have identical gene sequences). MLST on its own is sufficient for the identification or study of strains belonging to a well-circumscribed species, whose phylogeny is known. Phylogenetic analyses based on MLSA are essential for any taxon for which only 16S rRNA gene sequences are available in order to increase the taxonomic resolution, and thus gain support for the descriptions of species (or subspecies). On the other hand, MLST data provide information on how the alleles are assorted within each ST, and potential recombination events involved in the emergence of allelic variation(s) among different STs can be estimated by linkage disequilibrium analysis, and other statistical tests for recombination (see D'Alelio *et al.*, 2013, and references therein). Indexes of association (I_A) significantly higher than zero will indicate non-random association of the alleles, infrequent or no recombination events, and thus clonal population structure. In contrast, I_A values around zero are indicative of random allele association, frequent recombination, and thus panmictic population structure. This information, in combination with phylogeny based on MLSA, is essential for inferring the population structure of a species, or lineages within the species, in natural communities. Therefore, both methodologies are recommended, particularly for the study of potentially toxic cyanobacterial taxa.

Tanabe *et al.* (2007) applied a combination of MLST and MLSA on 164 isolates of *Microcystis aeruginosa* based on the sequences of seven housekeeping loci. Their results indicated a high level of genetic diversity and a clonal population structure for this species. While phylogenetic analyses of individual genes did not confidently resolve the phylogenetic relationships, using all of the seven concatenated sequences (2992 bp) provided a more resolved tree, with distinct clusters composed of toxic, nontoxic, and toxic plus nontoxic strains. In a subsequent study, expanding the MLST/MLSA analysis to 412 strains of *M. aeruginosa*, five intraspecific lineages were identified that appeared to be panmictic. However, the overall clonality of the species was again confirmed, probably owing to a reduced recombination rate among lineages (Tanabe and Watanabe, 2011). Similarly, Kurmayer *et al.* (2015) applied a combination of MLST/MLSA to study 138 toxic and nontoxic strains of *Planktothrix* species isolated from Europe, Russia, North America, and East Africa. The concatenation, alignment (2697 bp), and phylogenetic analysis of seven marker loci (three housekeeping genes and four intergenic spacer regions) permitted clear link between the phylogeny and groupings of genotypes with and without an active *mcy* gene cluster encoding the synthesis of microcystins. Three lineages were obtained, of which two were identified as *P. agardhii/rubescens* and the third one as *P. pseudagardhii*. Other examples of multi-locus (3–7 loci) based applications can be found in Acinas *et al.* (2009), Mazard *et al.* (2012), D'Alelio *et al.* (2013), Lara *et al.* (2013), and Gaget *et al.* (2015).

4.5.2   Genome-Based Extension of MLST and MLSA

More recently, whole genome sequences (WGS) can serve for retrieving gene sequences (or other loci) of the respective organism(s) and then performing *in silico* MLST or MLSA analyses by comparison with sequences already available in databases, without having to do any experimental work oneself. For well-characterized bacterial species, particularly pathogens, this can be done via MLST servers, which will identify the allelic profile(s) for the genome-derived set of sequences and, if new, can expand the relevant MLST database (Das *et al.*, 2014; Rosselló-Móra and Amann, 2015). In the study by Wu and Eisen (2008), WGS-derived housekeeping genes (31) were analyzed by MLSA for deducing the evolutionary relationships of a large number (578) of phylogenetically and phenotypically diverse bacteria. With respect to cyanobacteria, a similar approach allowed constructing phylogenetic trees inferred from 31 (Shih *et al.*, 2013; Komárek *et al.*, 2014) and 324 (Dagan *et al.*, 2013) concatenated protein families common to all the strains analyzed in the respective dataset. Based on their results and literature review, Komárek *et al.* (2014) revised the taxonomic classification of cyanobacterial genera and their arrangement into families and orders. This update can also be consulted on the AlgaeBase website (Guiry and Guiry, 2017).

4.6   Molecular Typing Methods Based on Gel Electrophoresis

"Genomic fingerprints" are banding patterns obtained from DNA fragments of different lengths separated by gel electrophoresis. Depending on the taxa analyzed, and the methods used to produce the DNA fragments, genomic fingerprints may help to discriminate between closely related organisms. In restriction fragment length polymorphism (RFLP), PCR products of specific genes/loci are digested by restriction enzyme(s) prior to electrophoresis (e.g. Bolch *et al.*, 1996; Palinska *et al.*, 2011). In the random amplified polymorphic DNA (RAPD) method, total genomic DNA is amplified by PCR with random short primers (generally 10-mers) that have a high probability to anneal frequently in the entire genome (Neilan *et al.*, 1995; Palinska *et al.*, 2011). However, the banding patterns are more complex than the ones obtained with RFLP and not easily reproducible within and between laboratories, and thus this method is no longer recommended. A more reproducible technique is Amplified Fragment Length Polymorphism (AFLP) (Janssen *et al.* 1996) that was applied by Oberholster *et al.* (2005) for examining the genetic diversity of *Microcystis aeruginosa* strains. Nowadays, the amplified genomic restriction fragments are labelled with fluorescent dyes (fluorophores), and electrophoresis and band detection are performed by capillary electrophoresis (e.g. Kütahya *et al.*, 2011). However, this method has been rarely used for cyanobacteria, probably because it requires axenic strains and DNA of high quality and high molecular weight. Genetic profiles of cyanobacteria have also been obtained by applying rep-PCR, performed with primers targeting noncoding repetitive sequences spread throughout the bacterial genomes (Lyra *et al.*, 2001; Piccini *et al.*, 2011), or HIP-PCR, in which the primers target a highly iterated palindrome which is overrepresented in cyanobacterial genomes (Smith *et al.*, 1998; Wilson *et al.*, 2005). These methods use higher annealing temperatures during PCR because the targets are well-defined sequences and result in more stable patterns than RAPD.

4.7 Denaturing Gradient Gel Electrophoresis (DGGE)

This method separates PCR products (or cloned DNA fragments) of identical (or very similar) lengths but differing by as little as a single base substitution in their nucleotide sequence according to their temperature-dependent dissociation ("melting") characteristics (Sheffield *et al.*, 1989). For this, electrophoresis is performed with acrylamide gels containing a linear ascending gradient of denaturing agents (urea and formamide), which has the same effect on DNA melting as a rise in temperature. During such gel electrophoresis conditions, DNA fragments migrate through the gel as double stranded molecules until they reach a concentration of the denaturing agents equivalent to the melting temperature (T_m) that will lead to strand dissociation in a region ("domain") lowest in percentage of guanine plus cytosine to total base content (mol % GC), and thus most sensitive to temperature. The "branching" or "bulge" formation of partly dissociated DNA fragments leads to an abrupt decrease of mobility, and thus retardation of migration. Differences in nucleotide sequence (and thus melting characteristics) between two DNA fragments slow migration at different concentrations of the denaturing agents, and thus permit their separation from each other. In order to distinguish nucleotide changes occurring in a DNA domain highest in GC content (having the highest T_m), and to prevent complete strand dissociation (at which sequence-dependent migration is no longer observed), a 40 bp G+C-rich sequence, called a GC clamp, is generally attached to the 5' or 3' ends of the amplification products prior to gel electrophoresis (Sheffield *et al.*, 1989). This method is particularly useful for the study of complex cyanobacterial communities and has been shown to distinguish between closely related cyanobacteria, including potentially toxic genera or species (Janse *et al.*, 2003; Boutte *et al.*, 2006). Complete details of DGGE are given in Chapter 9 and SOP 9.1.

4.8 Taxonomic and Molecular Databases

The online algal database AlgaeBase (Guiry and Guiry, 2017) provides current cyanobacterial taxonomy (Table 4.1) and information on synonyms, key references, habitat, and geographic distribution. Though this website is updated frequently by the site managers, it is based on a subjective selection of the most recent and relevant literature. Thus the website can provide very useful indications for the nomenclature of cyanobacteria, but cannot be considered the "ultimate" source for taxonomy. The classification and nomenclature of taxa should always be based on recent publications and consultation of the original literature. Another useful website for cyanobacterial genera and species, also listing rejected genera, is the CyanoDB archive (Komárek and Hauer, 2013) (Table 4.1).

The analysis of molecular markers is undertaken by comparing molecular sequences with those deposited in public databases. The INSDC (International Nucleotide Sequence

Database Collaboration) is an initiative that operates between EMBL-EBI (The European Bioinformatics Institute), NCBI (National Center for Biotechnology Information; Gen-Bank) and DDBJ (DNA Data Bank of Japan) (Table 4.1) (Cochrane *et al.*, 2016). The three centers from three continents host many molecular sequences. There is only partial control of sequence quality and taxonomic identification. Therefore, these data should be used with caution. BLAST data should not be used to measure similarities between sequences because ambiguities are counted as differences, whereas the ambiguous positions should be deleted from the analysis.

The Ribosomal Database Project (RDP), SILVA, Greengenes, and EzTaxon are slightly different databases that keep a curated alignment of 16S rRNA sequences (Table 4.1). These databases also allow matching and aligning sequences, and testing probes. leBIBI^{QBPP} (Quick Bioinformatic Phylogeny of Prokaryotes) is designed to reconstruct the phylogeny of archaea and bacteria, providing information on species identification using 16S rRNA and a selection of a few other genes (e.g. *rpo*B) (Table 4.1).

Genomes are now found in the Genome Online Database (GOLD) and EzGenome (Table 4.1). Though they are not of immediate use to identify toxic cyanobacteria, they are useful to improve the knowledge of evolution and design new molecular assays (e.g. based on the sequences of the gene clusters encoding cyanotoxin synthesis).

4.9 The Polyphasic Approach

Similar cyanobacterial morphotypes can differ extensively in physiology, ecology, and phylogeny (Bittencourt-Oliveira *et al.*, 2001; Willame *et al.*, 2005; Comte *et al.*, 2007; Lahr *et al.*, 2014). Conversely, different morphotypes, particularly among colonial forms, can genetically be almost identical (Palinska *et al.*, 1996; Otsuka *et al.*, 2001; Rajaniemi *et al.*, 2005; Lahr *et al.*, 2014). Furthermore, strains cultured and maintained for long periods under artificial conditions in the laboratory can exhibit uncommon or "distorted" features (Strunecký *et al.*, 2010; Komárek, 2014). In such cases, the organisms can no longer be readily identified by microscopy. Another example of morphological change is that of single-celled cultures of *Microcystis aeruginosa*, which, upon exposure to grazing by *Ochromonas*, respond actively by forming colonies (Yang *et al.*, 2008). For all these various reasons, a polyphasic approach is increasingly being applied, both in the botanical and bacteriological literature, for a better classification and understanding of already existing taxa or for the description of new ones (Komárek, 2016) (see SOP 4.2). This approach implies that the organisms are characterized based on their phenotypic characters (morphology, cellular differentiation, ultra-structure), physiology, ecology, and genetic profile (16S rRNA gene sequences, and/or other sequences of selected more discriminatory molecular markers) (Table 4.3; Fig. 4.4). The resulting combination of data allows distinguishing important (or diacritical) characters for identification, and to differentiate between phenotypic plasticity, resulting from environmental factors or culture conditions, and fundamental phenotypic distinctions caused by genetic differences (Fig. 4.5).

The application of the polyphasic approach has increased rapidly, especially during the last decade, and is a huge improvement compared with the exclusive use of phenotypic characters or selected single molecular markers (Table 4.3). However, its practical application

Table 4.3 *Examples of taxonomic identification of cyanobacteria using the polyphasic approach*

Taxa	Phenotypic characters	Physiology	Ecology	Genetic markers	Phylogenetic analyses	Notes (habitat)	Reference
Several genera	morphol, TEM (thylakoids)	—	—	16S rRNA	(Megalign, DNAStar)	Congruence between phenotype (morphology, thylakoid arrangements) and molecular markers.	Komárek and Kaštovský (2003)
51 strains of *Anabaena, Aphanizomenon, Nostoc, Trichormus* spp.	morphom, morphol	—	—	16S rRNA, *rpoB, rbcLX*	NJ, MP, ML (PHYLIP, PAUP)	Morphological analyses carried out at the time of isolation.	Rajaniemi et al. (2005)
Dolichospermum spp. (*Anabaena* spp.)	morphom, morphol	Effects of temp, light N, P	—	16S rRNA ITS	NJ, MP (Phylip)	Several planktic species.	Zapomělová et al. (2009)
Pseudanabaena rutilis-viridis	morphom, morphol, TEM	ELISA, PPIA, screening pigments	Field measurements (temp, light, chemistry)	16S rRNA ITS	ML, NJ (MEGA), Bayes (MrBayes)	Planktic.	Kling et al. (2012)
Sphaerospermopsis, Cylindrospermopsis, Raphidiopsis, Microcystis	morphom, morphol	—	—	*cpcBA*-IGS	ML (PAUP)	41 MCs variants detected in the environmental samples from Hartbeespoort Dam.	Ballot et al. (2014)

(Continued)

Table 4.3 (*Continued*)

Taxa	Phenotypic characters	Physiology	Ecology	Genetic markers	Phylogenetic analyses	Notes (habitat)	Reference
Reassessment of the cyanobacterial family Microchaetaceae	morphom, morphol	—	Habitat, biogeography	16S rRNA	Bayes (MrBayes), ML (PhyML), MP (SeaView)	Phylogenetic analyses showed that members of Microchaetaceae form several distant groups.	Hauer *et al.* (2014)
Cyanobacteria	morphol, TEM (reviewed)	—	General	31 conserved proteins	ML (RaxML)	Revision of the taxonomic classification of cyanobacterial genera, families and orders.	Komárek *et al.* (2014)
8 Chroococcales, 6 Oscillatoriales, 25 Nostocales	morphom, morphol	—	—	16S rRNA, *rpo*C1, *cpc*BA-IGS	NJ, MP, ML (MEGA)	Several Australian strains showed <95% similarity to available sequences in GenBank.	Lee *et al.* (2014)
Nodularia spp.	morphom, morphol	—	—	16S rRNA ITS	Bayes (MrBayes), ML (RaxML), MP (PAUP)	Soil and benthic habitats; problematic taxonomic resolution at the subgeneric level.	Řeháková *et al.*, (2014)

Planktothrix spp.	morphom, morphol	temperature and salinity tolerance, pigments	—	MLST using 16S rRNA, ITS, *gyrB*, *rpoC1*, *rpoB*	NJ, MP, ML (MEGA)	Axenic cultures; MALDI-TOF mass spectrometry revealed 13 chemotypes.	*Gaget et al.* (2015)
Cephalothrix	morphom, morphol TEM	—	Field measurements, habitat	16S rRNA, ITS, *rbcLX*, *rpoB*; Secondary structure of ITS	ML, NJ, Bayes (MEGA, MrBayes)	Planktic and benthic.	*Malone et al.* (2015)
Stigonema spp. and *Petalonema alatum*	morphom, morphol	—	Habitat	16S rRNA	Bayes (MrBayes), ML (PhyML), NJ (SeaView)	Single-cell and filament sequencing of cultivation-resistant terrestrial cyanobacteria with massive sheaths.	*Mareš et al.* (2015)
Dolichospermum lemmermannii	morphom, morphol	—	Field measurements, paleolimnology (eutrophication/ colonization history)	16S rRNA *rpoB*	ML (R, PhyML)	Natural and resurrected (akinete germination) planktic populations.	*Salmaso et al.* (2015a, 2015b)

(Continued)

Table 4.3 *(Continued)*

Taxa	Phenotypic characters	Physiology	Ecology	Genetic markers	Phylogenetic analyses	Notes (habitat)	Reference
Tychonema bourrellyi	morphom, morphol	—	Field measurements	16S rRNA rbclX	ML (MEGA, R, PhyML)	Planktic.	Shams et al. (2015), Salmaso et al. (2016)
17 cyanobacterial isolates	morphom, morphol, SEM	—	pH, conductivity, salinity, and water temperature	16S rRNA and 16S–23S Secondary structure of rRNA ITS	Bayes (MrBayes), ML (PhyML), MP (TNT). Topology differences (R)	Strains isolated from mats of 9 thermal springs in Greece; identification of 6 new taxa at the genus level.	Bravakos et al. (2016)
Snowella litoralis, *Woronichinia compacta*	morphom, morphol	Enzyme Inhibitory Activity	—	16S rRNA ITS	NJ, MP, ML (Phylip)	Example of controversial morphologic and phylogenetic results.	Häggqvist et al. (2016)

Abbreviations: morphom, morphometry; morphol, morphology; SEM, scanning electron microscopy; TEM, transmission electron microscopy; temp, temperature; N, nitrogen; P, phosphorus; NJ, Neighbor Joining; MP, Maximum Parsimony; ML, Maximum Likelihood, Bayes, Bayesian inference. Under phylogenetic analyses, the software used to estimate phylogenetic trees is reported within brackets.

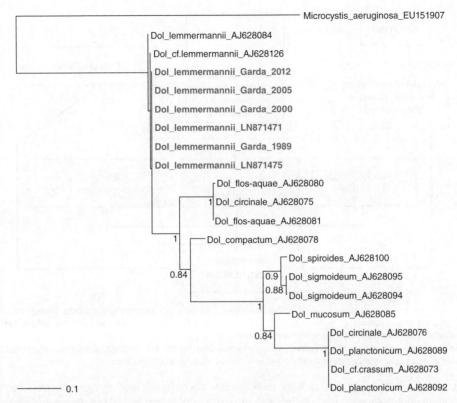

Figure 4.4 *Maximum likelihood (ML) rooted (M. aeruginosa) tree topology of Dolichospermum lemmermannii strains isolated from water samples collected in Lake Garda (LN871475, May 2014; LN871471, November 2013) and resurrected from akinetes isolated from the deep sediments of Lake Garda (1989–2012) (highlighted in blue). The other two D. lemmermannii strains were isolated from Finnish water bodies. The Dolichospermum species included in the analysis are identified by names and accession numbers. The tree is based on the alignment of the rpoB gene using R 3.3.0 and PhyML 3.1 (cf. SOP 4.2). The analysis of DNA substitution models indicated in the K80+G the best-fitting evolutionary model. Branch support aLRT-SH-like option values < 0.7 were not shown; the selection threshold for SH-like supports should be in the 0.8–0.9 range (Guindon et al. 2010). Modified from Salmaso et al. (2015a).*

for the "correct" nomenclature of taxa is not free of drawbacks. This is partly due to morphological similarity or environmentally induced variations (see above), but it is also the result of misidentified or poorly described organisms among those for which sequences are available in the INSDC archives (see the case study in Lee *et al.*, 2014). Furthermore, different authors have used different molecular markers for genetic distinctions (and thus the results cannot be readily compared), and the number of specific probes for several gene/loci is still limited to a relatively low number of well-studied taxa. Consequently, multiple loci analyses for the taxonomy of rarely observed, or unknown, cyanobacterial genera/species may only be possible after designing and testing new primers.

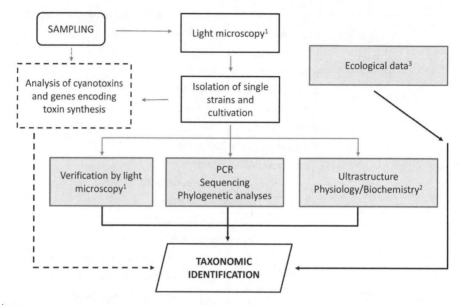

[1]Light microscopy: cell association (single, colony, filament bundle, etc.); cell morphology and cellular differentiation.
[2]Physiology and biochemistry: N_2-fixation, light, temperature and salinity optima (or tolerance) for growth, heterotrophic growth potential, etc.; pigment characterization.
[3]Ecological data: geography, type of habitat (water, terrestrial, etc.), salinity, pH, temperature, light intensity, mineral nutrients, and trophic status, cyanobacterial mass occurrence (water bloom, biofilm, etc.).

Figure 4.5 *Polyphasic approach for the detection and identification of cyanobacteria from environmental samples based on the analysis of strains obtained by culture-dependent methods. The gray boxes include the criteria used in the identification of cyanobacteria. The determination of secondary toxic (and nontoxic) metabolites and the analysis of genes encoding cyanotoxin synthesis (dashed box) add further information on the characteristics of species and their toxigenic potential.*

The polyphasic approach is generally performed on single axenic or non-axenic strains (Fig. 4.5) obtained by culture-dependent methods (e.g. SOP 4.2), and permitting phenotypic and physiological analyses after growth under different culture conditions. If culturing is not possible (or not desired), temporary filament (or colony) isolates from environmental samples (see SOPs 2.3, 5.4, 6.7) may also serve as the starting material, though the polyphasic characterization will then be limited mainly to structural (ultrastructural) features and sequence data (Lara *et al.*, 2013; Mareš *et al.*, 2015). Neither of these approaches is fully suitable for gaining complete insights into the overall taxonomic composition and biodiversity in environmental samples (but see below). In this regard, next-generation sequencing (NGS) or high-throughput sequencing based on the use of universal PCR primers to mass-amplify environmental DNA (Oulas *et al.*, 2015) has enormously increased the sequence coverage of 16S rRNA gene amplicons and, together with other associated metagenomic marker gene analyses, provides unprecedented possibilities for unveiling the diversity of prokaryotic (and eukaryotic) plankton communities (see Chapter 10). Nevertheless, for a robust taxonomic assignment of high-throughput cyanobacterial sequences, culture-independent metagenomic techniques require solid reference molecular databases (see Section 4.8). Their reliability, however, can only be assured by further updating the INSDC archives with more molecular sequences and corresponding correct taxonomic names resulting from polyphasic characterizations of cyanobacterial strains.

The insufficient number of cyanobacterial taxa represented in the current reference databases, together with other problematic issues (numerous sequences of organisms only known from environmental clones, unnamed OTUs within a taxon, or representatives that carry conflicting names, and/or lack specific epithets), is largely responsible for the rather high percentage of sequences that cannot be identified at the generic level (and even less so at the species level). This was, for example, shown in the metagenomic NGS studies based on 16S rRNA gene amplicons by Williams *et al.* (2016) on biological soil crusts (BSCs) of Western Europe, and by Pessi *et al.* (2016) on Antarctic lacustrine microbial mats. In addition, a comparison of the organisms identified by microscopy (after pre-cultivation of the samples) and the sequence analyses showed agreements between the two identification approaches but also discrepancies (the taxon either having been detected by microscopy but not by sequencing, or *vice versa*) (Williams *et al.*, 2016). In order to optimize the experimental conditions and bioinformatic tools ("pipelines") for their NGS sequence analyses, Pessi *et al.* (2016) first investigated artificial communities of cyanobacterial taxa already known from Antarctic habitats. In agreement with Williams *et al.* (2016), the latter authors concluded that several different methodologies should be tested and compared for faithfully assessing the true biodiversity in a given sampling site, as well as for distinguishing ecologically relevant cyanobacteria, even from allochthonous representatives. In this regard, metagenomic NGS studies can provide a lot of new insights with respect to niche specificity, biogeography, competition within a resident population, and other factors important for the more or less successful occupation over time and space of known (or yet to be identified) taxa in their natural environment.

4.10 Final Considerations

The correct identification of cyanobacteria is indispensable for every program aimed at the management of waters containing toxic species (or subspecies) and used for recreation, drinking water supply, and/or agricultural practices. Correct nomenclature provides a means for associating a taxon name with genotypic and ecological characteristics of the organisms occurring in water blooms (or other habitats), thus facilitating subsequent safety steps in management plans. Knowledge on the range of ecological adaptations of different cyanobacterial species (or subspecies) to different environments (Häggqvist *et al.*, 2017; Humbert and Fastner, 2017) is also helpful for management decisions on how best to counter the potential mass development of toxic representatives (Mantzouki *et al.*, 2016). Despite its limits, the polyphasic characterization is currently the best option available for the identification and characterization of cyanobacteria. In the future, in parallel with a better knowledge of ecological characteristics and geographic distribution, the application of genome-based analyses and the inclusion of metabolomic profiles will be instrumental for improving current identification methods and classification systems, as well as for drawing new conclusions on the ecological specificity and biogeography of cyanobacterial taxa. Progress is necessary, considering that the number of known cyanobacterial species is severely underestimated compared to the expected number of species yet to be described estimated from statistical models (Guiry, 2012; Nabout *et al.*, 2013). Overall, the knowledge of the number of toxic secondary metabolites and other bioactive compounds, including those of medical relevance, is expected to increase.

4.11   References

Abed, R.M.M. and Garcia-Pichel, F. (2001) Long-term compositional changes after transplant in a microbial mat cyanobacterial community revealed using a polyphasic approach, *Environmental Microbiology*, **3** 53–62, doi: 10.1046/j.1462-2920.2001.00159.x.

Abed, R.M.M., Garcia-Pichel, F., and Hernández-Mariné, M. (2002) Polyphasic characterization of benthic, moderately halophilic, moderately thermophilic cyanobacteria with very thin trichomes and the proposal of *Halomicronema excentricum* gen. nov., sp. nov, *Archives of Microbiology*, **177**, 361–370.

Acinas, S.G., Haverkamp, T.H.A., Huisman, J., and Stal, L.J. (2009) Phenotypic and genetic diversification of *Pseudanabaena* spp. (cyanobacteria), *ISME Journal*, **3** 31–46, doi: 10.1038/ismej.2008.78.

Ballot, A., Dadheech, P.K., Haande, S., and Krienitz, L. (2008) Morphological and phylogenetic analysis of *Anabaenopsis abijatae* and *Anabaenopsis elenkinii* (*Nostocales*, cyanobacteria) from tropical inland water bodies, *Microbial Ecology*, **55**, 608–18, doi: 10.1007/s00248-007-9304-4.

Ballot, A., Sandvik, M., Rundberget, T. *et al.* (2014) Diversity of cyanobacteria and cyanotoxins in Hartbeespoort Dam, South Africa, *Marine and Freshwater Research*, **65**, 175–189, doi: 10.1071/MF13153.

Baurain, D., Renquin, L., Grubisic, S., *et al.* (2002) Remarkable conservation of internally transcribed spacer sequences of *Arthrospira* ("*Spirulina*") (Cyanophyceae, Cyanobacteria) strains from four continents and of recent and 30-year-old dried samples from Africa, *Journal of Phycology*, **38**, 384–393, doi: 10.1046/j.1529-8817.2002.01010.x.

Benson, D.A., Cavanaugh, M., Clark, K., *et al.* (2013) GenBank, *Nucleic Acids Research*, **41**, D36–D42, doi: 10.1093/nar/gks1195.

Bernard, C., Ballot, A., Thomazeau, S., *et al.* (2017) Appendix: Cyanobacteria associated with the production of cyanotoxins, in J. Meriluoto, L. Spoof, and G.A. Codd, (eds), *Handbook on Cyanobacterial Monitoring and Cyanotoxin Analysis*, John Wiley & Sons, Ltd, Chichester, UK, 503–507.

Bittencourt-Oliveira, M.D.C., de Oliveira, M.C., and Bolch, C.J.S. (2001) Genetic variability of Brazilian strains of the *Microcystis aeruginosa* complex (Cyanobacteria/Cyanophyceae) using the phycocyanin intergenic spacer and flanking regions (*cpc*BA), *Journal of Phycology*, **37**, 810–818, doi: 10.1046/j.1529-8817.2001.00102.x.

Bolch, C.J.S., Blackburn, S.I., Neilan, B.A., and Grewe, P.M. (1996) Genetic characterization of strains of cyanobacteria using PCR-RFLP of the *cpc*BA intergenic spacer and flanking regions, *Journal of Phycology*, **32**, 445–451, doi: 10.1111/j.0022-3646.1996.00445.x.

Bolhuis, H., Severin, I., Confurius-Guns, V., *et al.* (2010) Horizontal transfer of the nitrogen fixation gene cluster in the cyanobacterium *Microcoleus chthonoplastes*, *ISME Journal*, **4**, 121–30.

Boone, D.R. and Castenholz, R.W. (2001) *Bergey's Manual of Systematic Bacteriology. Volume one: The Archaea and the Deeply Branching and Phototrophic Bacteria*, 2nd edn, Springer-Verlag, New York.

Boutte, C., Grubisic, S., Balthasart, P., and Wilmotte, A. (2006) Testing of primers for the study of cyanobacterial molecular diversity by DGGE, *Journal of Microbiological Methods*, **65**, 542–550, doi: 10.1016/j.mimet.2005.09.017.

Boyer, S.L., Flechtner, V.R., and Johansen, J.R. (2001) Is the 16S-23S rRNA Internal Transcribed Spacer region a good tool for use in molecular systematics and population genetics? A case study in cyanobacteria, *Molecular Biology and Evolution*, **18** (6), 1057–1069.

Bravakos, P., Kotoulas, G., Skaraki, K., *et al.* (2016) A polyphasic taxonomic approach in isolated strains of Cyanobacteria from thermal springs of Greece, *Molecular Phylogenetics and Evolution*, **98**, 147–60, doi: 10.1016/j.ympev.2016.02.009.

Briand, E., Gugger, M., Francois, J.C., *et al.* (2008) Temporal variations in the dynamics of potentially Microcystin-producing strains in a bloom-forming *Planktothrix agardhii* (cyanobacterium) population, *Applied and Environmental Microbiology*, **74**, 3839–3848, doi: 10.1128/AEM.02343-07.

Bryant, D.A. (1982) Phycoerythrocyanin and phycoerythrin: Properties and occurrence in cyanobacteria, *Journal of General Microbiology*, **128**, 835–844.

Case, R.J., Boucher, Y., Dahllöf, I., *et al.* (2007) Use of 16S rRNA and *rpo*B genes as molecular markers for microbial ecology studies, *Applied and Environmental Microbiology*, **73**, 278–88, doi: 10.1128/AEM.01177-06.

Catherine, A., Maloufi, S., Congestri, R., *et al.* (2017) Cyanobacterial samples: Preservation, enumeration and biovolume measurements, in J. Meriluoto, L. Spoof, and G.A. Codd, (eds), *Handbook on Cyanobacterial Monitoring and Cyanotoxin Analysis*, John Wiley & Sons, Ltd, Chichester, UK, 315–330.

Chen, M., Li, Y., Birch, D., and Willows, R.D. (2012) A cyanobacterium that contains chlorophyll *f*: A red-absorbing photopigment, *FEBS Letters*, **586**, 3249–3254.

Chisholm, S.W., Olson, R.J., Zettler, E.R., *et al.* (1988) A novel free-living prochlorophyte abundant in the oceanic euphotic zone, *Nature*, **334**, 340–343.

Christiansen, G., Molitor, C., Philmus, B., and Kurmayer, R. (2008) Non-toxic strains of cyanobacteria are the result of major gene deletion events induced by a transposable element, *Molecular Biology and Evolution*, **25**, 1695–1704.

Cochrane, G., Karsch-Mizrachi, I., Takagi, T., and International Nucleotide Sequence Database Collaboration (2016) The International Nucleotide Sequence Database Collaboration, *Nucleic Acids Research*, **44**, D48–D50, doi: 10.1093/nar/gkv1323.

Codd, G.A., Chorus, I., and Burch, M. (1999) Design of monitoring programmes, in I. Chorus, and J. Bartram (eds), *Cyanobacteria in Water: A guide to their public health consequences, monitoring and management*, Published on behalf of WHO by F and FN Spon, London, 313–328.

Cole, J.R., Wang, Q., Fish, J.A., *et al.* (2014) Ribosomal Database Project: Data and tools for high throughput rRNA analysis, *Nucleic Acids Research*, **42**, D633–D642, doi: 10.1093/nar/gkt1244.

Comte, K., Sabacká, M., Carré-Mlouka, A., *et al.* (2007) Relationships between the Arctic and the Antarctic cyanobacteria: Three *Phormidium*-like strains evaluated by a polyphasic approach, *FEMS Microbiology Ecology*, **59**, 366–376, doi: 10.1111/j.1574-6941. 2006.00257.x.

Crosbie, N.D., Pöckl, M., and Weisse, T. (2003) Dispersal and phylogenetic diversity of nonmarine picocyanobacteria, inferred from 16S rRNA gene and cpcBA-intergenic spacer sequence analyses, *Applied and Environmental Microbiology*, **69**, 5716–5721.

D'Alelio, D., Salmaso, N., and Gandolfi, A. (2013) Frequent recombination shapes the epidemic population structure of *Planktothrix* (Cyanoprokaryota) in Italian subalpine lakes, *Journal of Phycology*, **49**, 1107–1117, doi: 10.1111/jpy.12116.

Dagan, T., Roettger, M., Stucken, K., *et al.* (2013) Genomes of stigonematalean cyanobacteria (subsection V) and the evolution of oxygenic photosynthesis from prokaryotes to plastids, *Genome Biology and Evolution*, **5**, 31–44, doi: 10.1093/gbe/evs117.

Dall'Agnol, L.T., Ghilardi-Junior, R., Mcculloch, J.A., *et al.* (2012) Phylogenetic and gene trees of *Synechococcus*: Choice of the right marker to evaluate the population diversity in the Tucurui hydroelectric power station reservoir in Brazilian Amazonia, *Journal of Plankton Research*, **34**, 245–257, doi: 10.1093/plankt/fbr109.

Das, S., Dash, H.R., Mangwani, N., *et al.* (2014) Understanding molecular identification and polyphasic taxonomic approaches for genetic relatedness and phylogenetic relationships of microorganisms, *Journal of Microbiological Methods*, **103**, 80–100, doi: 10.1016/j.mimet.2014.05.013.

De Vos, P. and Trüper, H.G. (2000) Judicial Commission of the International Committee on Systematic Bacteriology. IXth International (IUMS) Congress of Bacteriology and Applied Microbiology. Minutes of the meetings, 14, 15 and 18 August (1999) Sydney, Australia, *International Journal of Systematic and Evolutionary Microbiology*, **50**, 2239–2244.

DeSantis, T.Z., Hugenholtz, P., Larsen, N., *et al.* (2006) Greengenes, a chimera-checked 16S rRNA gene database and workbench compatible with ARB, *Applied and Environmental Microbiology*, **72**, 5069–5072, doi: 10.1128/AEM.03006-05.

Dyble, J., Paerl, H.W., and Neilan, B.A. (2002) Genetic characterization of *Cylindrospermopsis raciborskii* (Cyanobacteria) isolates from diverse geographic origins based on *nifH* and *cpcBA*-IGS nucleotide sequence analysis, *Applied and Environmental Microbiology*, **68**, 2567–2571.

Edwards, U., Rogall, T., Blöcker, H., *et al.* (1989) Isolation and direct complete nucleotide determination of entire genes: Characterization of a gene coding for 16S ribosomal RNA, *Nucleic Acids Research*, **17**, 7843–7853, doi: 10.1093/nar/17.19.7843.

EN-15204 (2006) 2006, *Water Quality: Guidance standard on the enumeration of phytoplankton using inverted microscopy (Utermöhl technique)*, 46.

Engene, N. and Gerwick, W.H. (2011) Intra-genomic 16S rRNA gene heterogeneity in cyanobacterial genomes, *Fottea*, **11** 17–24.

European Bioinformatics Institute, (2014) *Annual Scientific Report*, European Molecular Biology Laboratory, Cambridge.

Fathalli, A., Jenhani, A.B.R., Moreira, C., *et al.* (2011) Molecular and phylogenetic characterization of potentially toxic cyanobacteria in Tunisian freshwaters, *Systematic and Applied Microbiology*, **34**, 303–310, doi: 10.1016/j.syapm.2010.12.003.

Felföldi, T., Somogyi, B., Márialigeti, K., and Vörös, L. (2009) Characterization of photoautotrophic picoplankton assemblages in turbid, alkaline lakes of the Carpathian Basin (Central Europe), *Journal of Limnology*, **68** (2), 385–395.

Flandrois, J.P., Perrière, G., and Gouy, M. (2015) leBIBIQBPP: A set of databases and a webtool for automatic phylogenetic analysis of prokaryotic sequences, *BMC Bioinformatics*, **16**, 251, doi: 10.1186/s12859-015-0692-z.

Gaget, V., Gribaldo, S., and Tandeau de Marsac, N. (2011) An *rpo*B signature sequence provides unique resolution for the molecular typing of cyanobacteria, *International Journal of Systematic and Evolutionary Microbiology*, **61**, 170–183, doi: 10.1099/ijs.0.019018-0.

Gaget, V., Welker, M., Rippka, R., and de Marsac, N.T. (2015) A polyphasic approach leading to the revision of the genus *Planktothrix* (Cyanobacteria) and its type species, *P. agardhii*, and proposal for integrating the emended valid botanical taxa, as well as three new species, *Planktothrix paucivesiculata* sp. nov.[ICNP], *Planktothrix tepida* sp. nov.[ICNP] and *Planktothrix serta* sp. nov.[ICNP], as genus and species names with nomenclatural standing under the ICNP, *Systematic and Applied Microbiology*, **38**, 141–158, doi: 10.1016/j.syapm.2015.02.004.

Garcia-Pichel, F. (2008) Molecular ecology and environmental genomics of cyanobacteria, in A. Herrero and E. Flores (eds), *The Cyanobacteria. Molecular Biology, Genomics and Evolution*, Caister Academic Press, Poole, UK, 59–87.

Gkelis, S., Rajaniemi, P., Vardaka, E., *et al.* (2005) *Limnothrix redekei* (Van Goor) Meffert (Cyanobacteria) strains from Lake Kastoria, Greece form a separate phylogenetic group, *Microbial Ecology*, **49**, 176–82, doi: 10.1007/s00248-003-2030-7.

Glazer, A.N. (1989) Light guides: Directional energy transfer in a photosynthetic antenna, *Journal of Biological Chemistry*, **264**, 1–4.

Gold-Morgan, M. and González-González, J. (2005) What is a species in cyanoprokaryotes?, *Algological Studies*, **117**, 209–222, doi: 10.1127/1864-1318/2005/0117-0209.

Greuter, D., Loy, A., Horn, M., and Rattei, T. (2016) probeBase-an online resource for rRNA-targeted oligonucleotide probes and primers: New features 2016, *Nucleic Acids Research*, **44**, D586–D589, doi: 10.1093/nar/gkv1232.

Grossman, A.R., Schaefer, M.R., Chiang, G.G., and Collier, J.L (1993) The phycobilisome: A light-harvesting complex responsive for environmental conditions, *Microbiological Reviews*, **57**, 725–749.

Gugger, M., Molica, R., Le Berre, B., *et al.* (2005) Genetic diversity of *Cylindrospermopsis* strains (Cyanobacteria) isolated from four continents, *Applied and Environmental Microbiology*, **71**, 1097–1100, doi: 10.1128/AEM.71.2.1097-1100.2005.

Guillard, R.R.L. and Sieracki, M.S. (2005) Counting cells in cultures with the light microscope, in R.A. Andersen (ed.), *Algal Culturing Techniques*, Elsevier Academic Press, Oxford, UK, 239–252.

Guindon, S., Dufayard, J.F., Lefort, V., *et al.* (2010) New algorithms and methods to estimate maximum-likelihood phylogenies: Assessing the performance of PhyML 3.0, *Systematic Biology*, **59**, 307–321, doi: 10.1093/sysbio/syq010.

Guiry, M.D. (2012) How many species of algae are there?, *Journal of Phycology*, **48**, 1057–1063, doi: 10.1111/j.1529-8817.2012.01222.x.

Guiry, M.D. and Guiry, G.M. (2017) *AlgaeBase*. World-wide electronic publication: National University of Ireland, Galway, http://www.algaebase.org.

Häggqvist, K., Akçaalan, R., Echenique-Subiabre, I., *et al.* (2017) Case studies of environmental sampling, detection and monitoring of potentially toxic cyanobacteria, in J. Meriluoto, L. Spoof, and G.A. Codd, (eds), *Handbook on Cyanobacterial Monitoring and Cyanotoxin Analysis*, John Wiley & Sons, Ltd, Chichester, UK, 70–83.

Häggqvist, K., Toruńska-Sitarz, A., Błaszczyk, A., *et al.* (2016) Morphologic, phylogenetic and chemical characterization of a brackish colonial picocyanobacterium (*Coelosphaeriaceae*) with bioactive properties, *Toxins*, **8**, 108, doi: 10.3390/toxins 8040108.

Hauer, T., Bohunická, M., Johansen, J.R., *et al.* (2014) Reassessment of the cyanobacterial family Microchaetaceae and establishment of new families Tolypothrichaceae and Godleyaceae, *Journal of Phycology*, **50**, 1089–100, doi: 10.1111/jpy.12241.

Hillebrand, H., Dürselen, C.D., Kirschtel, D., *et al.* (1999) Biovolume calculation for pelagic and benthic microalgae, *Journal of Phycology*, **35**, 403–424, doi: 10.1046/j.1529-8817.1999.3520403.x.

Hrouzek, P., Ventura, S., Lukesová, A., *et al.* (2005) Diversity of soil *Nostoc* strains: Phylogenetic and phenotypic variability, *Archives für Hydrobiologie/Supplement Algological Studies*, **117**, 251–264.

Humbert, J.F. and Fastner, J. (2017) Ecology of cyanobacteria, in J. Meriluoto, L. Spoof, and G.A. Codd, (eds), *Handbook on Cyanobacterial Monitoring and Cyanotoxin Analysis*, John Wiley & Sons, Ltd, Chichester, UK, 11–18.

Iteman, I., Rippka, R., Tandeau De Marsac, N., and Herdman, M. (2000) Comparison of conserved structural and regulatory domains within divergent 16S rRNA-23S rRNA spacer sequences of cyanobacteria, *Microbiology*, **146**, 1275–1286, doi: 10.1099/00221287-146-6-1275.

Janse, I., Kardinaal, W.E.A., Agterveld, M.K., *et al.* (2005) Contrasting microcystin production and cyanobacterial population dynamics in two *Planktothrix*-dominated freshwater lakes, *Environmental Microbiology*, **7** 1514–1524, doi: 10.1111/j.1462-2920.2005.00858.x.

Janse, I., Meima, M., Kardinaal, W.E.A., and Zwart, G. (2003) High-resolution differentiation of cyanobacteria by using rRNA-internal transcribed spacer denaturing gradient gel electrophoresis, *Applied and Environmental Microbiology*, **69**, 6634–6643, doi: 10.1128/AEM.69.11.6634-6643.2003.

Janssen, P., Coopman, R., Huys, G., *et al.* (1996) Evaluation of the DNA fingerprinting method AFLP as an new tool in bacterial taxonomy, *Microbiology*, **142**, 1881–1893, doi: 10.1099/13500872-142-7-1881.

Jasser, I. and Callieri, C. (2017) Picocyanobacteria: The smallest cell-size cyanobacteria, in J. Meriluoto, L. Spoof, and G.A. Codd, (eds), *Handbook on Cyanobacterial Monitoring and Cyanotoxin Analysis*, John Wiley & Sons, Ltd, Chichester, UK, 19–27.

Jasser, I., Karnkowska-Ishikawa, A., Kozlowska, E., *et al.* (2010) Composition of picocyanobacteria community in the Great Mazurian lakes: Isolation of phycoerythrin-rich and phycocyanin-rich ecotypes from the system: Comparison of two methods, *Polish Journal of Microbiology*, **59**, 21–31.

Johansen, J.R. and Casamatta, D.A. (2005) Recognizing cyanobacterial diversity through adoption of a new species paradigm, *Algological Studies*, **117**, 71–93, doi: 10.1127/1864-1318/2005/0117-0071.

Johansen, J.R., Kovacik, L., Casamatta, D.A., *et al.* (2011) Utility of 16S-23S ITS sequence and secondary structure for recognition of intrageneric and intergeneric limits within cyanobacterial taxa: *Leptolyngbya corticola* sp. nov (Pseudanabaenaceae, Cyanobacteria), *Nova Hedwigia*, **92** (3–4), 283–302, doi: 10.1127/0029-5035/2011/0092-0283.

Kim, O.S., Cho, Y.J., Lee, K., *et al.* (2012) Introducing EzTaxon-e: A prokaryotic 16S rRNA gene sequence database with phylotypes that represent uncultured species, *International Journal of Systematic and Evolutionary Microbiology*, **62**, 716–21, doi: 10.1099/ijs.0.038075-0.

Kling, H.J., Laughinghouse IV, H.D., Šmarda, J., *et al.* (2012) A new red colonial *Pseudanabaena* (Cyanoprokaryota, Oscillatoriales) from North American large lakes, *Fottea*, **12** (2), 327–339.

Kokociński, M., Stefaniak, K., Izydorczyk, K., *et al.* (2011) Temporal variation in microcystin production by *Planktothrix agardhii* (Gomont) Anagnostidis and Komárek (Cyanobacteria, Oscillatoriales) in a temperate lake, *Annales de Limnologie-International Journal of Limnology*, **47**, 363–371, doi: 10.1051/limn/2011046.

Komárek, J. (2013) Cyanoprokaryota, Part 3: Heterocytous Genera, in B. Büdel, G. Gärtner, L. Krienitz, and M. Schagerl (eds), *Süßwasserflora von Mitteleuropa, Band 19/3*, Springer Spektrum, Berlin.

Komárek, J. (2014) Modern classification of cyanobacteria, in N.K. Sharma, A.K. Rai, and L.J. Stal (eds), *Cyanobacteria: An economic perspective*, John Wiley and Sons, Ltd, Chichester, UK, 21–39.

Komárek, J. (2016) A polyphasic approach for the taxonomy of cyanobacteria: Principles and applications, *European Journal of Phycology*, **51**, 346–353, doi: 10.1080/09670262.2016.1163738.

Komárek, J. and Anagnostidis, K. (1999) Cyanoprokaryota: Part 1 (Chroococcales), in H. Ettl, J. Gerloff, H. Heyning, and D. Mollenhauer (eds), *Süßwasserflora von Mitteleuropa, Band 19/1*, Spektrum Akademischer Verlag, Heidelberg.

Komárek, J. and Anagnostidis, K. (2005) Cyanoprokaryota, Part 2: Oscillatoriales, in B. Büdel, G. Gärtner, L. Krienitz, and M. Schagerl (eds), *Süßwasserflora von Mitteleuropa, Band 19/2*, Elsevier Spektrum Akademischer Verlag, Munich, Germany.

Komárek, J. and Hauer, T. (2013) CyanoDB.cz: On-line database of cyanobacterial genera, World-wide electronic publication - Univ. of South Bohemia and Inst. of Botany AS CR, http://www.cyanodb.cz.

Komárek, J. and Kaštovský, J. (2003) Coincidences of structural and molecular characters in evolutionary lines of cyanobacteria, *Algological Studies*, **109**, 305–325.

Komárek, J., Kaštovský, J., Mareš, J., and Johansen, J.R. (2014) Taxonomic classification of cyanoprokaryotes (cyanobacterial genera) (2014) using a polyphasic approach, *Preslia*, **86**, 295–335.

Kosol, S., Schmidt, J., and Kurmayer, R. (2009) Variation in peptide net production and growth among strains of the toxic cyanobacterium *Planktothrix* spp., *European Journal of Phycology*, **44**, 49–62.

Kühl, M., Chen, M., and Larkum, A.W. (2007) Biology of the chlorophyll *d*-containing cyanobacterium *Acaryochloris marina*, in J Seckbach (ed.), *Algae and Cyanobacteria in Extreme Environments*, Springer, Netherlands, 101–123.

Kurmayer, R., Blom, J.F., Deng, L., and Pernthaler, J. (2015) Integrating phylogeny, geographic niche partitioning and secondary metabolite synthesis in bloom-forming *Planktothrix*, *ISME Journal*, **9** 909–921, doi: 10.1038/ismej.2014.189.

Kütahya, O.E., Starrenburg, M.J.C., Rademaker, J.L. *et al.* (2011). High-resolution amplified fragment length polymorphism typing of *Lactococcus lactis* strains enables identification of genetic markers for subspecies-related phenotypes, *Applied and Environmental Microbiology*, **77**, 5192–5198.

Lahr, D.J.G., Laughinghouse IV, H.D., Oliverio, A.M., *et al.* (2014) How discordant morphological and molecular evolution can revise our notions of biodiversity on Earth, *Bioessays*, **36**, 950–959.

Lapage, S.P., Sneath, P.H.A., Lessel, E.F., *et al.* (eds) (1992) International Code of Nomenclature of Bacteria *(1990 Revision)*, American Society for Microbiology, Washington DC.

Lara, Y., Lambion, A., Menzel, D., *et al.* (2013) A cultivation-independent approach for the genetic and cyanotoxin characterization of colonial cyanobacteria, *Aquatic Microbial Ecology*, **69**, 135–143, doi: 10.3354/ame01628.

Laughinghouse IV, H.D., Prá, D., Silva-Stenico, M.E., *et al.* (2012) Biomonitoring genotoxicity and cytotoxiciy of *Microcystis aeruginosa* (*Chroococcales*, cyanobacteria) using the *Allium cepa* test, *Science of theTotal Environment*, **432**, 180–188.

Lee, E., Ryan, U.M., Monis, P., *et al.* (2014) Polyphasic identification of cyanobacterial isolates from Australia, *Water Research*, **59**, 248–261, doi: 10.1016/j.watres.2014.04.023.

Lin, S., Wu, Z., Yu, G., *et al.* (2010) Genetic diversity and molecular phylogeny of *Planktothrix* (Oscillatoriales, Cyanobacteria) strains from China, *Harmful Algae*, **9**, 87–97, doi: 10.1016/j.hal.2009.08.004.

Linnaeus, C. (1753) *Species Plantarum*, **2** vols, Salvius, Stockholm.

Lyra, C., Suomalainen, S., Gugger, M., *et al.* (2001) Molecular characterization of planktic cyanobacteria of *Anabaena, Aphanizomenon, Microcystis and Planktothrix genera, International Journal of Systematic and Evolutionary Microbiology*, **51**, 513–526.

MacIsaac, E.A. and Stockner, J.G. (1993) Enumeration of phototrophic picoplankton by autofluorescence microscopy, in P.F. Kemp, B.F. Sherr, E.B., Sherr, and J.J. Cole (eds), *The Handbook of Methods in Aquatic Microbial Ecology*, CRC Press, Boca Raton, FL, 187–197.

Maiden, M.C., Bygraves, J.A., Feil, E., *et al.* (1998) Multilocus sequence typing: A portable approach to the identification of clones within populations of pathogenic microorganisms, *Proceedings of the National Academy of Sciences USA*, **95**, 3140–3145.

Malone, C.F.S., Rigonato, J., Laughinghouse IV, H.D., *et al.* (2015) *Cephalothrix* gen. nov. (Cyanobacteria): Towards an intraspecific phylogenetic evaluation by multilocus analyses, *International Journal of Systematic and Evolutionary Microbiology*, **65**, 2993–3007, doi: 10.1099/ijs.0.000369.

Mantzouki, E., Visser, P.M., Bormans, M., and Ibelings, B.W. (2016) Understanding the key ecological traits of cyanobacteria as a basis for their management and control in changing lakes, *Aquatic Ecology*, **50**, 333–350, doi: 10.1007/s10452-015-9526-3.

Mareš, J., Lara, Y., Dadáková, I., *et al.* (2015) Phylogenetic analysis of cultivation-resistant terrestrial cyanobacteria with massive sheaths (*Stigonema* spp. and *Petalonema alatum*, Nostocales, Cyanobacteria) using single-cell and filament sequencing of environmental samples, *Journal of Phycology*, **51**, 288–297.

Marie, D., Simon, N., and Vaulot, D. (2015) Phytoplankton cell counting by flow cytometry, in R.A. Andersen (ed.), *Algal Culturing Techniques*, Elsevier Academic Press, Oxford, UK, 253–285.

Mashima, J., Kodama, Y., Kosuge, T., *et al.* (2016) DNA data bank of Japan (DDBJ) progress report, *Nucleic Acids Research*, **44**, D51–D57, doi: 10.1093/nar/gkv1105.

Mazard, S., Ostrowski, M., Partensky, F., and Scanlan, D.J. (2012) Multi-locus sequence analysis, taxonomic resolution and biogeography of marine *Synechococcus*, *Environmental Microbiology*, **14** 372–386, doi: 10.1111/j.1462-2920.2011.02514.x.

McNeill, J., Barrie, F.R., Buck, W.R., *et al.* (eds) (2012) International Code of Nomenclature for Algae, Fungi and Plants (Melbourne code): Adopted by the Eighteenth International Botanical Congress Melbourne, Australia, July 2011, *Regnum Vegetabile, 154*, Koeltz Scientific Books.

Meriluoto, J. and Codd, G.A. (eds) (2005) *TOXIC: Cyanobacterial Monitoring and Cyanotoxin Analysis*, 1st edn, Åbo Akademi University Press, Turku, Finland.

Meriluoto, J., Spoof, L., and Codd, G.A. (eds) (2017) *Handbook on Cyanobacterial Monitoring and Cyanotoxin Analysis*, John Wiley & Sons, Ltd, Chichester, UK.

Moreira, C., Vasconcelos, V., and Antunes, A. (2013) Phylogeny and biogeography of cyanobacteria and their produced toxins, *Marine Drugs*, **11**, 4350–4369, doi: 10.3390/md11114350.

Nabout, J.C., da Silva Rocha, B., Carneiro, F.M., and Sant'Anna, C.L. (2013) How many species of Cyanobacteria are there? Using a discovery curve to predict the species number, *Biodiversity and Conservation*, **22**, 2907–2918, doi: 10.1007/s10531-013-0561-x.

Neilan, B., Jacobs, D., and Goodman, A. (1995) Genetic diversity and phylogeny of toxic cyanobacteria determined by DNA polymorphisms within the phycocyanin locus, *Applied and Environmental Microbiology*, **61**, 3875–3883.

Nübel, U., Garcia-Pichel, F., and Muyzer, G. (1997) PCR primers to amplify 16S rRNA genes from cyanobacteria, *Applied and Environmental Microbiology*, **63**, 3327–3332.

Oberholster, P.J., Botha, A.M., Muller, K., and Cloete, T.E. (2005) Assessment of the genetic diversity of geographically unrelated *Microcystis aeruginosa* strains using amplified fragment length polymorphisms (AFLPs), *African Journal of Biotechnology*, **4**, 389–399.

Ohki, K., Yamada, K., Kamiya, M., and Yoshikawa, S. (2012) Morphological, phylogenetic and physiological studies of pico-cyanobacteria isolated from the halocline of a saline meromictic lake, Lake Suigetsu, Japan, *Microbes and Environment*, **27**, 171–178.

Ong, L.J. and Glazer, A.N. (1991) Phycoerythrins of marine unicellular cyanobacteria: I: Bilin types and locations and energy transfer pathways in *Synechococcus* spp. Phycoerythrins, *Journal of Biological Chemistry*, **266**, 9515–9527.

Orcutt, K.M., Rasmussen, U., Webb, E.A., *et al.* (2002) Characterization of *Trichodesmium* spp. by genetic techniques, *Applied and Environmental Microbiology*, **68**, 2236–2245.

Oren, A. (2004) A proposal for further integration of the cyanobacteria under the Bacteriological Code, *International Journal of Systematic and Evolutionary Microbiology*, **54**, 1895–1902, doi: 10.1099/ijs.0.03008-0.

Oren, A. and Garrity, G.M. (2014) Proposal to change General Consideration 5 and Principle 2 of the International Code of Nomenclature of Prokaryotes, *International Journal of Systematic and Evolutionary Microbiology*, **64**, 309–310, doi: 10.1099/ijs.0.059568-0.

Ostermaier, V., Schanz, F., Köster, O., and Kurmayer, R. (2012) Stability of toxin gene proportion in red-pigmented populations of the cyanobacterium *Planktothrix* during 29 years of re-oligotrophication of Lake Zürich, *BMC Biology*, **10**, 100, doi: 10.1186/1741-7007-10-100.

Otsuka, S., Suda, S., Shibata, S., *et al.* (2001) A proposal for the unification of five species of the cyanobacterial genus *Microcystis* Kützing ex Lemmermann 1907 under the rules of the Bacteriological Code, *International Journal of Systematic and Evolutionary Microbiology*, **51**, 873–879, doi: 10.1099/00207713-51-3-873.

Oulas, A., Pavloudi, C., Polymenakou, P., *et al.* (2015) Metagenomics: Tools and insights for analyzing next-generation sequencing data derived from biodiversity studies, *Bioinformatics and Biology Insights*, **9**, 75–88, doi: 10.4137/BBI.S12462.

Overkamp, K.E., Gasper, R., Kock, K., *et al.* (2014) Insights into the biosynthesis and assembly of cryptophycean phycobiliproteins, *Journal of Biological Chemistry*, **289**, 26691–26707, doi: 10.1074/jbc.M114.591131.

Palenik, B. (1994) Cyanobacterial community structure as seen from RNA polymerase gene sequence analysis, *Applied and Environmental Microbiology*, **60**, 3212–3219.

Palinska, K.A., Deventer, B., Hariri, K., and Lotocka, M. (2011) A taxonomic study on *Phormidium*-group (Cyanobacteria) based on morphology, pigments, RAPD molecular markers and RFLP analysis of the 16S rRNA gene fragment, *Fottea*, **11**, 41–55, doi: 10.5507/fot.2011.006.

Palinska, K.A., Liesack, W., Rhiel, E., and Krumbein, W.E. (1996) Phenotype variability of identical genotypes: The need for a combined approach in cyanobacterial taxonomy demonstrated on *Merismopedia* –like isolates, *Archives of Microbiology*, **166**, 224–233, doi: 10.1007/s002030050378.

Pessi, I.S., Maalouf, P.D.C., Laughinghouse, H.D., *et al.* (2016) On the use of high-throughput sequencing for the study of cyanobacterial diversity in Antarctic aquatic mats, *Journal of Phycology*, **52**, 356–368, doi: 10.1111/jpy.12399.

Piccini, C., Aubriot, L., Fabre, A., *et al.* (2011) Genetic and eco-physiological differences of South American *Cylindrospermopsis raciborskii* isolates support the hypothesis of multiple ecotypes, *Harmful Algae*, **10**, 644–653, doi: 10.1016/j.hal.2011.04.016.

Piccin-Santos, V., Brandão, M.M., and Bittencourt-Oliveira, M.D.C. (2014) Phylogenetic study of *Geitlerinema* and *Microcystis* (Cyanobacteria) using PC-IGS and 16S-23S ITS as markers: Investigation of horizontal gene transfer, *Journal of Phycology*, **50**, 736–743, doi: 10.1111/jpy.12204.

Quast, C., Pruesse, E., Yilmaz, P., *et al.* (2013) The SILVA ribosomal RNA gene database project: Improved data processing and web-based tools, *Nucleic Acids Research*, **41**, D590–D596, doi: 10.1093/nar/gks1219.

Ragon, M., Benzerara, K., Moreira, D., *et al.* (2014) 16S rDNA-based analysis reveals cosmopolitan occurrence but limited diversity of two cyanobacterial lineages with contrasted patterns of intracellular carbonate mineralization, *Frontiers in Microbiology*, **5** 331, doi: 10.3389/fmicb.2014.00331.

Rajaniemi, P., Hrouzek, P., Kastovská, K., *et al.* (2005) Phylogenetic and morphological evaluation of the genera *Anabaena*, *Aphanizomenon*, *Trichormus* and *Nostoc*

(Nostocales, Cyanobacteria), *International Journal of Systematic and Evolutionary Microbiology*, **55**, 11–26, doi: 10.1099/ijs.0.63276-0.

Rajaniemi-Wacklin, P., Rantala, A., Mugnai, M.A., *et al.* (2005) Correspondence between phylogeny and morphology of *Snowella* spp. and *Woronichinia naegeliana*: Cyanobacteria commonly occurring in lakes, *Journal of Phycology*, **42**, 226–232, doi: 10.1111/j.1529-8817.2006.00179.x.

Rantala, A., Fewer, D.P., Hisbergues, M., *et al.* (2004) Phylogenetic evidence for the early evolution of microcystin synthesis, *Proceedings of the National Academy of Sciences USA*, **101**, 568–573, doi: 10.1073/pnas.0304489101.

Reddy, T.B.K., Thomas, A.D., Stamatis, D., *et al.* (2015) The Genomes OnLine Database (GOLD) v. 5: A metadata management system based on a four level (meta)genome project classification, *Nucleic Acids Research*, **43**, D1099–D1106, doi: 10.1093/nar/gku950.

Řeháková, K., Mareš, J., Lukešová, A., *et al.* (2014) *Nodularia* (Cyanobacteria, Nostocaceae): A phylogenetically uniform genus with variable phenotypes, *Phytotaxa*, **172**, 235, doi: 10.11646/phytotaxa.172.3.4.

Rippka, R., Deruelles, J., Waterbury, J.B., *et al.* (1979) Generic Assignments, strain histories and properties of pure cultures of cyanobacteria, *Journal of General Microbiology*, **111**, 1–61.

Rocap, G., Distel, D.L., Waterbury, J.B., and Chisholm, S.W. (2002) Resolution of *Prochlorococcus* and *Synechococcus* ecotypes by using 16S-23S ribosomal DNA internal transcribed spacer sequences, *Applied and Environmental Microbiology*, **68**, 1180–1191, doi: .10.1128/AEM.68.3.1180-1191.2002.

Rosselló-Móra, R. and Amann, R. (2015) Past and future species definitions for Bacteria and Archaea, *Systematic and Applied Microbiology*, **38**, 209–216.

Rudi, K., Skulberg, O.M., and Jakobsen, K.S. (1998) Evolution of cyanobacteria by exchange of genetic material among phyletically related strains, *Journal of Bacteriology*, **180**, 3453–3461.

Rudi, K., Skulberg, O.M., Larsen, F., and Jakobsen, K.S. (1997) Strain characterization and classification of oxyphotobacteria in clone cultures on the basis of 16S rRNA sequences from the variable regions V6, V7, and V8, *Applied and Environmental Microbiology*, **63**, 2593–2599.

Saker, M.L., Neilan, B.A., and Griffiths, D.J. (1999) Two morphological forms of *Cylindrospermopsis raciborskii* (Cyanobacteria) isolated from Solomon Dam, Palm Island, Queensland, *Journal of Phycology*, **35**, 599–606, doi: 10.1046/j.1529-8817.1999.3530599.x.

Salmaso, N., Bernard, C., Humbert, J.F., *et al.* (2017) Basic guide to detection and monitoring of potentially toxic cyanobacteria, in J. Meriluoto, L. Spoof, and G.A. Codd, (eds), *Handbook on Cyanobacterial Monitoring and Cyanotoxin Analysis*, John Wiley & Sons, Ltd, Chichester, UK, 46–69.

Salmaso, N., Boscaini, A., Capelli, C., *et al.* (2015a) Historical colonization patterns of *Dolichospermum lemmermannii* (Cyanobacteria) in a deep lake south of the Alps, *Advances in Oceanography and Limnology*, **6** 33–45, doi: 10.4081/aiol.2015.5456.

Salmaso, N., Capelli, C., Shams, S., and Cerasino, L. (2015b) Expansion of bloom-forming *Dolichospermum lemmermannii* (Nostocales, Cyanobacteria) to the deep lakes south of the Alps: Colonization patterns, driving forces and implications for water use, *Harmful Algae*, **50**, 76–87, doi: 10.1016/j.hal.2015.09.008.

Salmaso, N., Cerasino, L., Boscaini, A., and Capelli, C. (2016) Planktic *Tychonema* (Cyanobacteria) in the large lakes south of the Alps: Phylogenetic assessment and toxigenic potential, *FEMS Microbiology Ecology*, **92**, doi: 10.1093/femsec/fiw155.

Salmaso, N., Copetti, D., Cerasino, L., *et al.* (2014) Variability of microcystin cell quota in metapopulations of *Planktothrix rubescens*: Causes and implications for water management, *Toxicon*, **90**, 82–96, doi: 10.1016/j.toxicon.2014.07.022.

Shams, S., Capelli, C., Cerasino, L., *et al.* (2015) Anatoxin-a producing *Tychonema* (Cyanobacteria) in European waterbodies, *Water Research*, **69**, 68–79, doi: 10.1016/j.watres.2014.11.006.

Sheffield, V.C., Cox, D.R., Lerman, L.S., and Myers, R.M. (1989) Attachment of a 40-base-pair G+C-rich sequence (GC-clamp) to genomic DNA fragments by the polymerase chain reaction results in improved detection of single-base changes, *Proceedings of the National Academy of Sciences USA*, **86**, 232–236.

Shih, P.M., Wu, D., Latifi, A., *et al.* (2013) Improving the coverage of the cyanobacterial phylum using diversity-driven genome sequencing, *Proceedings of the National Academy of Sciences USA*, **110**, 1053–8, doi: 10.1073/pnas.1217107110.

Silva, P. (2016) *Index Nominum Algarum, University Herbarium*, University of California, Berkeley, URL http://ucjeps.berkeley.edu/CPD/, accessed January 1, 2016.

Smith, J., Parry, J., Day, J., and Smith, R. (1998) A PCR technique based on the Hipl interspersed repetitive sequence distinguishes cyanobacterial species and strains, *Microbiology*, **144**, 2791–2801.

Soares, M.C.S., Lürling, M., and Huszar, V.L.M. (2013) Growth and temperature-related phenotypic plasticity in the cyanobacterium *Cylindrospermopsis raciborskii, Phycological Research*, **61**, 61–67.

Stackebrandt, E. and Ebers, J. (2006) Taxonomic parameters revisited: Tarnished gold standards, *Microbiology Today*, **33**, 152–155.

Stackebrandt, E. and Goebel, B.M. (1994) Taxonomic note: A place for DNA–DNA reassociation and 16S rRNA sequence analysis in the present species definition in bacteriology, *International Journal of Systematic and Evolutionary Microbiology*, **44**, 846–849.

Staley, J.T., Bryant, M.P., Pfennig, N., and Holt, J.G. (1989) *Bergey's Manual of Systematic Bacteriology*, 31st edn, Williams & Wilkins, Baltimore.

Stanier, R.Y., Kunisawa, R., Mandel, M., and Cohen-Bazire, G. (1971) Purification and properties of unicellular blue-green algae (order Chroococcales), *Bacteriological Reviews*, **35**, 171–205.

Stanier, R.Y., Sistrom, W.R., Hansen, T.A., *et al.* (1978) Proposal to place the nomenclature of the Cyanobacteria (Blue-Green Algae) under the rules of the International Code of Nomenclature of Bacteria, *International Journal of Systematic and Evolutionary Microbiology*, **28**, 335–336, doi: 10.1099/00207713-28-2-335.

Steunou, A.S., Bhaya, D., Bateson, M.M., *et al.* (2006) *In situ* analysis of nitrogen fixation and metabolic switching in unicellular thermophilic cyanobacteria inhabiting hot spring microbial mats, *Proceedings of the National Academy of Sciences USA*, **103**, 2398–2403, doi: 10.1073/pnas.0507513103.

Strunecký, O., Elster, J., and Komárek, J. (2010) Phylogenetic relationships between geographically separate *Phormidium* cyanobacteria: Is there a link between North and South Polar regions?, *Polar Biology*, **33**, 1419–1428, doi: 10.1007/s00300-010-0834-8.

Suda, S., Watanabe, M.M., Otsuka, S., *et al.* (2002) Taxonomic revision of water-bloom-forming species of oscillatorioid cyanobacteria, *International Journal of Systematic and Evolutionary Microbiology*, **52**, 1577–1595.

Sun, D.L., Jiang, X., Wu, Q.L., and Zhoua, N.Y. (2013) Intragenomic heterogeneity of 16S rRNA genes causes overestimation of prokaryotic diversity, *Applied and Environmental Microbiology*, **79**, 5962–5969, doi: 10.1128/AEM.01282-13.

Sun, J. and Liu, D. (2003) Geometric models for calculating cell biovolume and surface area for phytoplankton, *Journal of Plankton Research*, **25**, 1331–1346, doi: 10.1093/plankt/fbg096.

Tan, W., Liu, Y., Wu, Z., *et al.* (2010) *cpc*BA-IGS as an effective marker to characterize *Microcystis wesenbergii* (Komárek) Komárek in Kondrateva (cyanobacteria), *Harmful Algae*, **9**, 607–612, doi: 10.1016/j.hal.2010.04.011.

Tanabe, Y. and Watanabe, M.M. (2011) Local expansion of a panmictic lineage of water bloom-forming cyanobacterium *Microcystis aeruginosa*, *PLOS ONE*, **6**, e17085, doi: 10.1371/journal.pone.0017085.

Tanabe, Y., Kasai, F., and Watanabe, M.M. (2007) Multilocus sequence typing (MLST) reveals high genetic diversity and clonal population structure of the toxic cyanobacterium *Microcystis aeruginosa*, *Microbiology*, **153**, 3695–3703, doi: 10.1099/mic.0.2007/010645-0.

Taton, A., Grubisic, S., Brambilla, E., *et al.* (2003) Cyanobacterial diversity in natural and artificial microbial mats of Lake Fryxell (McMurdo Dry Valleys, Antarctica): A morphological and molecular approach, *Applied and Environmental Microbiology*, **69**, 5157–5169, doi: 10.1128/AEM.69.9.5157-5169.2003.

Taton, A., Grubisic, S., Ertz, D., *et al.* (2006) Polyphasic study of Antarctic cyanobacterial strains, *Journal of Phycology*, **42**, 1257–1270.

Thomazeau, S., Houdan-Fourmont, A., Couté, A., *et al.* (2010) The contribution of sub-Saharan African strains to the phylogeny of cyanobacteria: Focusing on the Nostocaceae (Nostocales, Cyanobacteria), *Journal of Phycology*, **46**, 564–579, doi: 10.1111/j.1529-8817.2010.00836.x.

Thorne, S.W., Newcomb, E.H., and Osmond, C.B. (1977) Identification of chlorophyll *b* in extracts of prokaryotic algae by fluorescence spectroscopy, *Proceedings of the National Academy of Sciences USA*, **74**, 575–578.

Tian, R.M., Cai, L., Zhang, W.P., *et al.* (2015) Rare events of intragenus and intraspecies horizontal transfer of the 16S rRNA gene, *Genome Biology and Evolution*, **7**, 2310–2320, doi: 10.1093/gbe/evv143.

Ting, C.S., Rocap, G., King, J., and Chisholm, S.W. (2002) Cyanobacterial photosynthesis in the oceans: The origins and significance of divergent light-harvesting strategies, *Trends in Microbiology*, **10**, 134–142.

Toledo, G. and Palenik, B. (1997) *Synechococcus* diversity in the California current as seen by RNA polymerase (*rpo*C1) gene sequences of isolated strains, *Applied and Environmental Microbiology*, **63**, 4298–4303.

Tomitani, A., Knoll, A.H., Cavanaugh, C.M., and Ohno, T. (2006) The evolutionary diversification of cyanobacteria: Molecular-phylogenetic and paleontological perspectives, *Proceedings of the National Academy of Sciences USA*, **103**, 5442–5447, doi: 10.1073/pnas.0600999103.

Tsuji, T., Ohki, K., and Fujita, Y. (1986) Determination of photosynthetic pigment composition in an individual phytoplankton cell in seas and lakes using fluorescent microscopy; properties of the fluorescence emitted from picophytoplankton cells, *Marine Biology*, **93**, 343–349.

Utermöhl, H. (1931) Neue Wege in der quantitativen Erfassung des Planktons, *Verh. Internat. Verein. Limnol.*, **5**, 567–596.

Valério, E., Chambel, L., Paulino, S., *et al.* (2009) Molecular identification, typing and traceability of cyanobacteria from freshwater reservoirs, *Microbiology*, **155**, 642–656, doi: 10.1099/mic.0.022848-0.

Vandamme, P., Pot, B., Gillis, M., *et al.* (1996) Polyphasic taxonomy: A consensus approach to bacterial systematics, *Microbiological Reviews*, **60**, 407–438.

Via-Ordorika, L., Fastner, J., Kurmayer, R., *et al.* (2004) Distribution of microcystin-producing and non-microcystin-producing *Microcystis* sp. in European freshwater bodies: Detection of microcystins and microcystin genes in individual colonies, *Systematic and Applied Microbiology*, **27**, 592–602, doi: 10.1078/0723202041748163.

Waterbury, J.B. and Stanier, R.Y. (1978) Patterns of growth and development in pleurocapsalean cyanobacteria, *Microbiological Reviews*, **42**, 2–44.

Waterbury, J.B., Watson, S.W., Valois, F.W., and Franks, D.G. (1986) Biological and ecological characterization of the marine unicellular cyanobacterium *Synechococcus*, in T. Platt and W.K.W. Li, (eds), *Photosynthetic Picoplankton, Canadian Bulletin of Fisheries and Aquatic Sciences*, **214**, 71–120.

Wayne, L.G., Brenner, D.J., Colwell, R.R., *et al.* (1987) Report of the Ad Hoc Committee on Reconciliation of Approaches to Bacterial Systematics, *International Journal of Systematic and Evolutionary Microbiology*, **37**, 463–464, doi: 10.1099/00207713-37-4-463.

Whitton, B.A. and Potts, M. (2012) Introduction to the Cyanobacteria, in BA Whitton (ed.), *Ecology of Cyanobacteria II*, Springer, Dordrecht, 1–13.

Willame, R., Jurczak, T., Iffly, J.F., *et al.* (2005) Distribution of hepatotoxic cyanobacterial blooms in Belgium and Luxembourg, *Hydrobiologia*, **551**, 99–117, doi: 10.1007/s10750-005-4453-2.

Williams, L., Loewen-Schneider, K., Maier, S., and Büdel, B. (2016) Cyanobacterial diversity of western European biological soil crusts along a latitudinal gradient, *FEMS Microbiology Ecology*, **92**, 128–135, doi: 10.1093/femsec/fiw157.

Wilmotte, A. (1994) Molecular evolution and taxonomy of the Cyanobacteria, in D.A. Bryant (ed.), *The Molecular Biology of Cyanobacteria*, Springer, Berlin, 1–25.

Wilson, A.E., Sarnelle, O., Neilan, B.A., *et al.* (2005) Genetic variation of the bloom-forming cyanobacterium *Microcystis aeruginosa* within and among lakes: Implications for harmful algal blooms, *Applied and Environmental Microbiology*, **71**, 6126–6133, doi: 10.1128/AEM.71.10.6126-6133.2005.

Wood, S.A., Smith, F.M.J., Heath, M.W., *et al.* (2012) Within-mat variability in anatoxin-a and homoanatoxin-a production among benthic *Phormidium* (Cyanobacteria) strains, *Toxins*, **4** 900–912, doi: 10.3390/toxins4100900.

Wu, M. and Eisen, J.A. (2008) A simple, fast, and accurate method of phylogenomic inference, *Genome Biology*, **9**, R151, doi: 10.1186/gb-2008-9-10-r151.

Yang, Z., Kong, F., Shi, X., *et al.* (2008) Changes in the morphology and polysaccharide content of *Microcystis aeruginosa* (cyanobacteria) during flagellate grazing, *Journal of Phycology*, **44**, 716–720, doi: 10.1111/j.1529-8817.2008.00502.x.

Yerrapragada, S., Siefert, J.L., and Fox, G.E. (2009) Horizontal gene transfer in cyanobacterial signature genes, *Methods in Molecular Biology*, **532**, 339–366, doi: 10.1007/978-1-60327-853-9_20.

Yilmaz, P., Parfrey, L.W., Yarza, P., *et al.* (2014) The SILVA and "All-species Living Tree Project (LTP)" taxonomic frameworks, *Nucleic Acids Research*, **42**, D643–D648.

Zapomělová, E., Řeháková, K., Jezberová, J., and Komárková, J. (2009) Polyphasic characterization of eight planktonic *Anabaena* strains (Cyanobacteria) with reference to the variability of 61 *Anabaena* populations observed in the field, *Hydrobiologia*, **639**, 99–113, doi: 10.1007/s10750-009-0028-y.

SOP 4.1

Taxonomic Identification by Light Microscopy

Nico Salmaso[1]*, Rosmarie Rippka[2]‡, and Annick Wilmotte[3]

[1]*Research and Innovation Centre, Fondazione Edmund Mach – Istituto Agrario di S. Michele all'Adige, S. Michele all'Adige, Italy*
[2]*Institut Pasteur, Unité des Cyanobactéries, Centre National de la Recherche Scientifique (CNRS), Unité de Recherche Associé (URA) 2172, Paris, France*
[3]*InBios – Center for Protein Engineering, University of Liège, Liège, Belgium*

SOP 4.1.1   Introduction

The examination by microscopy is an essential and necessary step in the identification of the phenotypic characters of cyanobacteria, and a fast way to preliminarily check the presence of taxa belonging to potentially toxic genera or species (Lawton *et al.*, 1999). The level of taxonomic accuracy to be achieved will depend on the aims of the investigation. If a precise identification is required, or desired, microscopic observations must be confirmed by the genetic characterization of isolated strains (see SOP 4.2). This SOP reports a description of the procedures required for optimal identification of cyanobacterial samples by light microscopy. The collection of samples and their pre-treatment (e.g. concentration of individuals) are not considered in this SOP (see Chapter 2). A list of current species names and synonyms is given in the Appendix.

‡Retired.
*Corresponding author: nico.salmaso@fmach.it

SOP 4.1.2 Experimental

SOP 4.1.2.1 Materials

- Flat glass microscope slides (75 × 25 × *ca.* 1 mm).
- Glass cover slips (#1, 0.13 to 0.16 mm thick) allowing an effective cover glass thickness of around 0.17 mm; if the objectives are equipped with a correction collar, this requirement can be relaxed.
- Straight and angled tweezers.
- Mounting pins.
- Plastic and/or glass micropipettes.
- Laboratory filter paper and tissue paper.
- Immersion oil.
- Indian ink stain.
- Aqueous methylene blue solution (1–1.5% w/v)

SOP 4.1.2.2 Equipment

- Upright light compound microscope with phase contrast equipment, if possible. The optical components of the microscope should be aligned following the rules of Köhler illumination (Lacey, 1999; manufacturer instructions).
- Phase contrast objectives spanning a magnification range from 10× to 100× (e.g. 10×, 20×, 40×, 100× (oil immersion)). The 40× objective should have a numerical aperture greater than 0.70.
- Wide-field eyepieces (oculars) of at least 10× magnification, and a field of view of 18 or 20 mm diameter); one of the eyepieces must be equipped with a micrometer. It is assumed that the objectives with different magnification have been calibrated with a calibration slide (stage micrometer, 0.01 mm ruler).
- Digital microscope photo camera connected to a computer and a suitable software permitting the capture of images in the right format, inserting a scale onto them and (if desired) carry out measurements.

SOP 4.1.2.3 Procedure

Fresh samples should be observed on the day of collection, or the following day. Most aquatic samples containing cyanobacteria can be preserved in algal growth incubators at 15–20°C and low light (around 10–20 $\mu E\ m^{-2}\ s^{-1}$). Using a micropipette, place a small quantity (1 drop) of the sample onto the microscope slide. Larger and visible cyanobacterial filaments and colonies can first be gathered with a large-tipped plastic pipette, prior to placing them onto the microscope slide.

With the help of tweezers or mounting pins (or carefully with fingertips), place the cover glass diagonally onto the microscope slide until its edge touches the slide. Then gently lower the cover glass onto the sample to bring it into the horizontal position (Fig. 4.6). This will help prevent the formation of air bubbles.

The surface tension of the water will hold the cover glass in place. If there is too much water, the cover slip will float, and the surplus needs to be removed by placing a strip of

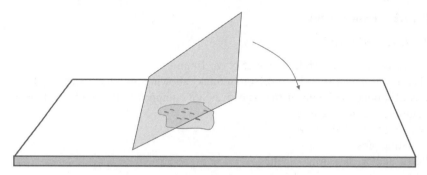

Figure 4.6 *Slide mount of an aqueous sample. To avoid the formation of air bubbles, after placing the sample on the slide, the cover glass should be carefully lowered at an angle onto the microscope slide with the aid of tweezers or mounting pins.*

filter paper or tissue along the edge of the cover slip. Conversely, if there is not enough water (or if the sample is drying), a minute amount of water can be carefully added on the edge of the cover slip with the micropipette (or wetted tweezers); the surface tension will drag the water beneath the cover slip.

In the case of mats or moist samples (e.g. cyanobacteria from wet soils), specimens can be gathered and placed onto the glass slide using tweezers or mounting pins. They are then covered with water of an osmotic potential compatible with the observed organisms (tap water is suitable for most freshwater species). If the cell material is too thick (and creates air bubbles), try to carefully press with a finger (or the extremity of a pencil) onto the cover slip to spread and flatten the material. Without this preparation, it will be impossible to have a sharp and precise view under the microscope.

Using the 10× objective, begin the observation by inspecting the whole sample for finding the right plane of observation, and examining large colonies/filaments. The search for smaller (width < 10 μm) cyanobacteria can then be performed by switching progressively to higher magnification objectives. Inspection of finer details (cell shape, differentiated reproductive cells, inclusion bodies, mucilage, sheaths, etc.), as well as measurements of cell dimensions, is generally done with 40× or 100× objectives permitting a higher resolution. The high magnification oil immersion objective (i.e. 100×) must be very close to the specimen, causing the risk of pressing the objective into the cover slip (potentially breaking it). If using this objective, place a small drop of immersion oil onto the center of the cover glass prior to very carefully positioning the objective into place. Use the immersion objective especially for observing thin filaments and small unicellular cyanobacteria (dimensions < 3 μm), or when internal cell structures need to be identified.

Bright field illumination is best for visualizing color differences of large (>5 μm) cyanobacteria, whereas phase contrast is required for small specimens (more faint in color), or if structural details are difficult to see under bright field illumination. After using the immersion oil objective, do not switch back into the position of the 40× objective, since the latter will pass and touch the drop of oil, leading to deterioration of optical performance and potentially damaging the objective (if oil leaks into the lens barrel). To avoid such accidents, it is advisable to place the 10× objective between the 40× and 100×.

If oil has accidentally touched the 40× objective, it must be wiped off immediately using several pieces of lens tissue paper until no more traces of oil can be detected on the tissue.

To make mucilage visible against a dark background (negative stain), add a drop of India ink onto one edge of the cover slip. Hold the edge of a strip of filter paper against the opposite side of the cover slip, which will allow the ink to be drawn into the preparation and thus distributed around the specimen. Conversely, cells or colonies can be stained by methylene blue (e.g. Gama *et al.*, 2012: Fig. 6).

The identification of genera/species should be made using the most updated and comprehensive taxonomic treatises (including dichotomous keys) available. The nomenclature and classification of particular taxa may further need to be updated based on more recent publications. At present, the identification of cyanobacteria should be made with the aid of the manuals by Komárek and Anagnostidis (1999, 2005) and Komárek (2013). Additional manuals (Whitton, 2011) and illustrated guides – such as those of Cronberg and Annadotter (2006), Joosten (2006) and Cirés and Quesada (2011) – are of great help in the identification of the main genera and species.

For each cyanobacterium identified, provide a photographic documentation. Digital images will be stored and cataloged in separate directories for future use.

SOP 4.1.2.4   Notes

Concavity glass slides (75 × 25 × *ca.* 1 mm) with one or two wells are especially useful for the observation of larger and thicker colonies and filaments (up to a few millimeters). They avoid the crushing and potential destruction of fragile organisms between the slide and the cover glass. However, given the greater depth of viewing the specimen layer, examination at higher magnifications would require long working distance objectives for optimal resolution. With upright microscopes, fresh (unfixed) samples can also be observed after depositing them into Sedgewick Rafter chambers.

Additional depth for specimens can also be obtained by firmly placing (with water or nail polish) two cover slips 1–1.5 cm apart onto the glass slide. The two cover slips will provide support for a third cover slip placed in between them, covering at its left and right sides the edges of the supporting cover slips. The gain of about 0.17 mm (i.e. the height of the cover slips) will create more depth for accommodating large specimens. However, this method is only useful for examinations with low magnification objectives.

The observation of specimens according to the procedures described in SOP 4.1.2.3 can also be made by using inverted microscopes. In this case, turn the glass slide upside down so that the cover slip will face the objectives. Be particularly careful to place the correct amount of sample onto the glass slide to ensure a firm adhesion of the cover slip by the surface tension of the water. Low volume (≤5 ml) Utermöhl chambers can also be used for observing unfixed samples. Their use has, however, the disadvantage of creating greater shaking of specimens, and the increased depth of the chamber will not permit precise viewing of organisms distributed in three dimensions because it is not suitable for regular working distance objectives.

Note that immersion oils are quite toxic chemicals. Therefore, clean carefully not only the high immersion oil objectives after their use but also any drops spilled onto the microscope, or elsewhere on the bench.

SOP 4.1.3 References

Cirés, G. and Quesada, A. (2011) Catálogo de Cianobacterias Planctónicas Potencialmente Tóxicas de las Aguas Continentales Españolas, Ministerio de Medio Ambiente y Medio Rural y Marino.

Cronberg, G. and Annadotter, H. (2006) *Manual of Aquatic Cyanobacteria. A Photo Guide and Synopsis of their Toxicology*, ISSHA and IOC of UNESCO, Copenhagen.

Gama Junior, W.A., De Paiva Azevedo, M.T., Komárková-Legnerová, J., and Sant'anna, C.L. (2012) A new species of *Lemmermanniella* (Cyanobacteria) from the Atlantic Rainforest, *Brazil, Revista Brasileira de Botanica*, **35**, 319–324.

Joosten, A.M.T. (2006) *Flora of the Blue–green Algae of the Netherlands: I: The non filamentous species of inland waters*, KNNV Publishing, Utrecht, The Netherlands.

Komárek, J. (2013) *Cyanoprokaryota: Part 3: Heterocytous Genera, in Süßwasserflora von Mitteleuropa, Band 19/3*, Springer Spektrum.

Komárek, J. and Anagnostidis, K. (1999) *Cyanoprokaryota Teil 1 (Chroococcales), in Süßwasserflora von Mitteleuropa, Band 19*, Gustav Fischer Verlag, Stuttgart.

Komárek, J. and Anagnostidis, K. (2005) *Cyanoprokaryota: Part 2: Oscillatoriales, in Süßwasserflora von Mitteleuropa, Band 19/2*, Springer Spektrum.

Lacey, A.J. (ed.) (1999) *Light Microscopy in Biology: A practical approach*, 2nd edn, Oxford University Press, Oxford.

Lawton, L., Marsalek, B., Padisák, J., and Chorus, I. (1999) Chapter 12: Determination of cyanobacteria in the laboratory, in I. Chorus and J. Bartram (eds), *Cyanobacteria in Water: A guide to their public health consequences, monitoring and management*, Published on behalf of WHO by F and FN Spon, London, 347–367.

Whitton, B.A. (2011) Cyanobacteria (Cyanophyta), in D.M. John, B.A., Whitton, and A.J. Brook (eds), *The Freshwater Algal Flora of the British Isles: An identification guide to freshwater and terrestrial algae*, 2nd edn, Cambridge University Press, Cambridge, 31–158.

SOP 4.2

Polyphasic Approach on Cyanobacterial Strains

Nico Salmaso[1]*, Camilla Capelli[1], Rosmarie Rippka[2‡], and Annick Wilmotte[3]

[1]*Research and Innovation Centre, Fondazione Edmund Mach – Istituto Agrario di S. Michele all'Adige, S. Michele all'Adige, Italy*
[2]*Institut Pasteur, Unité des Cyanobactéries, Centre National de la Recherche Scientifique (CNRS), Unité de Recherche Associé (URA) 2172, Paris, France*
[3]*InBios – Center for Protein Engineering, University of Liège, Liège, Belgium*

SOP 4.2.1   Introduction

Traditionally, the classification of cyanobacteria has been based on the analysis of morphological characteristics using light-microscopy observations. Nevertheless, owing to the morphological simplicity of some of these organisms, the variability of phenotypic traits within one species, and the dependence of certain characters on environmental factors, the taxonomic identification by microscopy alone can be quite problematic (e.g. Gugger *et al.*, 2002). On the other side, the molecular characterization requires the analysis of individual strains usually grown under standardized laboratory conditions, which often causes physiological and morphological changes of the organisms (Komárek, 2014). For this reason, the adoption of a combination of approaches for the identification and classification of cyanobacteria has been widely advocated. With the polyphasic approach, species are defined both by genotypic and phenotypic characterizations (Vandamme *et al.*, 1996; Komárek *et al.*, 2014). Despite this apparently simple statement, the identification of cyanobacteria at the species level with this approach can be achieved by different methods, such as DNA–DNA hybridizations (the basis for the bacterial species definition; ·

‡Retired.
*Corresponding author: nico.salmaso@fmach.it

Wayne *et al.*, 1987) or phylogenetic analyses, and selection of different combinations of phenotypic, ultrastructural, physiological, and ecological characters (e.g. Wilmotte and Herdman, 2001; Salmaso *et al.*, 2015). Moreover, the genotyping of cyanobacteria has, in the majority of cases, been performed on cultures of xenic unicyanobacterial strains and only rarely on axenic ones (e.g. Gaget *et al.*, 2015). A list of current species names and synonyms is given in the Appendix.

These considerations illustrate the difficulty of describing a detailed procedure valid for different investigation contexts and habitats (e.g. pelagic and benthic), and compatible with both the International Code of Nomenclature for algae, fungi, and plants (ICN), and the International Code of Nomenclature of Prokaryotes (ICNP).

The aim of this SOP is therefore to describe the procedures adopted in a polyphasic approach by focusing on the identification of the filamentous and colonial pelagic cyanobacteria, which are important for the management of toxic water blooms (Metcalf and Codd, 2012), and for benthic cyanobacteria that may also produce toxins (Gugger *et al.*, 2005). The notes at the end of the SOP will highlight the main alternatives to the standard procedure.

The procedure is applicable to axenic and non-axenic unicyanobacterial isolates. The analysis of environmental samples will not be considered here. The progress in NGS and metabarcoding techniques are rapidly changing the approach for the analysis of multi-species assemblages (Chapter 10).

SOP 4.2.2   Experimental

SOP 4.2.2.1   Materials

- Plankton net approx. 20 µm mesh size.
- Culture media, e.g. ASM-1 or Z8.
- Glass micropipettes.
- Petri dishes.
- Microwell plates.
- Culture flasks.
- Falcon tubes.
- Reagents for DNA extraction (see SOP 5.1).
- Primers for PCR amplifications of 16S rRNA and/or selected housekeeping genes.
- Reagents for PCR reactions and gel electrophoresis (see Chapter 6).
- Commercial PCR product purification kit.

SOP 4.2.2.2   Equipment

- Stereomicroscope.
- Growth chamber or culture room with controlled temperature and light intensity.
- Upright light compound microscope with phase contrast equipment and digital camera (see SOP 4.1).
- PCR equipment (electrophoresis equipment, thermocycler, UV transilluminator) (see Chapter 6).
- Taxonomical manuals and illustrated cyanobacterial guides (see SOP 4.1).

SOP 4.2.2.3 Procedure

Provide a description of the sampling site and measurements of the principal environmental variables (e.g. temperature, light, pH, conductivity, nutrients). This information is essential also for the choice of culture media and incubation conditions (Waterbury, 2006). The methods to be applied in the field are described, for example, in Wetzel and Likens (2000) (see Chapter 2).

SOP 4.2.2.3.1 Morphological Evaluation of Individuals in Natural Samples

On the same day or the day after the sampling, perform a first analysis of the cyanobacteria to be isolated by the microscopic observation of an aliquot of the fresh samples. Under the upright or inverted microscope (see SOP 4.1), the morphological properties to be analyzed include shape, size of the trichomes and cells, presence of sheaths, necridic cells, spirality, branching, morphology of apical cells, size and position of specialized cells such as akinetes and heterocytes, motility, as well as pigmentation. Perform a complete photographic documentation at different magnifications of the cyanobacteria to be isolated. Based on the above features, tentatively identify the organism(s) with the aid of taxonomic keys and illustrated guides, and give them the most appropriate name(s).

SOP 4.2.2.3.2 Isolation of Strains and Culturing

If the sample is too concentrated, dilute a small aliquot with culture medium. Conversely, if the sample is too diluted, concentrate a larger volume using clean plankton nets (or by membrane filtration), and then dilute again appropriately. Under a stereomicroscope close to a Bunsen burner, or in a laminar flow hood, pick up cyanobacterial specimens (filaments or colonies) from the samples using sterile glass micropipettes. Eject each single specimen into one drop of sterile growth medium deposited in a sterile Petri dish. From these initial drops, the individual specimens are then picked out again and transferred into a fresh drop of culture medium as described. This operation is repeated 3–4 times, and will eliminate as much as possible small cyanobacteria and contaminants not visible under the stereomicroscope. After these steps, transfer the filaments (or colonies) into microwell plates containing 3 mL culture medium. Place the plates into a growth chamber or a climate room. After initial growth, as assessed by visual inspection and by observation with the stereomicroscope, transfer the isolated strains to 30 mL culture medium and, upon successful growth, to 150 mL culture flasks. The choice of the culture medium, temperature, light, and photoperiod depends on the species and characteristics of their habitats. See Chapter 3 for a full description of methods that can be used for the isolation and cultivation. The diacritical features of the specimens grown under different culture conditions are further analyzed as described in SOP. 4.2.2.3.1.

SOP 4.2.2.3.3 Genotyping by PCR and Sequencing

DNA samples are obtained by extracting DNA from aliquots of the cultured strains after concentration by centrifugation (pellets) or filtration on filters (see SOPs in Chapter 5).

To assess the genetic distinctiveness of a strain, several genetic markers can be used. The 16S rRNA genes are the primary choice because of the wide availability of information in public sequence databases, and the many published examples of their value for phylogenetic analyses. Nevertheless, to better resolve or confirm clades in phylogenetic analyses, the

16S rRNA gene sequences should be complemented with data on other additional genetic markers (see Table 4.2).

A common protocol for the PCR amplification of the 16S rRNA gene (ca. 1430 bp) is based on the primers pA and 23S30R designed by Edwards *et al.* (1989) and Taton *et al.* (2003) respectively, and procedures described in Gkelis *et al.* (2005). The 23S30R primer is specific for cyanobacteria and ensures that the resulting amplicons correspond uniquely to this phylum, even if the cultures contain other bacteria.

pA   5′–AGAGTTTGATCCTGGCTCAG–3′
23S30R   5′–CTTCGCCTCTGTGTGCCTAGGT–3′

Add the isolated nucleic acids (ca. 10 ng) to a reaction mix containing 1× PCR buffer, 0.2 mM deoxynucleoside triphosphate mix, 0.2 µM of the two primers, 0.8 U of Taq DNA polymerase. Add one tube with pure H_2O as a negative control. The addition of tubes with positive controls, if necessary, should be done very carefully to avoid cross-contamination of samples (see Notes).

The cycling protocol includes a denaturation step at 94°C for 5 min; 30 cycles of DNA denaturation at 94°C for 45 sec, primer annealing at 57°C for 45 sec, and strand elongation at 68°C for 2 min; final elongation step at 68°C for 7 min.

Check and separate PCR products by 1% agarose gel electrophoresis, followed by staining with ethidium bromide or another convenient DNA stain. DNA sizes are evaluated with a commercial DNA ladder that is placed on the same gel. Visualize the gel under an UV transilluminator.

For DNA sequencing, clean the PCR products with a commercial PCR purification kit, or follow the protocols provided by internal or external (commercial) sequencing services. The 16S rRNA gene is sequenced in three parts with the internal primers 16S544R, 16S1092R, and 16S979F (Rajaniemi-Wacklin *et al.*, 2005; Hrouzek *et al.*, 2005):

16S544R   5′–ATTCCGGATAACGCTTGC–3′
16S1092R   5′–GCGCTCGTTGCGGGACTT–3′
16S979F   5′–CGATGCAACGCGAAGAAC–3′

After obtaining the chromatograms, trim the low-quality ends and assemble the forward and reverse sequences by using specific software (Tippmann, 2004). The examination of DNA sequencing chromatograms and the detection of low-quality zones (e.g. at the ends; Fig. 4.7a) can be improved by using both free or commercial software that, besides suggesting high-quality sequences (Fig. 4.7b), provide a final FASTA file to be used in the final sequence assembly. Nevertheless, it is necessary to control all the steps, as the software may make mistakes or provide controversial results (e.g. Fig. 4.7a).

16S rRNA analyses should be complemented with the analysis of at least one or more housekeeping genes or the internal transcribed spacer (ITS) between the 16S and 23S genes (see Table 4.2). If using the primers 16S27F and 23S30R for amplification, the ITS can be obtained and sequenced together with the 16S, whereas primers 16S1407F and 23S30R will yield the entire ITS (in between the end of the 16S rRNA and the beginning of the 23S rRNA genes (Taton *et al.*, 2003) (see also Table 4.2).

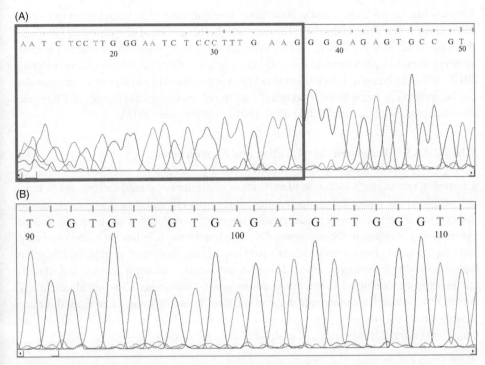

Figure 4.7 *Example of DNA sequencing chromatogram obtained using the internal primer 16S979F (16S rRNA). The partial sequence refers to the strain FEM_CPR107 (Planktothrix rubescens, accession number LT546527). (A) The rectangle indicates a zone of bad quality at the beginning of the sequence that should be deleted. These zones are characterized by multiple peaks that are difficult or impossible to distinguish. (B) Though the chromatogram has a little baseline noise, the actual peaks are easy to detect. The figures (modified) were obtained with the trace viewer Finch TV. The little grey vertical bars at the top of the chromatograms indicate the quality of the peaks. The quality of the sequence in (A) was determined with the help of Chromatogram Explorer Lite 4.0 (Heracle Biosoft SRL).*

SOP 4.2.2.3.4 *Testing Sequence Similarity and Phylogenetic Analyses*

Compare the DNA sequences with those deposited at the INSDC (the International Nucleotide Sequence Database Collaboration) public repositories (ENA, GenBank, and DDBJ) by using BLAST. If identical (or highly similar) sequences are not yet represented in the database, only more distantly related taxa stored in the database will show up, but will generally allow at least the assignment of the studied strain to a cyanobacterial order and/or family.

From an INSDC database, select a suitable number of closely related sequences suggested by the results of the BLAST query (Dereeper *et al.*, 2010), and/or belonging to the indicated species, genus, family, or order (for an update of taxonomic groupings, see Komárek *et al.* (2014). Alternatively, the tool "Seqmatch'" of the Ribosomal Database Project allows similar analyses, but will give trimmed, good-quality sequences (rdp.cme.msu.edu/seqmatch/). Seqmatch, however, does not replace a BLAST query, if the most recent sequence release is not yet available. Sequence alignments and phylogenetic analyses can be performed using standalone free software (e.g. Guindon and Gascuel, 2003; Gouy *et al.*, 2010; Hall, 2011; Kumar *et al.*, 2016), advanced programming languages (Paradis, 2012; R Core Team, 2016). Alternatively, Web-based and assisted pipelines allowing to build a phylogenetic tree starting from a set of raw sequences (Dereeper *et al.*, 2010, 2008) can be used. Information about the various types of phylogenetic analyses that can be applied (distance based, maximum parsimony, maximum likelihood, and Bayesian inference) are given in, for example, Lesk (2014) and Pevsner (2015).

SOP 4.2.2.3.5 *Taxonomic Identification and Classification*

Taking into consideration the observations made by microscopy (see SOP 4.1) and the genotyping results, the taxonomic identification of the strains should correspond as much as possible to the diacritical characteristics cited in the original taxa descriptions, and/or updated in a current and widely accepted taxonomic classification system (or other recent publications). At present, the taxonomy developed under the ICN has led to the most complete descriptions of cyanobacterial taxa but is in a dynamic state of change, and therefore highly provisional (Komárek *et al.*, 2014). In most cases, the nomenclature and classification schemes assembled in the books of the series "Süßwasserflora von Mitteleuropa" (see SOP 4.1), which also integrate some of the most recent literature, can be used for taxa identifications. Alternatively, the investigator may consult Web resources (e.g. AlgaeBase; Guiry and Guiry, 2017).

SOP 4.2.2.4 *Notes*

The isolation of filamentous and non-filamentous smaller cyanobacteria is generally easier by streaking on agar plates (see Chapter 3). The isolation of many, but not all, picocyanobacteria may require special protocols (see SOP 3.2). Axenization of strains (see SOP 3.3) is required for the validation of species according to the rules of the ICNP. An example for attempting the description of new species under the International Code of Nomenclature of Prokaryotes is the proposal by Gaget *et al.* (2015), which, however, was rebutted by Oren (2015).

The main risk of PCR amplifications is the contamination by amplicons of previous reactions as they may be very concentrated. Therefore, it is advisable to take multiple precautions, such as the use of sterile plasticware, cleaning of space with diluted bleach, and physical separations of laboratory steps, namely DNA extraction, PCR setup, and handling of PCR products. The micropipettes should be equipped with filter tips and dedicated to only one task. Negative controls should be made for each set of DNA extractions (one tube with water) and PCR reactions (one tube without DNA). Positive controls should be included with caution when conserved taxonomic markers are targeted, for example by using DNA from a species that is distinct (e.g. a marine one) and known to be absent in the sample under study.

The use of additives in the PCR mix has been described to solve problems with inhibitory substances, and up to 1 mg mL^{-1} bovine serum albumin (BSA) may be added (Taton *et al.*, 2003). Its inclusion is especially advisable with environmental samples as it increases PCR yields from low purity templates.

The analysis of ultrastructural, physiological, metabolic, and biochemical characteristics (including pigment analysis) can greatly aid in the classification and circumscription of cyanobacterial taxa, and may also be beneficial for the recognition of cryptic species and ecotypes. Nevertheless, it should be carefully taken into account that, when transferred to different culture conditions, strains isolated from ecologically distinct cyanobacterial populations in various habitats can change their morphological and biochemical characteristics, including toxicity (Gupta *et al.*, 2002; Amé and Wunderlin, 2005; Palinska and Surosz, 2014).

Many cyanobacteria with a specific ecology have not been cultured, or are difficult to maintain in culture (Waterbury, 2006). In such cases, alternative procedures for their genotypic characterization include DNA extraction and analysis of single cells, colonies, or filaments (see Chapters 2, 5, and 6 for recent applications, Lara *et al.*, 2013 and Mareš *et al.*, 2015).

The accuracy of the 16S rRNA gene sequences can be preliminarily assessed using RNAmmer (Lagesen *et al.*, 2007). In addition to the INSDC archives, the SILVA database can provide a comprehensive and quality checked dataset of aligned 16S rRNA sequences (Quast *et al.*, 2013).

The phylogenetic analyses based on the 16S rRNA should be complemented with sequence analyses of other housekeeping genes, which can provide a higher phylogenetic resolution between closely related organisms (at or below the species level) (Moreira *et al.*, 2013; Kurmayer *et al.*, 2015). The choice will depend on the availability of standardized protocols, designed PCR primers, and sequences deposited in the INSDC databases. Examples of genes and loci analyzed for taxonomic purposes are *rpo*B (Gaget *et al.*, 2011), *rbc*LX (Rudi *et al.*, 1998, Lin *et al.*, 2010), *rpo*C1 (D'Alelio *et al.*, 2013; Lin *et al.*, 2010). Another taxonomic marker with high resolution is the ITS linking the 16S and 23S ribosomal genes (e.g. Baurain *et al.*, 2002). However, its value should be carefully assessed for each species, since the presence of multiple and variable gene copies, especially in heterocytous species (Boyer *et al.*, 2001; Engene *et al.*, 2010), may lead to difficulties in taxonomic assignments.

Though being increasingly adopted, the implementation of the polyphasic approach is not free from difficulties (Lee *et al.*, 2014). Apart from the divergent taxonomic systems used by the sequence depositors, special attention should be given to incorrectly identified species in the INSDC databases, as the latter are not curated.

SOP 4.2.3   References

Amé, M.V. and Wunderlin, D.A. (2005) Effects of iron, ammonium and temperature on microcystin content by a natural concentrated *Microcystis aeruginosa* population, *Water, Air, and Soil Pollution*, **168**, 235–248.

Baurain, D., Renquin, L., Grubisic, S., *et al.* (2002) Remarkable conservation of internally transcribed spacer sequences of Arthrospira ("*Spirulina*") (Cyanophyceae, Cyanobacteria) strains from four continents and of recent and 30-year-old dried samples from Africa, *Journal of Phycology*, **38**, 384–393.

Boyer, S.L., Flechtner V.R., and Johansen, J.R. (2001) Is the 16S-23S rRNA internal transcribed spacer region a good tool for use in molecular systematics and population genetics? A case study in Cyanobacteria, *Molecular Biology and Evolution*, **18**, 1057–1069.

D'Alelio, D., Salmaso, N., and Gandolfi, A. (2013) Frequent recombination shapes the epidemic population structure of *Planktothrix* (Cyanoprokaryota) in Italian subalpine lakes, *Journal of Phycology*, **49**, 1107–1117.

Dereeper, A., Audic, S., Claverie, J.-M., and Blanc, G., (2010) BLAST-EXPLORER helps you building datasets for phylogenetic analysis. *BMC Evolutionary Biology*, **10**, 8.

Dereeper, A., Guignon, V., Blanc, G., *et al.* (2008) Phylogeny.fr: Robust phylogenetic analysis for the non-specialist, *Nucleic Acids Research*, **36**, W465–W469.

Edwards, U., Rogall, T., Blöcker, H., *et al.* (1989) Isolation and direct complete nucleotide determination of entire genes: Characterization of a gene coding for 16S ribosomal RNA, *Nucleic Acids Research*, **17**, 7843–7853.

Engene, N., Cameron Coates, R., and Gerwick, W.H. (2010) 16S rRNA gene heterogeneity in the filamentous marine cyanobacterial genus *Lyngbya*, *Journal of Phycology*, **46**, 591–601.

Gaget, V., Gribaldo, S., and Tandeau de Marsac, N. (2011) An *rpo*B signature sequence provides unique resolution for the molecular typing of cyanobacteria, *International Journal of Systematic and Evolutionary Microbiology*, **61**, 170–183.

Gaget, V., Welker, M., Rippka,R., and de Marsac, N.T. (2015) A polyphasic approach leading to the revision of the genus *Planktothrix* (Cyanobacteria) and its type species, *P. agardhii*, and proposal for integrating the emended valid botanical taxa, as well as three new species, *Planktothrix paucivesiculata* sp. nov.[ICNP], *Planktothrix tepida* sp. nov.[ICNP], and *Planktothrix serta* sp. nov.[ICNP], as genus and species names with nomenclatural standing under the ICNP, *Systematic and Applied Microbiology*, **38**, 141–158.

Gkelis, S., Rajaniemi, P., Vardaka, E., *et al.* (2005) *Limnothrix redekei* (Van Goor) Meffert (Cyanobacteria) strains from Lake Kastoria, Greece form a separate phylogenetic group, *Microbial Ecology*, **49**, 176–182.

Gouy, M., Guindon, S., and Gascuel, O. (2010) SeaView version 4: A multiplatform graphical user interface for sequence alignment and phylogenetic tree building, *Molecular Biology and Evolution*, **27**, 221–224.

Gugger, M., Lenoir, S., Berger, C., *et al.* (2005) First report in a river in France of the benthic cyanobacterium *Phormidium favosum* producing anatoxin-a associated with dog neurotoxicosis, *Toxicon*, **45**, 919–928.

Gugger, M., Lyra, C., Henriksen, P., *et al.* (2002) Phylogenetic comparison of the cyanobacterial genera *Anabaena* and *Aphanizomenon*, *International Journal of Systematic and Evolutionary Microbiology*, **52**, 1867–1880.

Guindon, S. and Gascuel, O. (2003) A simple, fast, and accurate algorithm to estimate large phylogenies by maximum likelihood, *Systematic Biology*, **52**, 696–704.

Guiry, M.D. and Guiry, G.M. (2017) *AlgaeBase: World-wide electronic publication* - National University of Ireland, Galway, http://www.algaebase.org.

Gupta, N., Bhaskar, A.S.B., and Rao, P.V.L. (2002) Growth characteristics and toxin production in batch cultures of *Anabaena flos-aquae*: Effects of culture media and duration, *World Journal of Microbiology and Biotechnology*, **18**, 29–35.

Hall, B.G. (2011) *Phylogenetic Trees Made Easy: A how-to manual*, 4th edn, Sinauer, USA.

Hrouzek, P., Ventura, S., Lukesová, A., *et al.* (2005) Diversity of soil *Nostoc* strains: Phylogenetic and phenotypic variability, *Archives für Hydrobiologie/Supplement Algological Studies*, **117**, 251–264.

Komárek, J. (2014) Modern classification of cyanobacteria, in N.K. Sharma, A.K. Rai, and L.J. Stal (eds), *Cyanobacteria: An economic perspective*, John Wiley & Sons, Ltd, Chichester, UK, 21–39.

Komárek, J., Kaštovský, J., Mareš, J., and Johansen, J.R. (2014) Taxonomic classification of cyanoprokaryotes (cyanobacterial genera) (2014) using a polyphasic approach, *Preslia*, **86**, 295–335.

Kumar, S., Stecher, G., and Tamura, K. (2016) MEGA7: Molecular Evolutionary Genetics Analysis Version 7.0 for Bigger Datasets, *Molecular Biology and Evolution*, **33**, 1870–1874.

Kurmayer, R., Blom, J.F., Ilona, L., and Pernthaler, J. (2015) Integrating phylogeny, geographic niche partitioning and secondary metabolite synthesis in bloom-forming *Planktothrix: The ISME Journal*, **9**, 909–921.

Lagesen, K., Hallin, P., Rødland, E.A., *et al.* (2007) RNAmmer: Consistent and rapid annotation of ribosomal RNA genes, *Nucleic Acids Research*, **35**, 3100–3108.

Lara, Y., Lambion, A., Menzel, D., *et al.* (2013) A cultivation-independent approach for the genetic and cyanotoxin characterization of colonial cyanobacteria, *Aquatic Microbial Ecology*, **69**, 135–143.

Lee, E., Ryan, U.M., Monis, P., *et al.* (2014) Polyphasic identification of cyanobacterial isolates from Australia, *Water Research*, **59**, 248–261.

Lesk, A.M. (2014) *Introduction to Bioinformatics*, 4th edn, Oxford University Press, Oxford, UK.

Lin, S., Wu, Z., Yu, G., *et al.* (2010) Genetic diversity and molecular phylogeny of *Planktothrix* (Oscillatoriales, Cyanobacteria) strains from China, *Harmful Algae*, **9**, 87–97.

Mareš, J., Lara, Y., Dadáková, I., *et al.* (2015) Phylogenetic analysis of cultivation-resistant terrestrial cyanobacteria with massive sheaths (*Stigonema* spp. and *Petalonema alatum* , Nostocales, Cyanobacteria) using single-cell and filament sequencing of environmental samples, *Journal of Phycology*, **51**, 288–297.

Metcalf, J.S. and Codd, G.A. (2012) Cyanotoxins, in BA Whitton (ed.), *Ecology of Cyanobacteria II: Their diversity in space and time*, Springer Netherlands, Dordrecht, 651–675.

Moreira, C., Vasconcelos, V., and Antunes, A. (2013) Phylogeny and biogeography of cyanobacteria and their produced toxins, *Marine Drugs*, **11**, 4350–4369.

Oren, A. (2015) Comments on: "A polyphasic approach leading to the revision of the genus *Planktothrix* (Cyanobacteria) and its type species, *P. agardhii*, and proposal for integrating the emended valid botanical taxa, as well as three new species, *Planktothrix paucivesiculata* sp. nov.(ICNP), *Planktothrix tepida* sp. nov.(ICNP), and *Planktothrix serta* sp. nov.(ICNP), as genus and species names with nomenclature standing under the ICNP," by Gaget, V., Welker, M., Rippka, R., and Tandeau de Marsac, N., *Systematic and Applied Microbiology*, **38**, 159–160.

Palinska, K.A. and Surosz, W. (2014) Taxonomy of cyanobacteria: A contribution to consensus approach, *Hydrobiologia*, **740**, 1–11.

Paradis, E. (2012) *Analysis of Phylogenetics and Evolution with R*, 2nd edn, Springer, New York.

Pevsner, J. (2015) *Bioinformatics and Functional Genomics*, 3rd edn, Wiley-Blackwell, Chichester, UK.

Quast, C., Pruesse, E., Yilmaz, P., *et al.* (2013) The SILVA ribosomal RNA gene database project: Improved data processing and web-based tools, *Nucleic Acids Research*, **41**, D590–D596.

R Core Team (2016) R: A language and environment for statistical computing. R Foundation for Statistical Computing, Vienna, https://www.R-project.org/.

Rajaniemi-Wacklin, P., Rantala, A., Mugnai, M.A., *et al.* (2005) Correspondence between phylogeny and morphology of *Snowella* spp. and *Woronichinia naegeliana*, cyanobacteria commonly occurring in lakes, *Journal of Phycology*, **42**, 226–232.

Rudi, K., Skulberg, O.M., and Jakobsen, K.S. (1998) Evolution of cyanobacteria by exchange of genetic material among phyletically related strains, *Journal of Bacteriology*, **180**, 3453–3461.

Salmaso, N., Capelli, C., Shams, S., and Cerasino, L. (2015) Expansion of bloom-forming *Dolichospermum lemmermannii* (Nostocales, Cyanobacteria) to the deep lakes south of the Alps: Colonization patterns, driving forces and implications for water use, *Harmful Algae*, **50**, 76–87.

Taton, A., Grubisic, S., Brambilla, E., *et al.* (2003) Cyanobacterial diversity in natural and artificial microbial mats of Lake Fryxell (McMurdo Dry Valleys, Antarctica): A morphological and molecular approach, *Applied and Environmental Microbiology*, **69**, 5157–5169.

Tippmann, H.F. (2004) Analysis for free: Comparing programs for sequence analysis, *Briefings in Bioinformatics*, **5**, 82–87.

Vandamme, P., Pot, B., Gillis, M., *et al.* (1996) Polyphasic taxonomy: A consensus approach to bacterial systematics, *Microbiological Reviews*, **60**, 407–438.

Waterbury, J.B. (2006) The Cyanobacteria: Isolation, purification and identification, in M. Dworki and S. Falkow, S. (eds), *The Prokaryotes: A handbook on the biology of bacteria: Volume 4: Bacteria: Firmicutes, cyanobacteria*, Springer, New York, 1053–1073.

Wayne, L.G., Brenner, D.J., Colwell, R.R., *et al.* (1987) Report of the *ad hoc* committee on reconciliation of approaches to bacterial systematics, *International Journal of Systematic and Evolutionary Microbiology*, **37**, 463–464.

Wetzel, R.G. and Likens, G. (2000) *Limnological Analyses*, 3rd edn, Springer, New York.

Wilmotte, A. and Herdman, M. (2001) Phylogenetic relationships among the Cyanobacteria based on 16S rRNA sequences, in D.R. Boone and R.W. Castenholz (eds), *Bergey's Manual of Systematic Bacteriology*, Springer, New York, 487–493.

5

Nucleic Acid Extraction

Elke Dittmann[1]*, Anne Rantala-Ylinen*[2], *Vitor Ramos*[3,4], *Vitor Vasconcelos*[3,4],
Guntram Christiansen[5,6], *and Rainer Kurmayer*[5]*

[1] *Institute of Biochemistry and Biology, University of Potsdam, Potsdam-Golm, Germany*
[2] *Department of Food and Environmental Sciences, Division of Microbiology and Biotechnology,*
University of Helsinki, Helsinki, Finland
[3] *Interdisciplinary Centre of Marine and Environmental Research (CIIMAR/CIMAR), University of*
Porto, Matosinhos, Portugal
[4] *Faculty of Sciences, University of Porto, Porto, Portugal*
[5] *Research Institute for Limnology, University of Innsbruck, Mondsee, Austria*
[6] *Miti Biosystems GmbH, Max F. Perutz Laboratories, Vienna, Austria*

5.1 Introduction

This chapter aims to provide basic knowledge on DNA and RNA extraction techniques for cyanobacterial laboratory strains and field samples. It includes general guidelines for qualitative and quantitative DNA and RNA extraction as well as recommendations for DNA and RNA storage. The success of any nucleic acid extraction technique depends on appropriate sampling strategies (see Chapter 2). Moreover, specific downstream applications require different qualities of nucleic acid purity and integrity. The selection of a nucleic acid extraction protocol thus has to depend on the question to be addressed, and the cyanobacterial genus to be analyzed. This chapter provides an overview of the critical steps that must be taken when extracting nucleic acid and outlines some example protocols. Qualitative or quantitative nucleic acid extraction from field samples is particularly challenging and will

*Corresponding authors: editt@uni-potsdam.de; rainer.kurmayer@uibk.ac.at

Molecular Tools for the Detection and Quantification of Toxigenic Cyanobacteria, First Edition.
Edited by Rainer Kurmayer, Kaarina Sivonen, Annick Wilmotte and Nico Salmaso.
© 2017 John Wiley & Sons Ltd. Published 2017 by John Wiley & Sons Ltd.

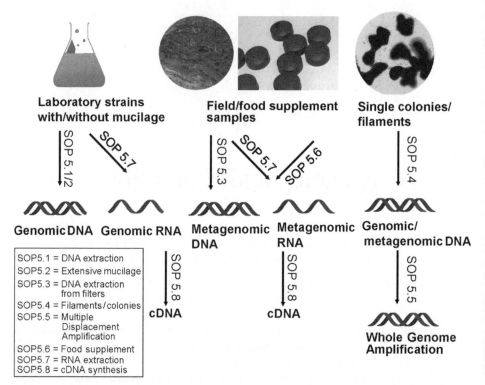

Figure 5.1 *Schematic overview of the protocols described in this chapter. (See color plate section for the color representation of this figure.)*

require method optimization or combinations of techniques for individual water samples. The chapter further includes a subchapter on single colony analysis and on whole genome amplification (WGA). See Fig. 5.1 for a schematic overview.

Nucleic acid extraction is a precondition for methods utilizing conventional PCR (Chapter 6), quantitative PCR (Chapter 7), DNA diagnostic microarrays and cDNA analysis (Chapter 8), DGGE (Chapter 9), and NGS techniques (Chapter 10). For PCR approaches, single colonies and filaments can be directly used thereby omitting the nucleic acid extraction step (this chapter). If the quantity of DNA is limited (e.g. for a whole genome analysis of a single colony or filament), a WGA technique may be necessary (this chapter).

The nucleic acid extraction protocols described here are used for (1) the detection of genes used for the taxonomic identification and phylogenetic classification of cyanobacteria, (2) the detection and quantification of genes (genotypes) indicative of toxin production, (3) the assignment of toxin production to different genera of cyanobacteria, (4) the analysis of the evolutionary processes determining the distribution of toxin genes among cyanobacteria, (5) the analysis of the transcriptional regulation of toxin synthesis genes by comparing the transcript amount under various environmental conditions, and (6) whole genome analysis of toxic cyanobacteria.

5.2 Specific Extraction Procedures and Storage

5.2.1 DNA Extraction from Laboratory Strains

For the isolation of DNA from laboratory strains, standard DNA extraction techniques typically include the following steps: (1) harvesting of cyanobacteria by centrifugation or filtration (see also Chapter 2), (2) washing of the pellet to remove contaminants, (3) homogenization and lysis of cells, (4) removal of proteins, (5) precipitation of nucleic acids, (6) resuspension of DNA pellet, and (7) removal of RNA (optional). Alternatively, DNA can be isolated using commercial DNA isolation kits that typically include anion-exchange columns for DNA binding and purification. Standard DNA isolation protocols (as exemplified in SOP 5.1) are more time-consuming but cheaper than kit-based DNA extraction techniques. In any case, techniques have to be validated for different cyanobacterial genera that differ in their cell surface characteristics and the presence of potential PCR contaminants and their lysis efficiency.

5.2.2 DNA Extraction from Field Samples

Cyanobacterial field samples show different levels of complexity that can complicate quantitative DNA extraction, in particular if toxic genotypes are underrepresented in mixtures obtained from benthic communities (Chapter 2). Moreover, cyanobacterial field samples typically contain a more pronounced sheath, and highly viscous DNA extracts may inhibit enzymes used for PCR amplifications. For quantitative analysis preferably fresh samples or samples stored in the freezer at –20° should be used (avoid freeze-drying or fixation). Owing to the different lysis efficiencies, extraction bias can be expected, such as to require a solid method optimization, in particular for (semi) quantitative approaches as those described in Chapters 7, 8, and 10. Principally, for complex environmental samples either standard DNA isolation approaches or kit-based approaches can be used. The efficiency of such techniques, however, has to be validated, and eventually different techniques have to be combined for quantitative DNA extraction. Environmental samples, in general, require more thorough washing and removal of contaminants and a more complete lysis of cells to minimize quantitative bias. Nevertheless, DNA isolation techniques should minimize mechanical stressing of DNA that can cause double-strand and single-strand breaks leading to bias in downstream technologies.

5.2.3 DNA Extraction from Food Supplements

DNA extraction from food supplements can be difficult as, typically, algae are processed by drying and may contain food additives that can influence the DNA extraction procedure. Pharmaceutical excipients with adsorbent properties, for example, can interfere with DNA extraction (Costa *et al.*, 2015). Because of the drying process and the high temperature treatment of food supplements, the extracted DNA is typically fragmented or even damaged (see SOP 5.6) (Ostermaier *et al.*, 2013). In order to estimate potential bias, it is important to document how samples have been processed (see Chapter 2) and apply DNA purification protocols, such as that described in SOP 5.2.

5.2.4   RNA Extraction from Laboratory Strains

RNA has a much shorter half-life than DNA. This is particularly true for mRNA, which can be easily degraded within minutes. For the assessment of the impact of environmental factors on toxin production, the cyanobacteria have to be immediately processed after sampling (or storage of the harvested pellet at –20°C). Note that even during a centrifugation or filtration step the factor light is changing, and this may have an impact on gene expression if the samples are not kept cool to minimize RNA degradation. SOP 5.7 exemplifies a standard laboratory protocol where cultures are harvested by gentle vacuum filtration and frozen immediately in liquid nitrogen. Standard protocols for RNA extraction include the following steps: (1) harvesting of cells, (2) washing of cells to remove contaminants, (3) homogenization and lysis of cells, (4) separation of RNA from proteins and genomic DNA using organic solvents and subsequent centrifugation, (5) precipitation of RNA, (6) resuspension of RNA, and (7) removal of contaminating DNA. RNA isolation can also be achieved by commercial kits but, similarly to DNA extraction, it requires additional method optimization steps depending on the cyanobacterial genus analyzed. All solutions used for RNA isolation should be free of RNases, and RNA dissolved in aqueous solution should be kept cool all the time. For downstream cDNA analysis any DNA contamination has to be effectively removed while maintaining the integrity of cyanobacterial RNA.

5.2.5   RNA Extraction from Field Samples

Standard routines for cyanobacterial sampling do not include immediate cooling of the samples and are thus not adequate for a quantitative RNA analysis. Any sampling for downstream RNA analysis has to be very fast with an exact monitoring of the environmental conditions. If the samples cannot be frozen immediately (e.g. with liquid nitrogen), they should be protected from RNA degradation by RNA storage and stabilization reagents such as RNAlater® (Thermo Fisher Scientific). The processing of the material should be otherwise the same as described above for laboratory samples and is typically based on commercials kits.

5.2.6   Single Colony and Filament Analysis

An increasing number of studies have analyzed toxigenic cyanobacteria directly from field colonies or from single filaments without nucleic acid extraction. Single colonies or filaments should be carefully isolated using the techniques described in SOP 2.3 and should not be fixed prior to analysis. Single colony analysis allows for a direct correlation of the genetic potential for toxin production and toxin production itself provided that it is combined with mass-spectrometric analytical techniques such as MALDI-TOF analysis. However, single colony analysis can be more error-prone than standard PCR after DNA extraction, owing to either inhibiting substances in the samples or limitations in the amount or the integrity of available DNA. The latter problem can be alleviated to some extent using a WGA technique. A protocol leading to single filament PCR is shown in SOP 5.4.

5.2.7 Whole Genome Amplification

The available amount of DNA obtained by lysis of single colonies and filaments is very low and often not sufficient for downstream applications. Detection of toxin biosynthesis genes by PCR (Chapters 6 and 7) might require an additional amplification step that increases the copy number of DNA. The most common technique for genome amplification is multiple displacement amplification (MDA) involving the binding of random short oligonucleotides and the polymerization of new DNA strands starting from each of these primers after strand displacement using the Phi 29 Polymerase. WGA can be performed using commercial kits following instruction of the manufacturers (see SOP 5.5).

5.2.8 Nucleic Acid Storage

For long-term storage it is important to keep the activity of nucleases as low as possible. This is typically achieved by storing DNA at –20°C and RNA at –80°C. Each freeze and thaw cycle causes mechanical stress for nucleic acids, in particular for high molecular weight DNA. Storage of DNase-free DNA in small aliquots at –20°C is thus more favorable than the repeated freezing/thawing of the same DNA sample. As RNA is much more sensitive to degradation, it should generally be stored at –20°C for shortterm or –80°C for longterm. It is important to keep the sample cool during the melting process, as RNases are highly active.

5.3 References

Costa, J., Amaral, J.S., Fernandes, T.J.R., *et al.* (2015) DNA extraction from plant food supplements: Influence of different pharmaceutical excipients, *Molecular and Cellular Probes*, **29**, 473–478.

Ostermaier, V., Christiansen, G., Schanz, F., and Kurmayer, R. (2013) Genetic variability of microcystin biosynthesis genes in *Planktothrix* as elucidated from samples preserved by heat desiccation during three decades, *PLOS ONE*, **8**, e80177.

SOP 5.1

Standard DNA Isolation Technique for Cyanobacteria

*Elke Dittmann**

Institute of Biochemistry and Biology, University of Potsdam, Potsdam-Golm, Germany

SOP 5.1.1   Introduction

This SOP describes a standard technique that can be used for DNA extraction from cyanobacterial laboratory strains without a commercial kit. The method was adapted from Franche and Damerval (1988) and combines osmotic shock treatment (to disrupt cell membranes), enzymatic treatment, and DNA extraction by phase separation.

SOP 5.1.2   Experimental

SOP 5.1.2.1   Materials

- *ca.* 200 mL cyanobacterial culture grown to mid- or late logarithmic growth phase (OD_{750} nm ≈ 0.1).
- TE buffer (10 mM Tris-HCl, 1 mM EDTA, pH 8.0).
- TES buffer (50 mM Tris-HCl, 100 mM EDTA, 25% (w/v) Sucrose pH 8.0).
- Lysozyme powder.
- Proteinase K solution (20 mg mL^{-1}), EDTA, SDS 20% (w/v).
- Phenol/chloroform/isoamyl alcohol (25:24:1, v/v/v), molecular biology grade, toxic (use laboratory gloves).
- Chloroform/isoamyl alcohol (24:1, v/v), molecular biology grade, toxic (use laboratory gloves).

---

*Corresponding author: editt@uni-potsdam.de

- Isopropanol (analysis grade).
- 70% (v/v) ethanol (analysis grade).
- Eppendorf tubes (2.0, 1.5 mL).

SOP 5.1.2.2   Equipment

- Table centrifuge with cooling function for Eppendorf tubes (4°C).
- Thermostat for Eppendorf tubes (37°C, 60°C).
- Fume hood for work with solvents.
- Ice bath.
- Adjustable mechanical pipettes (100–1000 μL, 20–200 μL, 1–10 μL), filter tips.
- Gel electrophoresis and gel imaging.
- NanoDrop™ (Thermo Fisher Scientific).

SOP 5.1.3   Procedure

1. Harvest cells from 200 mL liquid culture by centrifugation at 4.000 × g, 10 min.
2. Alternatively: pulverize filter with cyanobacterial material from field using liquid nitrogen in a mortar.
3. Wash pellet twice using TE buffer and centrifuge at 4.000 × g for 10 min.
4. Re-suspend in 0.5 mL TES buffer and incubate on ice at 4°C for 1 h.
5. Add lysozyme to a final concentration of 2 mg mL^{-1} and incubate at 37°C for 1 h.
6. Add EDTA, proteinase K and SDS (to a final concentration of 0.05 M, 50 μg mL^{-1}, and 2% (w/v), respectively) and incubate at 60°C for 1 h (or overnight).
7. Add one volume phenol/chloroform/isoamyl alcohol (25:24:1), mix well, and centrifuge for 10 min at 4°C at 4.000 × g. Take the upper aqueous fraction and transfer it to a new tube and repeat the step.
8. Add one volume chloroform/isoamyl alcohol (24:1) to the aqueous fraction in as obtained in the previous step and centrifuge as before. Take the upper aqueous fraction and transfer it to a new tube.
9. Add 1 volume isopropanol and incubate for 1 h at room temperature. Centrifuge for 10 min at 6.000 × g at 4°C. Remove the supernatant from precipitated DNA.
10. Add 500 μL 70% ethanol, wash, and centrifuge at 4.000 × g for 10 min.
11. Air-dry pellet and re-suspend it in 50–100 μL TE buffer depending on size and solubility of pellet.
12. Add 1 μL RNase and incubate for 1 h at 37°C (optional).

SOP 5.1.4   Notes

Phenol/chloroform/isoamyl alcohol (25:24:1, v/v/v) are available as ready-to-use solutions which is recommended because phenol and chloroform are both toxic and hazardous. For the phenol/chloroform extraction step safe-lock Eppendorf tubes are recommended.

The quantity of the DNA should be measured using a NanoDrop™ photometer at 260 nm (OD = 1 at 260 nm is equivalent to 50 μg mL^{-1} double-strand DNA). The quality of the DNA should be tested on a 0.8% (w/v) agarose gel (load at least 0.2 μg to check integrity). DNA can be stained using ethidium bromide (toxic) or nonhazardous stains such as Midori Green following standard conditions (Sambrook and Russell, 2001). The majority of DNA

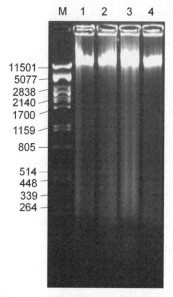

Figure 5.2 *Genomic DNA extracted from* Chamaesiphon *strain PCC 6505 and separated by agarose gel electrophoresis (0.8%, w/v) using ethidium bromide staining. Bands 1, 2, 3, and 4 were loaded with DNA extracts from different culture flasks (1 µg of DNA). M, λPstI marker (in bp).*

will run in the 20–40 kbp (double-stranded DNA) range, owing to mechanical shearing (Fig. 5.2).

SOP 5.1.5   References

Franche, C. and Damerval, T. (1988) Tests on *nif* probes and DNA hybridizations, *Methods in Enzymology*, **167**, 803–808.

Sambrook, J. and Russell, D. (2001) *Molecular Cloning: A laboratory manual*, 3rd edn, Cold Spring Harbor Laboratory Press, Cold Spring Harbor, New York.

SOP 5.2

DNA Isolation Protocol for Cyanobacteria with Extensive Mucilage

Guntram Christiansen[1,2], Elisabeth Entfellner[1], and Rainer Kurmayer[1]

[1]*Research Institute for Limnology, University of Innsbruck, Mondsee, Austria*
[2]*Miti Biosystems GmbH, Max F. Perutz Laboratories, Vienna, Austria*

SOP 5.2.1   Introduction

This SOP describes a protocol that can be used for DNA extraction from cyanobacterial laboratory strains which show extensive mucilage production indicated by high viscosity of the cell culture. The high proportion of polysaccharides can inhibit subsequent PCR amplification (see Chapters 6 and 7).

SOP 5.2.2   Experimental

SOP 5.2.2.1   Materials

- 50 mL cyanobacterial culture grown to mid- or late log phase.
- TE buffer (10 mM Tris-HCl, 1 mM EDTA, pH 8.0).
- Proteinase K (600–1000 U mL^{-1}).
- SDS 20% (w/v).
- Phenol/chloroform/isoamyl alcohol (25:24:1, v/v/v), toxic solution (to be handled with laboratory gloves).
- Chloroform/isoamyl alcohol (24:1, v/v), toxic solution (to be handled with laboratory gloves).
- 39% polyethylene glycol (PEG) 8000 (w/v), 30 mM MgCl$_2$.
- 70% (v/v) ethanol.
- Safe-lock Eppendorf tubes (2.0 mL).
- Eppendorf tube (1.5 mL).

SOP 5.2.2.2   Equipment (Instruments)

- Cell disruption bomb (nitrogen pressure 200 bar); or pistil, mortar (and liquid nitrogen).
- Table centrifuge with cooling function for Eppendorf tubes (4°C).
- Thermostat for Eppendorf tubes (37°C, 60°C).
- Fume hood for work with solvents.
- Ice bath.
- Adjustable mechanical pipettes (100–1000 μL, 20–200 μL, 1–10 μL), filter tips.

SOP 5.2.3   Procedure

1. Harvest cells by centrifugation (10000 × g, 10 min).
2. Wash pellet twice with TE buffer and centrifuge (10000 × g, 10 min).
3a. Transfer washed cell pellet either to precooled mortar and grind until a greenish power is obtained, collect powder with a spatula into a safe-lock Eppendorf tube (2.0 mL), add TE buffer up to 900 μL total volume

   or

3b. Transfer pellet with 1 mL of back flow to the cell disruption bomb and apply 100 bar. Open valve carefully and collect suspension at the valve in an Eppendorf tube. Repeat cell disruption three times. Adjust volume to 900 μL with TE buffer.
4. Add 100 μL of 20% SDS and vortex.
5. Add 5 μL of proteinase K (600–1000 U mL^{-1}) and vortex.
6. Incubate at 65°C for one hour (check color: should turn from green to brown).
7. Add 1 volume phenol/chloroform/isoamyl alcohol (under fume hood) and mix vigorously. Centrifuge (10000 × g, 10 min).
8. Transfer aqueous supernatant to a new Eppendorf tube, without disturbing interphase or lower phase.
9. Repeat steps 7 and 8.
10. Add one volume chloroform/isoamyl alcohol to supernatant and mix vigorously. Centrifuge (10000 × g 10 min).
11. Transfer aqueous supernatant to 1.5 mL Eppendorf tube and add ½ volume of 39% PEG 8000, 30 mM MgCl$_2$ (final concentration 13% PEG 8000, 10 mM MgCl$_2$). Mix vigorously and incubate at room temperature for 15 min.
12. Centrifuge (10000 × g, 30 min). Carefully remove the supernatant without disturbing the white pellet.
13. Add 1 mL 70% EtOH and vortex until pellet is detached from the bottom of the tube. Centrifuge (10000 × g, 5 min, 4°C).
14. Carefully remove supernatant without disturbing the pellet. Repeat step 13.
15. After finally having removed residual EtOH, let pellet air dry by ventilation in the opened Eppendorf tube.
16. Re-suspend in 50–200 μL TE buffer depending on size and solubility of pellet.
17. Quantify DNA concentration as described in SOP 5.1.

SOP 5.2.4   Notes

Phenol/chloroform/isoamyl alcohol (25:24:1, v/v/v) are available as ready-to-use solutions, which is recommended because phenol and chloroform are both toxic and hazardous. For the phenol/chloroform extraction step, safe-lock Eppendorf tubes are recommended.

By using PEG 8000, DNA fragments will be selectively precipitated while long-chained polysaccharides remain in solution (Paithankar and Prasad, 1991). This precipitation protocol is not suitable for recovery of short DNA fragments (<150 bp).

Alternatively to PEG 8000, in order to purify the DNA from polysaccharides, the aqueous phase is mixed with two volume parts of cetyltrimethylammonium bromide (CTAB) precipitation solution (15 mM CTAB, 40 mM NaCl, pH 7.0), incubated for 1 h at 4°C and pelleted (Reinard, 2008; Entfellner *et al.*, 2017). The DNA pellet is then dissolved in 1.2 M NaCl solution and extracted again with chloroform/isoamyl alcohol (24:1, v/v) and washed using EtOH.

SOP 5.2.5   References

Entfellner, E., Frei, M., Christiansen, G. *et al.* (2017). Evolution of anabaenopeptin peptide structural variability in the cyanobacterium *Planktothrix*. *Frontiers in Microbiology*, **8** (219).

Paithankar, K.R. and Prasad, K.S. (1991) Precipitation of DNA by polyethylene glycol and ethanol, *Nucleic Acids Research*, **19**, 1346–1349.

Reinard, T. (2008) *Molekularbiologische Methoden*, Eugen Ulmer KG, Stuttgart.

SOP 5.3

Quantitative DNA Isolation from Filters

*Rainer Kurmayer**

Research Institute for Limnology, University of Innsbruck, Mondsee, Austria

---

*Corresponding author: rainer.kurmayer@uibk.ac.at

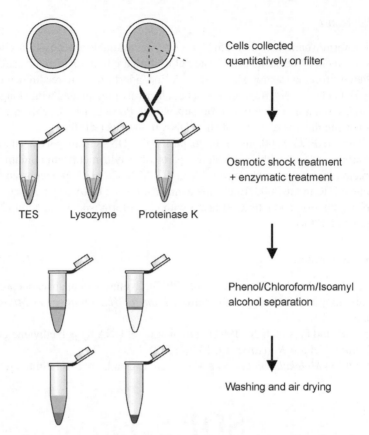

Figure 5.3 *Flow diagram showing steps of DNA extraction from cells collected on glass fiber filters (see SOP 5.3). (See color plate section for the color representation of this figure.)*

SOP 5.3.1 Introduction

In order to quantify toxin genes by qPCR in absolute numbers (e.g. gene copies, cell number equivalents) (see Chapter 7) it is necessary to extract DNA from water samples quantitatively. This protocol describes how DNA is quantitatively extracted from cells harvested on glass fiber filters (see Chapter 2) following a protocol published by Franche and Damerval (1988). The modified protocol has been published in Kurmayer *et al.* (2003). See Fig. 5.3 for a flow diagram of the various steps.

SOP 5.3.2 Experimental

SOP 5.3.2.1 Materials

- TES buffer (50 mM Tris-HCl, 100 mM EDTA, 25% (w/v) sucrose pH 8.0).
- SDS 20% (w/v).
- Phenol/chloroform/isoamyl alcohol (25:24:1, v/v/v), toxic solution (to be handled with laboratory gloves).

- Chloroform/isoamyl alcohol (24:1, v/v), toxic solution (to be handled with laboratory gloves).
- TE buffer (10 mM Tris-HCl, 1 mM EDTA, pH 8.0).
- Lysozyme, proteinase K.
- 2 mL safe-lock Eppendorf tubes.
- Eppendorf tube (1.5 mL).

SOP 5.3.2.2   Equipment (Instruments)

- Table centrifuge with cooling function for Eppendorf tubes (4°C).
- Thermostat for Eppendorf tubes (37°C, 60°C).
- Fume hood for work with solvents.
- Ice bath.
- Adjustable mechanical pipettes (100–1000 μL, 20–200 μL, 1–10 μL), filter tips.

SOP 5.3.3   Procedure

1. Cut the frozen filter (e.g. glass fiber filter GF/C containing harvested cells; see Chapter 2) into pieces and keep on ice.
2. Suspend the filter in 0.5 mL TES (using a 2 mL safe-lock Eppendorf tube) and incubate on ice for 2 h. If the filter is > 2.5 cm in diameter (e.g. 4.7 cm), split it in half and aliquot it into two Eppendorf tubes, and the DNA extract is combined at Step 12.
3. Add lysozyme (one tip of a spatula ≈ 5 mg mL^{-1} final concentration) and incubate for 1 h at 37°C.
4. Add proteinase K (final concentration 50 μg mL^{-1}) and SDS (final concentration 2% = 55 μL SDS (20%, w/v)) and incubate for 1 h at 60°C. Smashing of the filter assists mechanical destruction of the cells.
5. Add 1 mL of phenol/chloroform/isoamyl alcohol (25:24:1, v/v/v) and vortex (use filter tips to protect the pipette against the solvents).
6. Centrifuge 10 min 10000 × g, and suck clear supernatant (water phase with DNA) with a pipette and put it into a new tube; do not disturb the white interphase (proteins). At this step, the final recovered volume should be around 0.2–0.4 mL.
7. Add 1 mL phenol/chloroform/isoamyl alcohol to the supernatant once again, vortex, centrifuge and suck the water phase again.
8. Add 1 mL chloroform/isoamyl alcohol (24:1, v/v) to the water phase collected in Step 7 and centrifuge.
9. Suck the water phase for a last time and add 2.5 volumes of ethanol (–20 °C) (*ca.* 1.3 mL) and precipitate DNA on ice (0°C, 1 h).
10. Centrifuge 20 min at 10000 × g, keep the pellet (with DNA), and discard the supernatant (the pellet is strong and the reaction tubes can be easily turned around and remove the last drop by tipping the top of the Eppendorf tube on paper).
11. Wash with 0.5 mL 70% (v/v) ethanol (–20°C) and centrifuge 20 min at 10000 × g (4°C), then remove the supernatant with a pipette, the pellet is looser now.
12. Air dry the pellet, re-suspend the pellet in 100 μL of TE buffer.
13. Store the dissolved DNA at –20°C.

SOP 5.3.4   Notes

In general glass fiber filters (GF/C type) have been found to be easier to handle than membrane filters (non-dissolvable in phenol/chloroform). Owing to their more rigid nature, the membrane filters do not form a pellet after centrifugation and also disturb the phase separation. Membrane filters must not dissolve in phenol/chloroform.

Owing to the effective phase separation, all lipophilic and hydrophobic molecules will be effectively removed. However, in the water phase, in addition to nucleic acids, other abundant water-soluble molecules, such as phycobilins (phycoerythrin, phycocyanin) and intra- and extracellular carbohydrates (e.g. from the mucilage), will accumulate as well. Subsequent DNA purification (see SOP 5.2) will inevitably lead to the loss of DNA, and thus for a quantitative DNA extraction protocol it is not recommended. In contrast, dilution of the DNA extracts (>1:100) typically results in a negligible influence of the contaminants, potentially interfering with qPCR amplification.

Since phenol and chloroform are both toxic solvents, it is important to pipette the solvents using filter tips (to protect the mechanical pipettes) and use a fume hood. The phenol/chloroform waste has to be collected. A ready-to-use mixture of chloroform/phenol/isoamyl alcohol is offered commercially.

When compared with commercially available DNA extraction kits, this protocol has been found more effective and reliable (i.e. resulting in higher DNA template amounts as revealed by qPCR) (Schober and Kurmayer, 2006). The cell numbers as estimated by qPCR (see Chapter 7) and estimated by counting using a microscope have been shown to correlate with high correlation coefficients (Kurmayer and Kutzenberger, 2003; Ostermaier and Kurmayer, 2010).

SOP 5.3.5   References

Franche, C. and Damerval, T. (1988) Tests on *nif* probes and DNA hybridizations, *Methods in Enzymology*, **167**, 803–808.

Kurmayer, R., Christiansen, G., and Chorus, I. (2003) The abundance of microcystin-producing genotypes correlates positively with colony size in *Microcystis* sp. and determines its microcystin net production in Lake Wannsee, *Applied and Environmental Microbiology*, **69**, 787–795.

Kurmayer, R. and Kutzenberger, T. (2003) Application of real-time PCR for quantification of microcystin genotypes in a population of the toxic cyanobacterium *Microcystis* sp., *Applied and Environmental Microbiology*, **69**, 6723–6730.

Ostermaier, V. and Kurmayer, R. (2010) Application of real-time PCR to estimate toxin production by the cyanobacterium *Planktothrix* sp., *Applied and Environmental Microbiology*, **76**, 3495–3502.

Schober, E. and Kurmayer, R. (2006) Evaluation of different DNA sampling techniques for the application of the real-time PCR method for the quantification of cyanobacteria in water, *Letters in Applied Microbiology*, **42**, 412–417.

SOP 5.4

Genomic DNA Extraction from Single Filaments/Colonies for Multiple PCR Analyses

Guntram Christiansen[1,2], Chen Qin[1,3], and Rainer Kurmayer[1]*

[1]*Research Institute for Limnology, University of Innsbruck, Mondsee, Austria*

[2]*Miti Biosystems GmbH, Max F. Perutz Laboratories, Vienna, Austria*

[3]*College of Natural Resources and Environment, Northwest A & F University, Yangling, P. R. China*

SOP 5.4.1   Introduction

This SOP describes a protocol that is useful for the provision of genomic DNA from individual cyanobacterial filaments/colonies directly isolated from the field (Fig. 5.4). The obtained DNA quantities are sufficient for multiple individual PCR experiments.

SOP 5.4.2   Experimental

SOP 5.4.2.1   Materials

- Eppendorf tubes (0.5 mL).
- Plastic Petri dishes.
- Forceps.
- BG11 medium (Rippka, 1988).
- Sample dilution buffer (Phire Plant Direct PCR Kit, Thermo Fisher Scientific).
- Milli-Q® (Merk Millipore) water.
- 10% H_2O_2 containing water for rinsing the Sonifier® cell disruptor tip.

*Corresponding author: rainer.kurmayer@uibk.ac.at

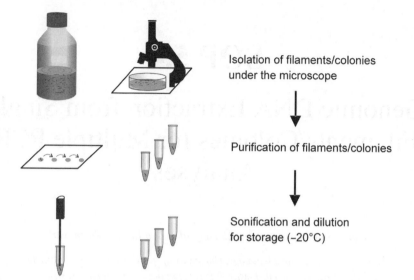

Figure 5.4 *Flow diagram showing steps of individual filament/colony isolation, purification by washing in medium and DNA extraction using sonification.*

SOP 5.4.2.2   Equipment

- Stereomicroscope (4–40× magnification) with working space to handle forceps.
- Sonifier® cell disruptor equipped with a microtip (e.g. Digital Sonifier® 450, Branson, Danbury, Connecticut, USA).
- Adjustable mechanical pipettes (10–100 μL, 20–200 μL)

SOP 5.4.3   Procedure

1. Dilute field sample with BG11 so that approximately five filaments/colonies are visible in one drop (see SOP 2.3 for details).
2. Isolate all single filaments/colonies from one drop, so that random selection is assured.
3. Pick individual filaments/colonies with a forceps, a needle, or a glass micropipette (see SOP 3.1) and wash by serial transfer between three fresh drops of BG11 medium.
4. Transfer the so washed single filament/colony to 0.5 mL Eppendorf tube containing 10 μL of Milli-Q® water. Be sure that the filament/colony is successfully transferred by visual inspection under the Stereomicroscope. Store sample at –20°C. Prevent sublimation.
5. Submerse the clean horn tip of the ultrasonification device into the Eppendorf tube containing the filament/colony.
6. Apply low force (output 15%) for a short pulse (Digital Sonifier® 450, Branson, 1 sec).
7. Add 40 μL of sample dilution buffer (Phire Plant Direct PCR kit) and mix.
8. Use 1 μL of the sample as template DNA for PCR experiment (see SOP 6.7).
9. In between sample sonification, rinse the microtip twice with 10% H_2O_2 containing water, to prevent cross-contamination.

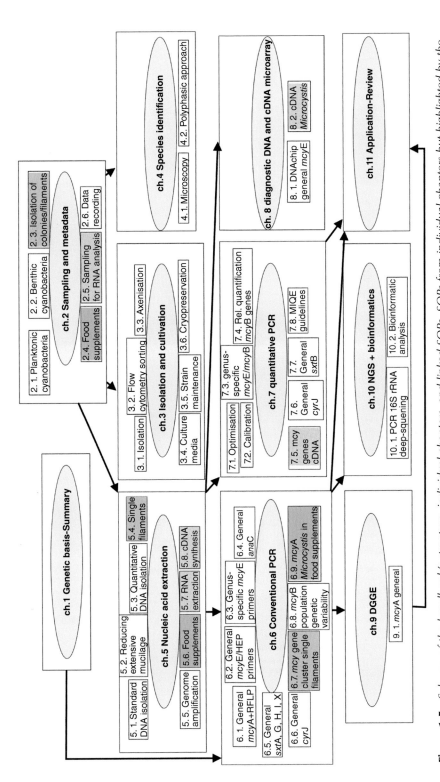

Figure 1.5 *Scheme of the handbook's structure, individual chapters, and linked SOPs. SOPs from individual chapters but highlighted by the same background are directly related.*

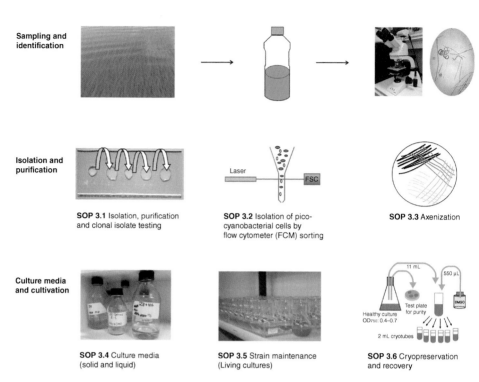

Figure 3.1 *Schematic overview of protocols (see SOPs 3.1–3.6) described in this chapter.*

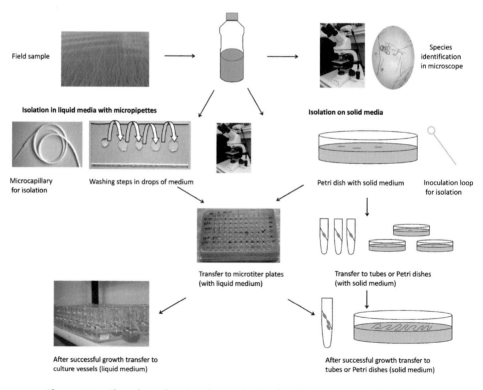

Figure 3.2 *Flowchart showing the methods of isolation presented in SOP 3.1.*

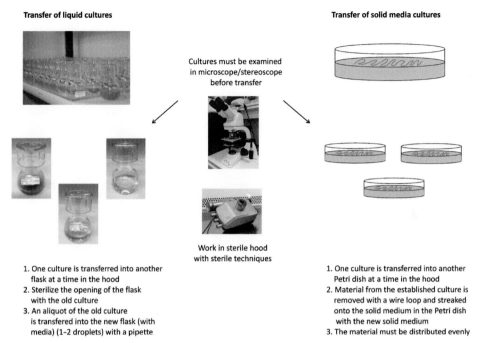

Transfer of liquid cultures

Transfer of solid media cultures

Cultures must be examined in microscope/stereoscope before transfer

Work in sterile hood with sterile techniques

1. One culture is transferred into another flask at a time in the hood
2. Sterilize the opening of the flask with the old culture
3. An aliquot of the old culture is transfered into the new flask (with media) (1-2 droplets) with a pipette

1. One culture is transferred into another Petri dish at a time in the hood
2. Material from the established culture is removed with a wire loop and streaked onto the solid medium in the Petri dish with the new solid medium
3. The material must be distributed evenly

Figure 3.11 *Flowchart showing the methods for maintenance of living cultures presented in SOP 3.5.*

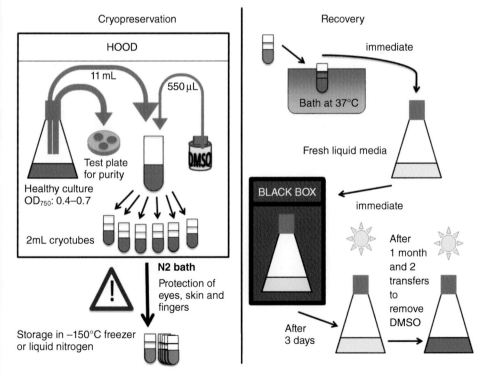

Cryopreservation

HOOD

11 mL

550 µL

DMSO

Test plate for purity

Healthy culture OD$_{750}$: 0.4–0.7

2mL cryotubes

N2 bath
Protection of eyes, skin and fingers

Storage in –150°C freezer or liquid nitrogen

Recovery

immediate

Bath at 37°C

Fresh liquid media

BLACK BOX

immediate

After 1 month and 2 transfers to remove DMSO

After 3 days

Figure 3.14 *Flowchart showing the methods for cryopreservation and recovery presented in SOP 3.6.*

Figure 4.1 *(A) Bloom of* Microcystis aeruginosa *in a lake of the Institute of Botany of São Paulo (photo: A. Tucci). (B) Yellow coloration from a bloom of* Cylindrospermopsis raciborskii *(photo: H.D. Laughinghouse IV). (C) Bloom of* Planktothrix rubescens; *Lake Ledro (photo: A. Boscaini). (D) Bloom of* Planktothrix rubescens, *showing a particular and enlarged section of Lake Ledro (photo: A. Boscaini). (E) Bloom of* Dolichospermum lemmermannii; *Lake Garda (photo: N. Salmaso). (F) Bundles of* Aphanizomenon flos-aquae *looking like larch needles in Bergnappweiher, a small pond used for bathing in Bavaria (photo: K. Teubner). (G) Emus crossing a mixed bloom of cyanobacteria and euglenophytes at the Zoo of Rio Grande so Sul (photo: H.D. Laughinghouse IV and V.R. Werner). (H)* Microcystis aeruginosa, *Lake Garda, net sample, magnification 100× (photo: N. Salmaso). (I)* Cylindrospermopsis raciborskii, *scale bar 10 μM (photo: V.R. Werner). (J)* Planktothrix rubescens, *Lake Como, scale bar 10 μM (photo: A. Boscaini). (K)* Dolichospermum lemmermannii *with attached vorticellids, Lake Garda, scale bar 40 μM (photo: S. Shams). (L) Cluster of akinetes of* Dolichospermum lemmermannii, *Lake Garda, scale bar 40 μM (photo: N. Salmaso). (M) Aggregate of* Aphanizomenon flos-aquae *filaments, Lake Winnipeg; flakes are between 3 and up to 10–20 mm (photo: H.J. Kling). (N)* Dolichospermum crassum *(planktic), scale bar 50 μM (photo: H.D. Laughinghouse IV). (O)* Tychonema bourrellyi, *Lake Garda, scale bar 20 μM (photo: N. Salmaso). Source: Courtesy of Tucci, Boscaini, Teubner, Werner, Shams, Kling.*

(G)   (H)   (I)

(J)   (K)   (L)

(M)   (N)   (O)

Figure 4.1   (Continued)

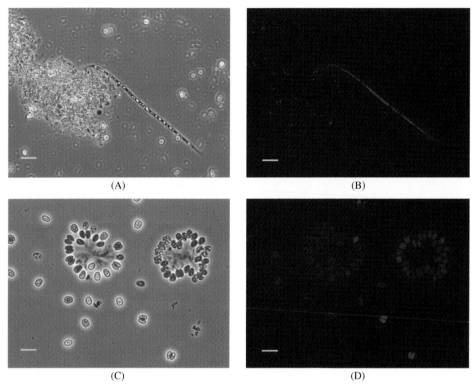

(A)

(B)

(C)

(D)

Figure 4.2 *Identification of cyanobacteria in a bloom sample from a freshwater lake (Etagnac, Poitou-Charentes, France, 10/08/2000) by phase contrast (A, C) and autofluorescence (B, D) microscopy. A green filter (Zeiss 15, excitation band path 546/12 nm, beam splitter 580 nm, long path emission 590 nm) was used for fluorescence detection. Bar markers: 10 μM. In (B) note the red fluorescence of "picocyanobacteria" of various different cell dimensions, in addition to the filamentous cyanobacterium of unknown taxonomic position. The presence of the unicellular cyanobacteria cannot be differentiated from other bacteria within and around the mineral precipitate and cellular debris in the corresponding phase contrast image (A). In the absence of the autofluorescence image (B), even the identification of the filamentous cyanobacterium containing large light refractile inclusions (most likely polyalkanoate) would have been difficult, given that morphologically similar non-cyanobacterial taxa exist. The phase contrast image (C) shows a colony of* Woronichinia, *and some individual cells of this cyanobacterial genus. Note the doublets of lengthwise dividing cells and light refractile gas vesicle clusters ("aerotopes") typical of this planktonic taxon. Gas vesicle collapse resulting from the pressure exerted by the cover slip gives some of the cells a darker appearance. Note also the mucilaginous appendages interconnecting the* Woronichinia *cells within the colony, and associated thin rod-shaped cells. From the corresponding autofluorescent image (D), it is clear that in this field of view* Woronichinia *is the only representative of the cyanobacterial lineage. Differences in the intensity of red fluorescence may be attributable to partial loss of phycocyanin, and/or the fact that not all* Woronichinia *cells are located in the same plane of observation (photos: R. Rippka).*

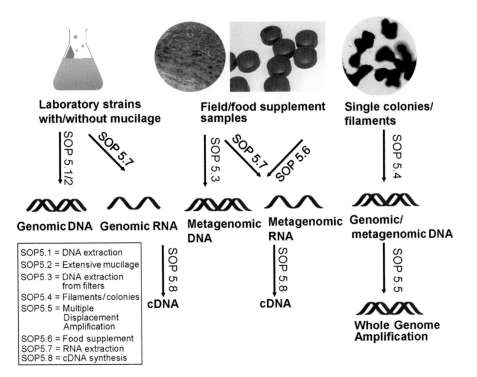

Laboratory strains with/without mucilage

SOP 5.1/2

SOP 5.7

Genomic DNA   **Genomic RNA**

Field/food supplement samples

SOP 5.3

SOP 5.7

SOP 5.6

Metagenomic DNA   **Metagenomic RNA**

Single colonies/ filaments

SOP 5.4

Genomic/ metagenomic DNA

SOP 5.8

cDNA

SOP 5.8

cDNA

SOP 5.5

Whole Genome Amplification

SOP5.1 = DNA extraction
SOP5.2 = Extensive mucilage
SOP5.3 = DNA extraction from filters
SOP5.4 = Filaments/colonies
SOP5.5 = Multiple Displacement Amplification
SOP5.6 = Food supplement
SOP5.7 = RNA extraction
SOP5.8 = cDNA synthesis

Figure 5.1   *Schematic overview of the protocols described in this chapter.*

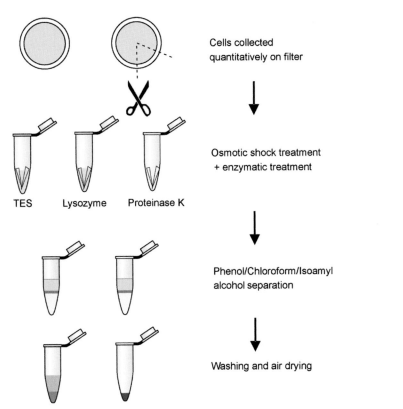

Cells collected quantitatively on filter

Osmotic shock treatment + enzymatic treatment

TES   Lysozyme   Proteinase K

Phenol/Chloroform/Isoamyl alcohol separation

Washing and air drying

Figure 5.3   *Flow diagram showing steps of DNA extraction from cells collected on glass fiber filters (see SOP 5.3).*

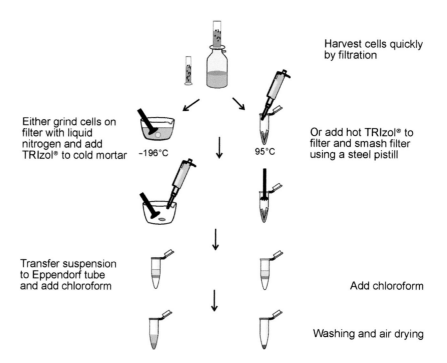

Harvest cells quickly
by filtration

Either grind cells on
filter with liquid
nitrogen and add
TRIzol® to cold mortar   −196°C

95°C   Or add hot TRIzol® to
filter and smash filter
using a steel pistill

Transfer suspension
to Eppendorf tube
and add chloroform

Add chloroform

Washing and air drying

Figure 5.6 *Flow diagram showing steps of RNA extraction from cells collected on filters.*

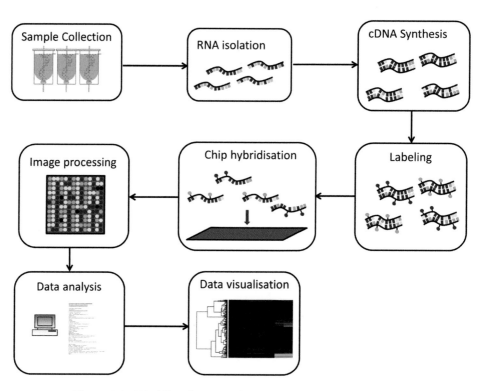

Sample Collection

RNA isolation

cDNA Synthesis

Image processing

Chip hybridisation

Labeling

Data analysis

Data visualisation

Figure 8.1 *Workflow for two color DNA microarray experiments.*

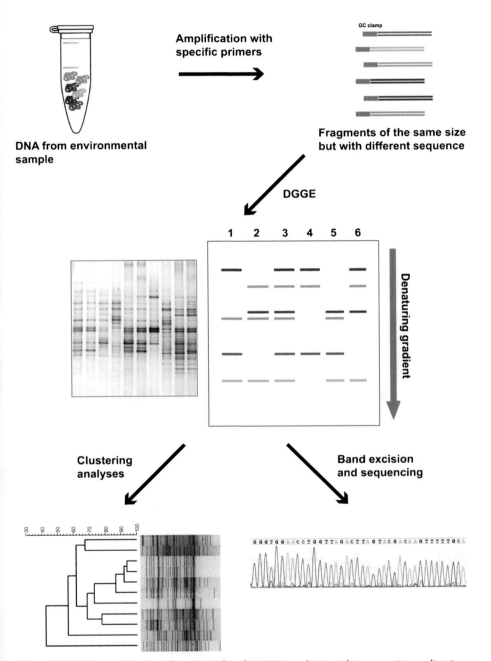

Figure 9.1 *Scheme showing the principle of DGGE analysis with two main applications: clustering analyses and sequencing of separated bands.*

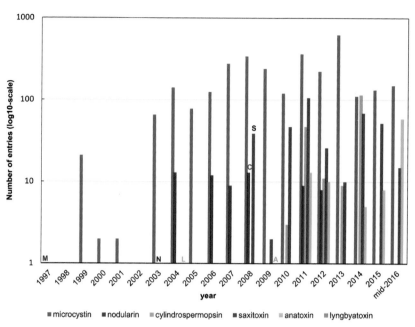

Figure 11.1 *Evolution of the number of new sequences, deposited in the GenBank nucleotide database, for genes involved in the biosynthesis of cyanotoxins. Example of a Boolean search string used: (((microcystin[Title]) OR mcy*[Title]) AND cyanobacter*[Organism]) AND ("1997"[Publication Date] : "1997"[Publication Date]). The first (capitalized) letter of each toxin name indicates the release date of the first annotated gene sequence for this toxin (see text to an accession number and publication reference).*

SOP 5.4.4   Notes

This protocol is useful for the genetic analysis of individual cyanobacterial fila-
ments/colonies directly isolated from field samples and thus bypassing potential
cultivation biases. The possibility to perform multiple PCRs amplifying 3.5 kbp of the *mcy*
gene cluster without interruption allows in-depth analysis of entire gene clusters encoding
toxin synthesis (Chen *et al.*, 2016, and SOP 6.7). Pilot experiments showed that using the
Phire Hot start II DNA polymerase (Thermo Fisher Scientific) following manufacturer's
instructions, PCR products ranging from 500 bp to 9 kbp could be obtained from one
single filament.

SOP 5.4.5   References

Chen, Q., Christiansen, G., Deng, L., and Kurmayer, R. (2016) Emergence of nontoxic
    mutants as revealed by single filament analysis in bloom-forming cyanobacteria of the
    genus *Planktothrix*, *BMC Microbiology*, **16**, 1–12.
Rippka, R. (1988) Isolation and purification of cyanobacteria, *Methods in Enzymology*, **167**,
    3–27.

SOP 5.5

Whole Genome Amplification Using Bacteriophage Phi29 DNA Polymerase

Guntram Christiansen[1,2] *and Rainer Kurmayer*[1]

[1]*Research Institute for Limnology, University of Innsbruck, Mondsee, Austria*
[2]*Miti Biosystems GmbH, Max F. Perutz Laboratories, Vienna, Austria*

SOP 5.5.1   Introduction

This SOP describes a protocol that can be used for quasi-random amplification of whole
genomic DNA starting with minute amounts. It is useful to obtain sufficient template DNA
for a large number of subsequent PCR analyses. The method was adapted from Dean *et al.*
(2001).

SOP 5.5.2   Experimental

SOP 5.5.2.1   Materials

- Single filaments/colonies (in Eppendorf Tubes 0.5 mL).
- Phi 29 DNA polymerase (10 U μL^{-1}) (Thermo Fisher Scientific) supplied with 10 × reaction.
- Buffer.
- Deoxynucleotide triphosphates, dNTPs (10 mM each).
- Exonuclease resistant random hexamer primer (500 µM), 5′-NNNNnnn-3′.
- 20× concentrated pyrophosphatase (0.1 U μL^{-1}).
- Milli-Q® water, DNase-free.
- Eppendorf (0.2 mL) PCR tubes.

SOP 5.5.2.2   Equipment

- Stereomicroscope (4–40 × magnification).
- PCR thermal cycler.
- Table centrifuge (PCR product purification).
- Adjustable mechanical pipettes (100–1000 µL, 20–200 µL, 1–10 µL), filter tips.

SOP 5.5.3   Procedure

1. Prepare isolated single filaments/colonies as described in SOP 2.3. Check that the filament/colony is successfully transferred by visual inspection under the stereomicroscope. Store sample at –20°C. Prevent sublimation.
2. Prepare a Phi polymerase mastermix (10 µL) containing: 6.26 µL Milli-Q® water, 1.0 µL dNTPs (10 mM each), 0.5 µL primer (500 µM), 1.0 µL Phi polymerase buffer (10×), 0.04 µL pyrophosphatase, 0.2 µL phi-polymerase.
3. Add the filament (or 1.0 µL of DNA in control samples) to the phi polymerase mastermix and incubate at 30°C overnight (12–16 h).
4. Stop reaction using PCR product purification (using anion-exchange columns).
5. Dilute 1:10 to 1:100 with Milli-Q® water and use 1 µL for PCR experiment

SOP 5.5.4   Notes

Instead of using single enzymes commercial kits are available bypassing master mix preparations. As the multiple displacement amplification (MDA) works on circular molecules, any sonification or other mechanical disruption treatment is not recommended.

Using Phi 29 DNA polymerase and MDA (Dean *et al.*, 2001) produces consistent quality and quantity of DNA templates for DNA sequencing. Phi 29 DNA polymerase has an associated 3′–5′ exonuclease proofreading activity and a reported error rate of 5×10^{-6}, about 100-fold lower than that of Taq DNA polymerase.

WGA promises great potential, although the reliability in the fidelity of obtained sequences still needs to be confirmed. Nevertheless, the absence/presence of genes of interest (e.g. *mcy* genes) can be analyzed.

SOP 5.5.5   Reference

Dean, F.B., Nelson, J.R., Giesler, T.L., and Lasken, R.S. (2001) Rapid amplification of plasmid and phage DNA using Phi 29 DNA polymerase and multiply-primed rolling circle amplification, *Genome Research*, **11**, 1095–1099.

SOP 5.6

DNA Extraction from Food Supplements

Vitor Ramos[1,2], *Cristiana Moreira*[1], *and Vitor Vasconcelos*[1,2*]

[1]*Interdisciplinary Centre of Marine and Environmental Research (CIIMAR/CIMAR), University of Porto, Matosinhos, Portugal*
[2]*Faculty of Sciences, University of Porto, Porto, Portugal*

SOP 5.6.1   Introduction

After sampling coverage being adequately considered (see SOP 2.4), it is necessary to extract and test the suitability of DNA for the subsequent PCR-based detection of cyanobacterial contaminants (i.e. testing the purity of the sample) and/or test the potential toxigenicity of the sample (for health risk assessment). The method presented here was partially adapted from Rohland and Hofreiter (2007).

This SOP is linked to SOP 2.4 and SOP 6.9, and all in combination are primarily intended to help design an effective plan for early-warning prediction for food supplements quality and food safety assurance. See also Chapter 11 (Section 11.3.5) for additional rationale for these approaches. (An outline is given in Fig. 5.5.)

SOP 5.6.2   Experimental

SOP 5.6.2.1   Materials

- Subsamples, including spiked/internal control (see SOP 2.4).
- Microcentrifuge tubes (1.5 mL), sterile.
- Storage buffer (150 mM NaCl; 1 mM EDTA; 10 mM Tris-HCl, pH 8.0); sterile.
- Milli-Q® water; sterile.
- Glass beads (425–600 μm, acid washed).

---

*Corresponding author: vmvascon@fc.up.pt

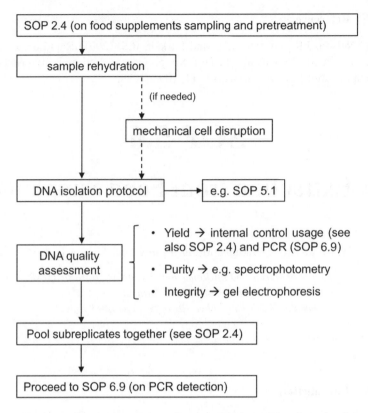

Figure 5.5 *Flow diagram showing steps related to DNA extraction from food supplements.*

- Filter tips.
- Commercial kit for DNA extraction (optional).

SOP 5.6.2.2 Equipment

- Vortex mixer.
- (DNA/RNA) spectrophotometer.
- Agarose gel electrophoresis equipment.
- Adjustable mechanical pipettes (100–1000 μL, 20–200 μL, 1–10 μL).

SOP 5.6.3 Procedure

1. Prior to start with the DNA extraction protocol, it is advisable to rehydrate the samples.
2. The rehydration can be achieved by using either sterile, nuclease-free water (Milli-Q®) or a resuspension/storage buffer such as TE buffer (store the pellet at –20°C). To ensure that the volume added exceeds the expected rehydration ratio, add 20 × (in volume, μL) the weight of the sample. As an example, for 50 mg of biomass add 1 mL of water/buffer. Mix by vortexing, briefly spin down and incubate at room temperature for 5 min.

3. In some cases, a mechanical cell disruption can be desirable in addition to the chemical lysis (e.g. when binders or fillers are used during the production of supplements). If this additional step is thought to be needed, add 0.3–0.5 g of micro glass beads after the rehydration step. An additional volume of water/buffer might be necessary. Vortex vigorously (5 min). Do a short-spin and transfer the supernatant to a new, clean 1.5 mL tube.
4. To recover disrupted cells and cellular debris, subject samples to a medium-speed centrifugation ($8000 - 10000 \times g$) for 10 min. Carefully discard the upper aqueous layer and proceed with the DNA isolation protocol following SOP 5.1, starting at Step 3 (alternatively, use a commercial kit and follow the manufacturer's instructions).
5. Check for DNA quality (see notes) and pool together the DNA from each replicate.
6. To continue with further procedures for food supplements analysis see SOP 6.9.

SOP 5.6.4   Notes

The rehydration of pre-treated samples (from SOP 2.4) prior to starting the DNA isolation protocol will prevent the consumption of the first reagent to be added, which typically is the "lysis buffer" in commercial kits.

For mechanical cell disruption, higher yields may be obtained and less time required if a bead mill homogenizer (bead beater) is available.

Nucleic acids recovered from processed samples, such as the food supplements, may be degraded or contain unspecified compounds that might inhibit PCR. Thus, after the extraction it is important to assess the suitability of the DNA for PCR amplification. Three aspects need to be addressed simultaneously:

- DNA yield: the adsorption of DNA to clay-like substances (such as those used as binders and fillers during supplement production) and to tube walls (to circumvent this problem, "DNA low binding" tubes can be used) can seriously reduce the yield of DNA extraction. Since the bulk DNA is supposed to be mainly from the edible algae itself, it is important to guarantee a satisfactory extraction yield that allows the recovering of genomic DNA from organisms underrepresented in the sample (see SOP 6.9 for additional comments).
- DNA purity: concentration is usually measured using UV absorption with a spectrophotometer (such as NanoDrop™, Thermo Fisher Scientific). Moreover, a very common method to assess the purity of a DNA preparation is to use the ratios of the absorbance at 260 and 280 nm (A260/280) and at 260 and 230 nm (A260/230). Pure DNA is supposed to have A260/280 and A260/230 of around 1.8 (however, it can be slightly higher). Lower values (typically < 1.5) indicate the presence of contaminants (e.g. proteins, salts, solvents) within the sample.
- DNA integrity: to make an evaluation of possible DNA degradation, load and run 1–5 μL of total DNA on an agarose gel (0.8–1%), using an appropriate DNA ladder. An intact total DNA will reveal a single high molecular weight band. On the other hand, degraded DNA appears as a smear in the gel lane, extending down from the high molecular weight band toward the bottom of the gel. If the DNA is significantly degraded (i.e. totally in the form of fragments with low molecular weight), only a more or less extended smear can be seen, which appears in the lower part of the gel.

Samples to be tested should not be limited to blue–green algae food supplements, but also to supplements derived from eukaryotic microalgae such as the widely merchandised *Chlorella* spp. (Görs *et al.*, 2010). Open ponds commonly used for the cultivation of these microalgae are as susceptible to contamination with cyanobacteria (and other organisms) as those used for the cultivation of edible cyanobacteria.

SOP 5.6.5   References

Görs, M., Schumann, R., Hepperle, D., and Karsten, U. (2010) Quality analysis of commercial *Chlorella* products used as dietary supplement in human nutrition, *Journal of Applied Phycology*, **22**, 265–276.

Rohland, N. and Hofreiter, M. (2007) Comparison and optimization of ancient DNA extraction, *Biotechniques*, **42**, 343–352.

SOP 5.7

RNA Extraction from Cyanobacteria

Guntram Christiansen[1,2] *and Rainer Kurmayer*[1]*

[1]*Research Institute for Limnology, University of Innsbruck, Mondsee, Austria*
[2]*Miti Biosystems GmbH, Max F. Perutz Laboratories, Vienna, Austria*

SOP 5.7.1   Introduction

This SOP describes a protocol that can be used for RNA extraction from cyanobacterial laboratory strains and field samples. This protocol has been used by Kaebernick *et al.* (2000) and by Christiansen *et al.* (2008). (A flow diagram on the various steps is given in Fig. 5.6.)

SOP 5.7.2   Experimental

SOP 5.7.2.1   Materials

- 50–100 mL cyanobacterial culture grown to early logarithmic growth phase ($OD_{880} \approx$ 0.1 (5–cm cuvette).)

---

*Corresponding author: rainer.kurmayer@uibk.ac.at

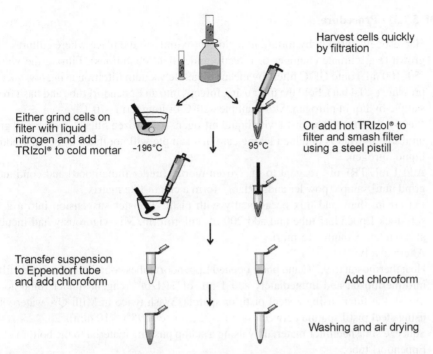

Harvest cells quickly
by filtration

Either grind cells on
filter with liquid
nitrogen and add
TRIzol® to cold mortar   −196°C

95°C   Or add hot TRIzol® to
filter and smash filter
using a steel pistill

Transfer suspension
to Eppendorf tube
and add chloroform

Add chloroform

Washing and air drying

Figure 5.6 *Flow diagram showing steps of RNA extraction from cells collected on filters. (See color plate section for the color representation of this figure.)*

- Glass fiber filters (diameter 25 mm) type GF/C (Whatman).
- Liquid nitrogen.
- Forceps.
- TRIzol® reagent (Thermo Fisher Scientific), toxic solution (to be handled with laboratory gloves).
- Chloroform, toxic solution (to be handled with laboratory gloves).
- Isopropanol.
- 75% (v/v) ethanol.
- Safe-lock Eppendorf tubes (2.0 mL).
- 1.5 mL Eppendorf tubes.
- DEPC (toxic) treated Milli-Q® water.

SOP 5.7.2.2   Equipment

- Vacuum filtration unit.
- Pistil, mortar (liquid nitrogen).
- Table centrifuge.
- Thermostat.
- Fume hood.
- Adjustable mechanical pipettes (100–1000 µL, 20–200 µL, 1–10 µL), filter tips.

SOP 5.7.3   Procedure

1. Harvest cells quickly by filtration in close proximity to the place where cultures were grown (e.g. climate chamber or directly under field conditions). Filtrate the volume (50–100 mL) onto GF/C filters by means of gentle vacuum filtration using low vacuum pressure (<0.4 bar). Fold the filter using forceps into an Eppendorf tube and flash freeze sample by liquid nitrogen. Store sample –20°C or long term –70°C.

2a. Precool the mortar and pistil with liquid nitrogen. Add frozen filter to mortar and grind until fine powder is obtained taking care that sample will not thaw, by serially adding liquid nitrogen.

3a. Add 1 mL TRIzol® reagent to the frozen mortar (under fume hood) and continue to grind until sample powder and TRIzol® form a cream-like matrix.

4a. Let cream thaw and mix occasionally with pistil. Transfer suspension into a 2 mL safe-lock Eppendorf tube and add 200 μL chloroform. Mix vigorously and incubate at room temperature (15 min).

    Alternatively:

2b. Heat thermostat to 95°C and place opened Eppendorf tubes containing the frozen filters into thermostat and immediately add 1 mL of TRIzol® reagent (under fume hood). Smash the filters using a steel pistil or spatula. Wash twice in Milli-Q® water when using steel pistil/spatula between samples. Incubate at 95°C (10 min).

3b. Squeeze smashed filter material by using a stamp pushing material to the bottom of the Eppendorf tube.

4b. Add 200 μL of chloroform, mix vigorously and incubate at room temperature (15 min).

5. Centrifuge at 4°C for 15 min and transfer supernatant without disturbing interphase or pelleted filter material. Expected volume is 600 μL aqueous supernatant.

6. Add 420 μL isopropanol and mix vigorously and incubate at room temperature (15 min). Centrifuge at 4°C for 15 min. After this step a white pellet will be visible.

7. Wash pellet twice with 75% (v/v) ice cold EtOH and centrifuge in between at 4°C for 5 min.

8. Air-dry pellet by ventilation in the opened Eppendorf tube.

9. Add 50 μL of DEPC treated Milli-Q® water using filter tips, quantify by spectrophotometry (OD_{260} 1 = 40 μg mL^{-1} RNA); store at –70°C.

SOP 5.7.4   Notes

While the use of the mortar and pistil is probably more effective in RNA extraction, the number of samples that can be processed is relatively low. In contrast, by incubating the filters in parallel using hot TRIzol® reagent (95°C) under the fume hood a larger number of samples (e.g. obtained during experiments) can be processed.

Owing to potential short half-lives of certain mRNAs, harvest cultures directly in the culture room and preserve harvested samples immediately in liquid nitrogen (sample processing times are optimally < 90 sec). This is why centrifugation steps are not recommended. Alternatively, the addition of fixatives like RNAlater® (Thermo Fisher Scientific) can be used instead.

RNases are ubiquitous and thus RNA degradation risks can be minimized by using gloves, barrier tips, and DEPC treated water. RNA extraction quality control can be performed by RNA gel electrophoresis visualizing rRNA (16S and 23S). If both RNAs are visible in a 1:2 ratio and no digestion signals (distinct additional bands) appear, high-quality RNA can be ascertained.

SOP 5.7.5 References

Christiansen, G., Molitor, C., Philmus, B., and Kurmayer, R. (2008) Nontoxic strains of cyanobacteria are the result of major gene deletion events induced by a transposable element, *Molecular Biology and Evolution*, **25**, 1695–1704.

Kaebernick, M., Neilan, B.A., Börner, T., and Dittmann, E. (2000) Light and the transcriptional response of the microcystin biosynthesis gene cluster, *Applied and Environmental Microbiology*, **66** (8), 3387–3392.

SOP 5.8

cDNA Synthesis

Guntram Christiansen[1,2] and Rainer Kurmayer[1]*

[1]*Research Institute for Limnology, University of Innsbruck, Mondsee, Austria*
[2]*Miti Biosystems GmbH, Max F. Perutz Laboratories, Vienna, Austria*

SOP 5.8.1 Introduction

This SOP describes the production of cDNA from extracted mRNA (see SOP 5.7). After treatment with DNase, the RNA is purified using RNeasy® MinElute® columns (Qiagen). For cDNA production 1 µg of total RNA is used for reverse transcription using the RevertAid H minus First strand cDNA Synthesis Kit (Thermo Fisher Scientific). See Fig. 5.7 for an overview.

SOP 5.8.2 Experimental

SOP 5.8.2.1 Materials

- Gloves, filter tips, RNase-free reaction tubes (1.5 mL).
- DEPC treated Milli-Q® water (RNase-free).
- DNase I (Fermentas, Thermo Fisher Scientific).

---

*Corresponding author: rainer.kurmayer@uibk.ac.at

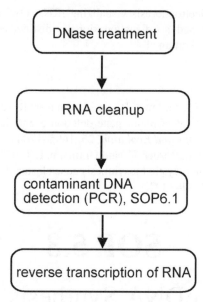

Figure 5.7 *Flow diagram showing steps of cDNA synthesis in SOP 5.8 from RNA extracted as described in SOP 5.7.*

- DNase I buffer (Fermentas, Thermo Fisher Scientific)
- Beta-Mercaptoethanol (β-ME) (molecular biology grade).
- RNeasy MinElute Cleanup Kit (Qiagen).
- RevertAid H minus First strand cDNA Synthesis Kit (Thermo Fisher Scientific).

SOP 5.8.2.2   Equipment

- Table centrifuge.
- Adjustable mechanical pipettes (100–1000 μL, 20–200 μL, 1–10 μL), filter tips.

SOP 5.8.3   Procedure

SOP 5.8.3.1   DNase Treatment

1. Digestion of DNA: use 1 U of DNase I for 1 μg of extracted RNA. For 10 μL reaction volume add extracted RNA (and water), 1 μL DNase I buffer (10 ×) and 1 μL DNase I and incubate at 37°C (30 min).

SOP 5.8.3.2   RNA Cleanup

2. Add beta-Mercaptoethanol to RLT-buffer (RNeasy MiniElute Cleanup Kit, Qiagen): 1 mL RLT-buffer + 10 μL of β-ME.
3. Inactivate the DNase (95°C, 5 min) and add 5 μL of 25 mM EDTA.
4. Add 45 μL of DEPC water + 5 μL of 25 mM EDTA, vortex.

5. Adjust the sample to 100 μL; add 350 μL of RLT-buffer and vortex, add 250 μL of EtOH and vortex, finally transfer the sample to one silica based membrane column (RNeasy MinElute Spin column, Qiagen) and place into 2.0 mL collection tube (according to manufacturer's instructions).
6. Centrifuge (15 sec, 8000 × g) and discard the flow through.
7. Add 500 μL of 80% EtOH and centrifuge (8000 g, 2 min) discard the flow through (washing step).
8. Place the RNeasy MinElute spin column into a new 2.0 mL collection tube, open the lid, and centrifuge (full speed, 5 min).
9. Place the RNeasy MinElute Spin column into a new 1.5 mL reaction tube, add 16 μL of DEPC treated water into the center of the spin column, centrifuge (1 min, full speed) to elute the RNA.

SOP 5.8.3.3   Contaminant DNA detection

10. The presence of DNA should be tested using nucleotide primers amplifying a short DNA fragment of chromosomal DNA (see SOP 6.1) from 1 μL of RNA template during 40 cycles; if the PCR test is positive the efficiency of the DNA digestion step 1) needs to be increased.

SOP 5.8.3.4   Reverse Transcription of RNA

11. 1 μg of RNA is reverse transcribed following the manufacturer's instructions (Thermo Fisher Scientific): add RNase-free water (variable), 1 μg of RNA, and 1 μL of random hexamer primer resulting in total of 12 μL and incubate (50 min, 70°C).
12. Add 4 μL of reaction buffer (5×), 1 μL of ribonuclease inhibitor, 2 μL of dNTP mix (provided by the manufacturer), incubate 37°C (5 min).
13. Add 1 μL of RevertAid H minus Reverse transcriptase resulting in a total volume of 20 μL, incubate at 42°C (60 min).
14. Dilute 1:10 (20+180 μL) and store at −20°C.

SOP 5.8.4   Notes

Using this RNA cleanup protocol and the RNA extraction (see SOP 5.7) and sampling (see SOP 2.5) procedure as described, on average 700 ng μL^{-1} are obtained, resulting in approx. 10 μg of RNA in total.

To avoid contamination with RNase the same cautions apply as under SOP 5.7: use gloves, barrier tips, and DEPC-treated water.

SOP 5.8.5   References

Qiagen (2010) *RNeasy® MinElute® Cleanup Handbook: For RNA cleanup and concentration with small elution volumes*. User guide.
Thermo Fisher Scientific (2016) *Thermo Scientific RevertAid H Minus First Strand cDNA Synthesis Kit*, User guide.

6

Conventional PCR

Elke Dittmann[1]*, Anne Rantala-Ylinen[2], Kaarina Sivonen[2], Ilona Gągała[3],
Joanna Mankiewicz-Boczek[3,4], Samuel Cirés[5], Andreas Ballot[6], Guntram Christiansen[7,8],
Rainer Kurmayer[7]*, Vitor Ramos[9,10], Vitor Vasconcelos[9,10], and Martin Saker[9,11]

[1]*Institute of Biochemistry and Biology, University of Potsdam, Potsdam-Golm, Germany*
[2]*Department of Food and Environmental Sciences, Division of Microbiology and Biotechnology,*
University of Helsinki, Helsinki, Finland
[3]*European Regional Centre for Ecohydrology of the Polish Academy of Sciences, Łódź, Poland*
[4]*Department of Applied Ecology, Faculty of Biology and Environmental Protection, University of*
Lodz, Łódź, Poland
[5]*Department of Biology, Autonomous University of Madrid, Madrid, Spain*
[6]*Norwegian Institute for Water Research, Oslo, Norway*
[7]*Research Institute for Limnology, University of Innsbruck, Mondsee, Austria*
[8]*Miti Biosystems GmbH, Max F. Perutz Laboratories, Vienna, Austria*
[9]*Interdisciplinary Centre of Marine and Environmental Research (CIIMAR/CIMAR), University of*
Porto, Matosinhos, Portugal
[10]*Faculty of Sciences, University of Porto, Porto, Portugal*
[11]*Alpha Environmental Solutions, Dubai, United Arab Emirates*

6.1 Introduction

This chapter gives an overview about PCR strategies aimed to detect and differentiate toxigenic cyanobacteria. The overarching principle of any PCR-based detection of

*Corresponding authors: editt@uni-potsdam.de; rainer.kurmayer@uibk.ac.at

Molecular Tools for the Detection and Quantification of Toxigenic Cyanobacteria, First Edition.
Edited by Rainer Kurmayer, Kaarina Sivonen, Annick Wilmotte and Nico Salmaso.
© 2017 John Wiley & Sons Ltd. Published 2017 by John Wiley & Sons Ltd.

toxigenic cyanobacteria is the fact that the production of cyanobacterial toxins by individual strains is typically linked with the presence of the corresponding biosynthesis genes, whereas the functional genes are absent in non-toxic strains (see below for special notes). Biosynthesis genes have been assigned for the toxins microcystin (MC), nodularin (NOD), cylindrospermopsin (CYL), anatoxin (ANA), and saxitoxin (STX) (Dittmann *et al.*, 2013). For MC and NOD, PCR-based strategies allow for an assignment of the producing genus (Rantala *et al.*, 2004). The increasing wealth of sequences of toxin biosynthesis genes from laboratory strains and field studies enables the design of more reliable and specific primers. This chapter offers some guidelines for the selection of primers for PCR approaches and discusses possible bias of the techniques and troubleshooting as well as downstream applications.

Conventional PCR is a precondition for the detection and differentiation of toxic cyanobacteria (this chapter) and DNA diagnostic microarrays (Chapter 8). Eventually, single colonies and filaments can be directly used thereby omitting the nucleic acid extraction step (Chapters 5 and 6). The PCR-based detection of different toxin biosynthesis genes in complex environmental samples can be combined in multiplex-approaches. The methods can be used for studying the dynamics of toxigenic cyanobacteria and for early-warning approaches. Conventional PCR can be combined with downstream applications such as sequencing, cloning, and restriction fragment length polymorphism (RFLP) analysis (this chapter).

The conventional PCR (polymerase chain reaction) described here can be used for (1) the detection of genes (genotypes) indicative of toxin production, (2) the assignment of toxin production to different genera of cyanobacteria, (3) the analysis of the phylogenetic dependence on the distribution of toxin genes among cyanobacteria, (4) a combined detection of different toxin biosynthesis genes, and (5) the discovery of new toxin-producing genera. Moreover, the PCR techniques described here can be adopted, in connection with the use of specific primers, to detect a variety of molecular markers (16S rRNA, housekeeping genes) used for the identification and phylogenetic classification of cyanobacteria (see Chapter 4).

6.2 Principle of PCR and Available Enzymes

PCR is a technique that allows the creation of multiple copies of specific gene fragments (Sambrook and Russell, 2001). The technique can be used to identify small DNA fragments of interest, which serve as templates for the design of primers. Polymerases are the driving enzymes for PCR reactions. They will synthesize a complementary strand to a single stranded DNA fragment, provided that it has a double-stranded starting point. This initial double strand is formed between a specific primer of approximately 20 nucleotides in length and the complementary DNA strand it binds to. The amplification is the result of multiple cycles of: (1) denaturation to generate single strands, (2) annealing that allows primer binding to the complementary DNA copies, and (3) elongation to synthesize the complementary DNA strands. As the denaturation step typically requires temperatures above 90°C, polymerases need to be thermostable. The most common enzyme applied for PCR is the

Taq polymerase isolated from the thermophilic bacterium *Thermus aquaticus*. Alternative thermostable enzymes include the Pfu polymerase from the thermophilic archaebacterium *Pyrococcus furiosus*. Different thermostable polymerases show different frequencies of random sequence errors ranging from 10^{-6} (high fidelity) to 10^{-4} (low fidelity). Information on the fidelity of specific polymerases is usually provided by the manufacturer of kits. There are numerous kits available from different companies that can be utilized for PCR approaches. Some of them are more robust even when substances inhibiting the Taq polymerase are present in the sample or when the amount of DNA is limited. PCR analysis of field samples and single colonies and filaments may require the testing of different enzymes that can significantly differ in their amplification efficiency and specificity.

6.2.1 Primer Development

The most critical step for the reliable detection of toxic cyanobacteria is the selection of appropriate primers. The responsible gene clusters for the synthesis of MC, NOD, CYL, ANA, and STX are all very large and include giant genes encoding multi-enzymes of the non-ribosomal peptide synthetase (NRPS) and/or polyketide synthase (PKS) classes. Whereas this gives broad opportunities for the design of primers, the underlying genes show considerable variation between different cyanobacterial genera and even within individual strains of the same genus. Typically, there are highly variable gene regions and regions that are highly conserved. For a robust PCR detection strategy it is important to avoid variable regions that can cause strong bias in toxic cyanobacteria detection. Many studies have uncovered variable regions for MC and NOD biosynthesis genes, whereas the database is much smaller for CYL, ANA, and STX biosynthesis genes. For MC and NOD gene synthesis detection, it may thus be appropriate to use primers available and tested in the literature (and listed in Table 6.1), whereas primer design for CYL, ANA, and STX genes may still require further optimization.

The selection of primers will also depend on the downstream applications. If the goal is to reliably detect a certain biosynthesis gene, it will be appropriate to design primers for highly conserved gene regions. If the aim is to further differentiate between different toxin-producing genera by downstream technologies such as sequencing, RFLP, or diagnostic microarray, the selected gene fragment should also contain variable, genus-specific DNA regions. For MC and NOD, the most commonly used primer pairs listed in Table 6.1 combine both features. If the goal is to simultaneously detect different toxin genes in a multiplex approach, it is essential to adjust the primers for the different toxin biosynthesis genes to a similar melting temperature and optimize their length. For single colony/single filament analysis selected gene fragments (and therefore the distance of primer binding sites) should be rather short as high molecular weight DNA may not be released from the lysing cells (but see SOP 5.4). For the self-design of a primer pair for toxin biosynthesis gene detection, researchers need to be experienced both with the rationale (Sambrook and Russell, 2001; Basu, 2015) and with the software used for nucleotide acid comparison (e.g. BLAST) and for primer design (e.g. Primer3; Untergasser *et al.*, 2012). Otherwise, they should preferably utilize primers published in the literature.

Table 6.1 Commonly used primers for the detection and differentiation of cyanotoxin biosynthesis genes and for general detection of cyanobacterial DNA

Primer (Name, Sequence)	Toxin genes/Genera	Suitable for	Reference
mcyA-Cd 1R (5′–AAAAGTGTTTTATTAGCGGCTCAT–3′) mcyA-Cd 1F (5′–AAAATTAAAAGCCGTATCAAA–3′)	MC, NOD/*Microcystis*, *Planktothrix*, *Anabaena*, *Nostoc*, *Nodularia*	Detection and differentiation of MC biosynthesis genes, RFLP analysis, phylogenetic analysis	Hisbergues et al. (2003); Rantala et al. (2004)
HEPF (5′–TTTGGGGTTAACTTTTTGGGCATAGTC–3′) HEPR (5′–AATTCTTGAGGCTGTAAATCGGGTTT–3′)	MC/NOD/ *Microcystis*, *Planktothrix*, *Anabaena*, *Nostoc*, *Nodularia*	Detection and differentiation of MC and NOD biosynthesis genes	Jungblut and Neilan (2006)
mcyDF (5′–GATCCGATTGAATTAGAAAG–3′) mcyDR (5′–GTATTCCCCAAGATTGCC–3′)	MC, NOD/*Microcystis*, *Planktothrix*, *Anabaena*, *Nostoc*, *Nodularia*	Detection and differentiation of MC biosynthesis genes, phylogenetic analysis	Rantala et al. (2004)
mcyE-F2 (5′–GAAATTTGTGTAGAAGGTGC–3′) mcyE-R4 (5′–AATTCTAAAGCCCAAAGACG–3′)	MC, NOD/*Microcystis*, *Planktothrix*, *Anabaena*, *Nostoc*, *Nodularia*	Detection and differentiation of MC biosynthesis genes, phylogenetic analysis, microarray analysis	Rantala et al. (2004)
mcyE-R1 (5′–ATAGGATGTTTAGAGAGAATTTTTCCC–3′) mcyE-S1 (5′–GGGACGAAAAGATAATCAAGTTAAGG–3′)	MC, NOD/*Microcystis*, *Planktothrix*, *Anabaena*, *Nostoc*, *Nodularia*	Detection and differentiation of MC biosynthesis genes, phylogenetic analysis	Mankiewicz-Boczek et al. (2015)
mcyE-F2 (5′–GAAATTTGTGTAGAAGGTGC–3′) mcyE-12R (5′–CAATCTCGGTATAGCGGC–3′)	MC/*Anabaena*	Identification of MC biosynthesis genes specific for the genus *Anabaena*	Vaitomaa et al. (2003)

Primer	Target	Purpose	Reference
ndaF8452 (5'–GTGATTGAATTTCTTGGTCG–3') ndaF8640 (5'–GGAAATTTCTATGTCTGACTCAG–3')	NOD/*Nodularia*	Identification of NOD biosynthesis genes	Koskenniemi et al. (2007)
anaC-gen F (5'–TCTGGTATTCAGTCCCCTCTAT–3'), anaC-gen R (5'–CCCAATAGCCTGTCATCAA–3')	ANA/*Anabaena*, *Aphanizomenon*, *Oscillatoria*	Detection and differentiation of ANA biosynthesis genes	Rantala-Ylinen et al. (2011)
cyrJ cynsulfF (5'–ACTTCTCTCCTTTCCCTATC–3') cylnamR (5'–GAGTGAAAATGCCGTAGAACTTG–3')	CYL/*Cylindrospermopsis*, *Aphanizomenon*	Detection and differentiation of CYL biosynthesis genes	Mihali et al. (2008)
sxtAf (5'–GCGTACATCCAAGCTGGACTCG–3') sxtAr (5'–GTAGTCCAGCTAAGGCACTTGC–3')	*Aphanizomenon*, *Anabaenopsis*, *Dolichospermum crassum*, *Chrysosporum*, *Cuspidothrix*, *Cylindrospermum* sp., *Geitlerinema* sp., *Phormidium* sp. and *Scytonema* sp.	Detection and differentiation of STX producing cyanobacteria	Ballot et al. (2010)
PCβF (5'–GGCTGCTTGTTTACGCGACA–3') PCαR (5'–CCAGTACCACCAGCAACTAA–3')	PC-IGS[1] All cyanobacteria	Detection and assignment of cyanobacterial DNA	Neilan et al. (1995)

[1]PC-IGS, intergenic spacer region of the phycocyanin genes cpcB and cpcA.

6.2.2 Setup of PCR Conditions for DNA and Single Colony Analysis

The PCR conditions applied for the amplification of the desired gene fragments depend on the following criteria: (1) the specific polymerase used for the reaction, (2) the melting temperature of the utilized primers, (3) the length of the selected gene fragment, and (4) the nature of the template (isolated DNA or single colony/filament). PCR kits provided by manufacturers typically provide protocols suggesting the appropriate time and temperature for the denaturation and elongation steps that are specific for different PCR enzymes. The annealing temperature in a PCR protocol is set slightly below the melting temperature of the primers. At this temperature the primer binds with high specificity, whereas lower temperatures open the door for unspecific binding of primers to undesired DNA regions and thus false-positive results. Adjusting the annealing temperature can thus be part of the optimization process of a PCR reaction. At lower temperatures even gene fragments are detected that show slight deviations in their primer binding sites, whereas close to the melting temperature the primer binding is very stringent (i.e. does not tolerate nucleotide deviations in the template DNA fragment). For inexperienced researchers it is thus recommendable to apply PCR conditions published in the literature. It is important to note that any contamination can cause false-positive results. All PCR solutions should therefore be kept in small aliquots and DNA tubes should only be opened for a short time. For single colony or single filament analysis different protocols have to be used (see SOP 6.7).

6.2.3 Gel Electrophoresis and Documentation

The results of PCR reactions are analyzed on agarose gels that are stained with fluorescing intercalating DNA agents allowing for visualization under UV light (typically Midori Green or the more toxic ethidium bromide (EtBr), which is hazardous to human health). The concentration of the agarose gel depends on the size of the gene fragments to be analyzed. Fragments of 500–3000 bp can be analyzed on 2.0–0.8% agarose gels. Along with the samples to analyze, each agarose gel should be loaded with a DNA size marker (DNA ladder) that can be purchased from companies. Moreover, for each PCR reaction, the positive and the negative control should be loaded on the same gel. The gels can be analyzed under UV light, typically on imaging systems (UV transilluminators) that are equipped with cameras that enable documentation of gels.

6.2.4 Troubleshooting of PCR Results

PCR reactions are prone to errors and can yield false-positive or false-negative results. This is particularly true for field samples that may contain extremely low amounts of, or poor-quality, DNA or PCR-inhibiting substances. Thus, for the interpretation of the PCR results different types of controls should be included. Samples without DNA can serve as negative controls to exclude DNA contaminations in the PCR reagents. DNA of known cyanotoxin-producing laboratory strains should be included as positive control. For a list of strains to produce reference material see Table S6.1 in the Appendix. For quality assessment of field DNA an "entry PCR" should be used that yields PCR products whenever cyanobacterial DNA is present. An established marker for the reliable detection of cyanobacterial DNA is the PC-IGS region (see Table 6.1). Possible inhibitory substances in DNA samples

Table 6.2   *Troubleshooting guide for the interpretation of PCR results*

Phenomenon	Possible Reason	Possible Solution
No PCR bands, not even in positive control	PCR reaction not successful Enzyme inactive Missing or inappropriate reagents Annealing temperature too high	Repeat PCR reaction and carefully check inclusion of all reagents Repeat PCR reaction at lower annealing temperature
PCR bands everywhere, even in negative control	Contamination with template DNA	Check and re-prepare all reagents
PCR bands only in positive control	PCR reaction successful No toxin genes present in field samples PCR-inhibiting substances in field samples	Repeat PCR reaction with lower concentrations of template material (higher dilution) Spike with positive control DNA to test for inhibitors
PCR products not at expected size	Unspecific PCR products	Repeat PCR with higher annealing temperature
PCR bands only for standard marker gene, not for toxin genes	PCR successful No toxin genes present *or* Toxin genes more difficult to amplify (e.g. owing to larger size of fragment)	Eventually design new primers for toxin biosynthesis genes Repeat with higher concentrations of template material
PCR bands for toxin genes but not for marker genes	PCR successful Marker genes possibly more difficult to amplify (e.g. owing to larger size of fragment)	Eventually design new primers for marker genes Repeat with higher concentrations of template material

can be detected by using a "spiking" control with positive control DNA. Please see Table 6.2 as a troubleshooting guide for the interpretation of PCR results.

6.2.5   PCR Product Downstream Processing (RFLP, Cloning, Sequencing)

Whereas plain detection of PCR products in samples can give qualitative information about the presence of toxin biosynthesis genes, differentiation of toxin producers requires additional downstream technologies. PCR products from pure laboratory strains can be directly sequenced using in-house sequencing platform or services offered by commercial sequencing companies. Single PCR primers can then directly be used for the sequencing of the double strand from one side. Protocols are typically provided by the companies and depend on the individual sequencing technology and enzymes applied. PCR products from mixed samples (i.e. any products obtained from field samples usually contain mixtures of PCR fragments with variations in sequences). Such PCR products cannot be directly sequenced but have to be separated first.

A simple possibility for a separation is cloning where single PCR fragments are cloned into individual *E. coli* hosts (Sambrook and Russell, 2001). Cloning of PCR fragments relies on the fact that PCR products produced using Taq polymerase contain 5' overhanging adenine (A) nucleotides. Several companies provide kits with cloning vectors carrying 3' thymine nucleotides that can build complement pairs with the PCR products that can be directly ligated. Single *E. coli* cells carrying these vectors are selected using antibiotic resistance markers. The manufacturers provide the corresponding protocols. Plasmid vectors from positive *E. coli* cells can be isolated and sequenced. Sequence information of clone libraries can provide information on the diversity of sequences present and represents a semi-quantitative approach. A simple technique to differentiate between different toxin producers without sequencing is RFLP analysis. Here, a PCR product obtained from a mixture of strains is digested using different sequence-specific restriction enzymes. Alternatively, the PCR product mixtures can be analyzed by deep amplicon sequencing (see Chapter 10).

6.3   Special Notes

It is important to note that the correlation of the presence of genes and the production of toxins is not without exception. A number of studies have shown the presence of mutants possessing genes (e.g. for MC biosynthesis but lacking production of MC). For a full interpretation of PCR results from the field, a combination with chemical analytics, preferentially on the single colony/single filament level is needed (e.g. Via-Ordorika *et al.*, 2004).

6.4   References

Ballot, A., Fastner, J., and Wiedner, C. (2010) Paralytic shellfish poisoning toxin-producing cyanobacterium *Aphanizomenon gracile* in northeast Germany, *Applied and Environmental Microbiology*, **76**, 1173–1180.

Basu, C. (ed.) (2015) *PCR Primer Design*, 2nd ed., Humana Press, Springer, New York.

Dittmann, E., Fewer, D.P., and Neilan, B.A. (2013) Cyanobacterial toxins: Biosynthetic routes and evolutionary roots, *FEMS Microbiology Reviews*, **37**, 23–43.

Hisbergues, M., Christiansen, G., Rouhiainen, L., *et al.* (2003) PCR-based identification of microcystin-producing genotypes of different cyanobacterial genera, *Archives in Microbiology*, **180**, 402–410.

Jungblut, A.D. and Neilan, B.A. (2006) Molecular identification and evolution of the cyclic peptide hepatotoxins, microcystin and nodularin, synthetase genes in three orders of cyanobacteria, *Archives in Microbiology*, **185**, 107–114.

Koskenniemi, K., Lyra, C., Rajaniemi-Wacklin, P., *et al.* (2007) Quantitative real-time PCR detection of toxic *Nodularia* cyanobacteria in the Baltic Sea, *Applied and Environmental Microbiology*, **73**, 2173–2179.

Mankiewicz-Boczek, J., Gągała, I., Jurczak, T., *et al.* (2015) Incidence of microcystin-producing cyanobacteria in Lake Tana, the largest waterbody in Ethiopia, *African Journal of Ecology*, **53**, 54–63.

Mihali, T.K., Kellmann, R., Muenchhoff, J., *et al.* (2008) Characterization of the gene cluster responsible for cylindrospermopsin biosynthesis, *Applied and Environmental Microbiology*, **74**, 716–722.

Neilan, B.A., Jacobs, D., and Goodman, A.E. (1995) Genetic diversity and phylogeny of toxic cyanobacteria determined by DNA polymorphisms within the phycocyanin locus, *Applied and Environmental Microbiology*, **61**, 3875–3883.

Rantala, A., Fewer, D.P., Hisbergues, M., *et al.* (2004) Phylogenetic evidence for the early evolution of microcystin synthesis, *Proceedings of the National Academy of Science USA*, **101**, 568–573.

Rantala-Ylinen, A., Kana, S.E.C., Wang, H., *et al.* (2011) Anatoxin-a synthetase gene cluster of the cyanobacterium *Anabaena* sp. strain 37 and molecular methods to detect potential producers, *Applied and Environmental Microbiology*, **77**, 7271–7278.

Sambrook, J. and Russell, D. (2001) *Molecular Cloning: A laboratory manual*, 3rd edn, Cold Spring Harbor Laboratory Press, Cold Spring Harbor, New York.

Untergasser, A., Cutcutache, I., Koressaar, T., *et al.* (2012) Primer3: New capabilities and interfaces. *Nucleic Acids Research*, **40** (15), e115.

Vaitomaa, J., Rantala, A., Halinen, K., *et al.* (2003) Quantitative real-time PCR for determination of microcystin synthetase E copy numbers for *Microcystis* and *Anabaena* in lakes, *Applied and Environmental Microbiology*, **69**, 7289–7297.

Via-Ordorika, L., Fastner, J., Kurmayer, R., *et al.* (2004) Distribution of microcystin-producing and non-microcystin-producing *Microcystis* sp. in European freshwater bodies: Detection of microcystins and microcystin genes in individual colonies, *Systematic and Applied Microbiology*, **27**, 592–603.

SOP 6.1

PCR Detection of Microcystin Biosynthesis Genes Combined with RFLP Differentiation of the Producing Genus

*Elke Dittmann**

Institute of Biochemistry and Biology, University of Potsdam, Potsdam-Golm, Germany

SOP 6.1.1   Introduction

This SOP describes a simple PCR-based detection technique for microcystin (MC) biosynthesis genes using the marker gene *mcy*A (291 bp) and the subdifferentiation by RFLP analysis. The technique was initially published by Hisbergues *et al.* (2003). See Fig. 6.1 for an overview. Reference strains are given in Table S6.1 (Appendix).

SOP 6.1.2   Experimental

SOP 6.1.2.1   Materials

- PCR reagents (or PCR cycling kit).
- Primers, *mcy*A-Cd 1R (5′–AAAAGTGTTTTATTAGCGGCTCAT–3′); *mcy*A-Cd 1F (5′–AAAATTAAAAGCCGTATCAAA–3′).
- Isolated nucleic acids of laboratory strains or field samples.
- Isolated nucleic acid of an MC-producing strain.
- Sterile Milli-Q® water.

*Corresponding author: editt@uni-potsdam.de

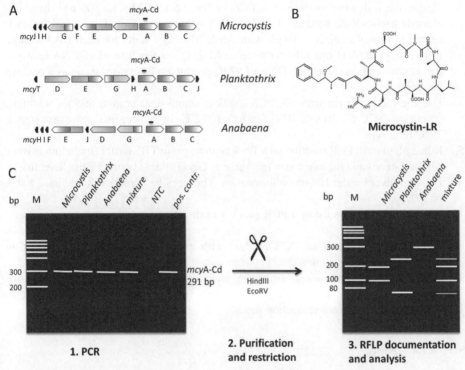

Figure 6.1 *Overview of PCR detection of microcystin biosynthesis genes combined with RFLP differentiation of individual genera: (A) Schematic representation of* mcy *operons in three genera of cyanobacteria. The detected gene fragment is indicated with a black line; (B) Structure of microcystin; (C) Standard workflow of procedure. NTC = non-template control.*

- TBE buffer (Tris-borate-EDTA, 89 mM TRIS Base, 89 mM boric acid, 2 mM EDTA, pH 8.0).
- EtBr.
- PCR product purification kit.
- Restriction enzymes: HindIII, EcoRV.
- EtBr (or Midori Green) for DNA staining.

SOP 6.1.2.2   Equipment

- Thermocycler.
- Electrophoresis equipment.
- UV transilluminator with documentation.
- Adjustable mechanical pipettes (100–1000 µL, 20–200 µL, 1–10 µL).

SOP 6.1.3   Procedure

1. Adjust the amount of isolated DNA to approximately 10 ng µL^{-1}.
2. Adjust the amount of primer to 10 pmol µL^{-1}.

3. Pipette the following mixture: 1 × PCR buffer, 4.5 mM MgCl$_2$, 300 μM deoxynucleotide triphosphate mixture, 1 U Taq DNA polymerase, 1 μL DNA, 1 μL of each primer (10 pmol); adjust to 10 μL with sterile Milli-Q$^®$ water for each DNA sample to be analyzed. Add one tube with sterile Milli-Q$^®$ water instead of DNA as negative control and one tube with DNA of a MC-producing laboratory strain as a positive control.

4. Use the following program for PCR cycling: initial denaturation at 95°C (3 min); 35 cycles: 94°C for 30 sec, 59°C for 30 sec, 72°C for 30 sec; final extension step at 72°C for 5 min.

5. Run 2 μL of each PCR reaction on a 1.5% agarose gel in TBE buffer (including EtBr or Midori Green) and include a size marker, e.g. GeneRuler (Thermo Fisher Scientific).

6. Photograph gel under UV transillumination. The amplified product should have a size of 291 bp.

7. Purify PCR products using a PCR product purification kit following manufacturer's instructions.

8. Incubate 5 μL of purified PCR product with either HindIII or EvoRV restriction enzymes at 37°C for 30 min.

9. Load onto a 1.5% agarose gel as in Step 5. Include a size marker and the positive control DNA.

10. Document and compare restriction pattern.

SOP 6.1.4 Notes

This PCR was developed based on the observation that the condensation domain (Cd) within the *mcy*A gene was found relatively conserved across *mcy* genes from different genera (Hisbergues *et al.*, 2003). Please note that the PCR protocol may be subject to changes depending on the PCR kit used and the recommendations of the manufacturer.

SOP 6.1.5 Reference

Hisbergues, M., Christiansen, G., Rouhiainen, L., *et al.* (2003) PCR-based identification of microcystin-producing genotypes of different cyanobacterial genera, *Archives in Microbiology*, **180**, 402–410.

SOP 6.2

PCR Detection of Microcystin and Nodularin Biosynthesis Genes in the Cyanobacterial Orders Oscillatoriales, Chroococcales, Stigonematales, and Nostocales

Elke Dittmann[1]*, Joanna Mankiewicz-Boczek[2,3], and Ilona Gągała[2]

[1] *Institute of Biochemistry and Biology, University of Potsdam, Potsdam-Golm, Germany*
[2] *European Regional Centre for Ecohydrology of the Polish Academy of Sciences, Łódź, Poland*
[3] *Department of Applied Ecology, Faculty of Biology and Environmental Protection, University of Lodz, Łódź, Poland*

SOP 6.2.1 Introduction

This SOP describes a simple PCR-based detection technique for microcystin (MC) and nodularin (NOD) biosynthesis genes using the marker genes *mcy*E or *nda*F (472 bp, 809–812 bp, and 405 bp amplification products for HEPF/R, *mcy*E-F2/R4, and *mcy*ER1/S1, respectively) in four orders of cyanobacteria. The technique was initially published by Jungblut and Neilan (2006), Rantala *et al.* (2004), and Mankiewicz-Boczek *et al.* (2015a, b). The PCR fragments can be assigned to the producing genus using sequencing and phylogenetic analysis. The methods are suitable for laboratory strain analysis and population studies. See Fig. 6.2 for an overview. Reference strains are given in Table S6.1 (Appendix).

SOP 6.2.2 Experimental

SOP 6.2.2.1 Materials

- PCR reagents (or PCR cycling kit).
- Primers:

---

*Corresponding author: editt@uni-potsdam.de

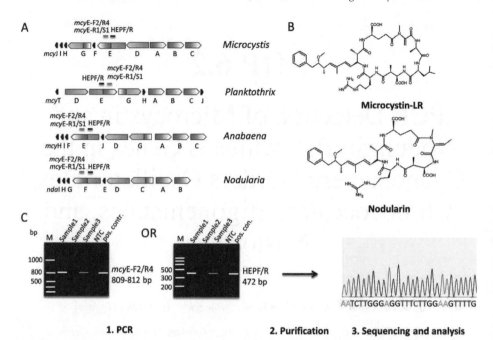

Figure 6.2 *Overview of PCR detection of microcystin and nodularin biosynthesis genes in the cyanobacterial orders Oscillatoriales, Chroococcales, Stigonematales, and Nostocales. (A) Schematic representation of* mcy *and* nda *operons in four genera of cyanobacteria. The detected gene fragments are indicated with a black and gray line, respectively. (B) Structure of microcystin and nodularin. (C) Standard workflow of procedure. NTC = non-template control.*

- ○ HEPF (5′–TTTGGGGTTAACTTTTTTGGGCATAGTC–3′)
- ○ HEPR (5′–AATTCTTGAGGCTGTAAATCGGGTTT–3′)
- ○ or
- ○ *mcy*E-F2 (5′–GAAATTTGTGTAGAAGGTGC–3′)
- ○ *mcy*E-R4 (5′–AATTCTAAAGCCCAAAGACG–3′)
- ○ or
- ○ *mcy*E-R1 (5′–ATAGGATGTTTAGAGAGAATTTTTTCCC–3′)
- ○ *mcy*E-S1 (5′–GGGACGAAAAGATAATCAAGTTAAGG–3′).
- Isolated nucleic acids of laboratory strains or field samples.
- Isolated nucleic acid of an MC-producing strain.
- Sterile Milli-Q® water.
- 0.5× TAE buffer (stock 50× Tris-Acetic acid-EDTA, pH 8.0).
- Midori Green or EtBr.

SOP 6.2.2.2   Equipment

- Thermocycler.
- Electrophoresis equipment.

- UV transilluminator with documentation.
- Adjustable mechanical pipettes (100–1000 µL, 20–200 µL, 1–10 µL)

SOP 6.2.3   Procedure

1. Adjust the amount of isolated DNA to approximately 10 ng µL^{-1}.
2. Adjust the amount of primer to 10 pmol µL^{-1} (for *mcy*E or *nda*F) or to 5 pmol µL^{-1} (for *mcy*ER1/S1).
3. Pipette the following mixture:

   for *mcy*E or *nda*F: (F2/R4, HEPF/R)

   1 × PCR buffer, 2.5 mM MgCl$_2$, 200 µM deoxynucleotide triphosphate mixture, 0.2 U Taq DNA polymerase, 1 µL DNA, 1 µL of each primer (final concentration 1.0 µM); adjust to 10 µL with sterile Milli-Q® water for each DNA sample to be analyzed. Add one tube with sterile Milli-Q® water instead of DNA as negative control (NTC = non-template control) and one tube with DNA of a MC-producing laboratory strain as a positive control;

   for *mcy*E (R1/S1):

   1 × PCR buffer, 3 mM MgCl$_2$, 200 µM deoxynucleotide triphosphate mixture, 0.1 mg mL^{-1} BSA (recommended if working with environmental samples), 1 U Taq DNA polymerase, 1 µL DNA sample, 1 µL of each primer (final concentration 0.25 µM); adjust to 20 µL with sterile Milli-Q® water for each DNA sample to be analyzed. Add one tube with sterile Milli-Q® water instead of DNA as negative control (NTC) and one tube with DNA of a MC-producing laboratory strain as a positive control.
4. Use the following program for PCR cycling:

   for *mcy*E or *nda*F: (HEPF/R, F2/R4):

   initial denaturation at 95°C (5 min), 35 cycles: 94°C for 30 sec, 59°C for 30 sec, 72°C for 1 min, and a final extension step at 72°C (10 min);

   for *mcy*E (R1/S1):

   initial denaturation at 92°C (2 min), 35 cycles: 92°C for 20 sec, 52°C for 30 sec, 72°C for 1 min, and a final extension step at 72°C (5 min).
5. Run 2 µL of each PCR reaction on a 1.5% agarose gel in TBE buffer (including EtBr or Midori Green) and include a size marker, e.g. GeneRuler (Thermo Fisher Scientific).
6. Photograph gel under UV transillumination (see Fig. 6.3 for an example).
7. (Optional) purify PCR products using a PCR product purification kit following manufacturer's instructions.
8. (Optional) subject the PCR product to in-house or commercial sequencing facilities, and perform phylogenetic analysis.

SOP 6.2.4   Notes

Please note that the PCR protocol may be subject to changes depending on the PCR kit used and the recommendations of the manufacturer.

The exact PCR product size can vary, which will not be revealed by gel electrophoresis but by sequencing.

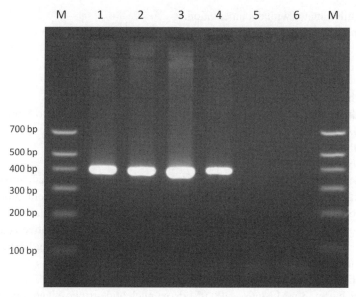

Figure 6.3   *PCR amplification of the* mcyE *gene fragment (*mcyE-R1/mcyE-S1; 405 bp) from environmental samples. (1) Dominance of* Microcystis aeruginosa *in Sulejów Reservoir (Central Poland). (2) Co-occurrence of* M. aeruginosa *and* Aphanizomenon flos-aque *in Jeziorsko Reservoir (Central Poland). (3) Dominance of* Dolichospermum sp. *in Wielgie Lake (Western Poland). (4) Toxic strain* Microcystis aeruginosa *PCC7806. (5) Strain* Cylindrospermopsis raciborskii *CS505 (CSIRO, Australia) lacking* mcy *genes. (6) Negative control with sterile Milli-Q® water. (M) DNA marker M100-700 (DNA Gdańsk, Poland) (1.5% agarose gel, EtBr 4 μg/mL, 15 μL of PCR product on gel).*

When direct sequencing of the amplicon yields multiple peaks in the electropherogram, cloning before sequencing according to standard techniques is required (Sambrook and Russel, 2001).

SOP 6.2.5   References

Jungblut, A.D. and Neilan, B.A. (2006) Molecular identification and evolution of the cyclic peptide hepatotoxins, microcystin and nodularin, synthetase genes in three orders of cyanobacteria, *Archives in Microbiology*, **185**, 107–114.

Mankiewicz-Boczek, J., Gągała, I., Jurczak, T., *et al.* (2015a) Incidence of microcystin-producing cyanobacteria in Lake Tana, the largest waterbody in Ethiopia, *African Journal of Ecology*, **53**, 54–63.

Mankiewicz-Boczek, J., Gągała, I., Jurczak T., *et al.* (2015b) Bacteria homologous to *Aeromonas* capable of microcystin degradation, *Open Life Sciences*, **10**, 119–129.

Rantala, A., Fewer, D.P., Hisbergues, M., *et al.* (2004) Phylogenetic evidence for the early evolution of microcystin synthesis, *Proceedings of the National Academy of Science USA*, **101**, 568–573.

Sambrook, J. and Russell, D. (2001) *Molecular Cloning: A laboratory manual*, 3rd edn, Cold Spring Harbor Laboratory Press, Cold Spring Harbor, New York.

SOP 6.3

Genus-Specific PCR Detection of Microcystin Biosynthesis Genes in *Anabaena*/*Nodularia* and *Microcystis* and *Planktothrix*, Respectively

*Anne Rantala-Ylinen and Kaarina Sivonen**

Department of Food and Environmental Sciences, Division of Microbiology and Biotechnology, University of Helsinki, Helsinki, Finland

SOP 6.3.1   Introduction

This SOP describes a PCR method for the genus-specific detection of microcystin (*mcy*) and nodularin (*nda*) biosynthesis genes in *Anabaena*/*Nodularia* and *Microcystis* and *Planktothrix*, respectively. Specificity of the method is based on the reverse primer that was designed to recognize a sequence found only in the *mcy*E gene sequence variants of *Anabaena*/*Nodularia* and *Microcystis* and *Planktothrix* strains, respectively. The specific primers are used in combination with a forward primer that recognizes a *mcy*E gene sequence found in several microcystin producer genera. The specificity of the method was tested with several microcystin-producing and non-producing strains of the genera *Anabaena*, *Microcystis*, *Planktothrix* (Vaitomaa *et al.*, 2003) and later used to detect MC-producing *Anabaena* in lake water samples (Rantala *et al.*, 2006). See Fig. 6.4 for an overview. Reference strains are given in Table S6.1 (Appendix).

SOP 6.3.2   Experimental

SOP 6.3.2.1   Materials

- DyNAzyme II DNA polymerase, 10× optimized DyNAzyme buffer (Thermo Fisher Scientific), and deoxynucleotides, e.g., dNTP Mix 10 mM each (Thermo Fisher Scientific).

---

*Corresponding author: kaarina.sivonen@helsinki.fi

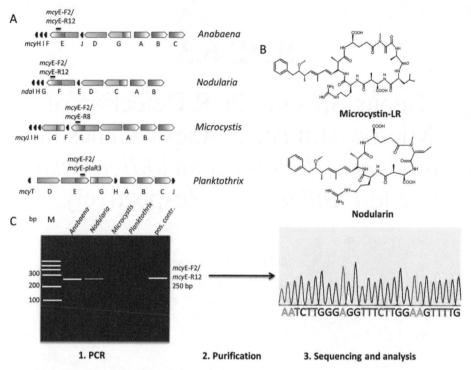

Figure 6.4 *Overview of PCR detection of genus-specific microcystin and nodularin biosynthesis genes. (A) Schematic representation of* mcy *and* nda *operons. The detected gene fragment is indicated with a black line. (B) Structure of microcystin and nodularin. (C) Standard workflow of procedure.*

- Primers (stock solutions 100 pmol μL^{-1}):
    ○ *mcy*E-F2 (5′–GAAATTTGTGTAGAAGGTGC–3′)
    ○ *mcy*E-12R (5′–CAATCTCGGTATAGCGGC–3′) (*Anabaena/Nodularia*)
    ○ *mcy*E-R8 (5′–CAATGGGAGCATAACGAG–3′) (*Microcystis*)
    ○ *mcy*E-plaR3 (5′–CTCAATCTGAGGATAACGAT–3′) (*Planktothrix*).
- Isolated nucleic acids of laboratory strains or field samples.
- Isolated nucleic acids MC-producing strains.
- Sterile Milli-Q® water.
- 0.5× TAE buffer (stock 50× Tris-Acetic acid-EDTA, pH 8.0).
- Agarose (LE, analytical grade).
- Midori Green or EtBr.

SOP 6.3.2.2 Equipment

- Thermocycler.
- Electrophoresis equipment.
- UV transilluminator with documentation.
- Adjustable mechanical pipettes (100–1000 μL, 20–200 μL, 1–10 μL).

SOP 6.3.3 Procedure

1. Adjust the amount of isolated DNA of laboratory strains to approximately 10 ng μL^{-1} and that of environmental samples to approximately 30 ng μL^{-1}.
2. Adjust the amount of primers to 10 pmol μL^{-1}.
3. Pipette the following mixture: For each DNA sample to be analyzed, mix 2 μL of 10× PCR buffer (final concentration 1×), 0.5 μL of dNTP mix (final concentration 250 μM), 1 μL of each primer (final concentration 0.5 pmol μL^{-1}), 0.5 μL (=1 U) DyNAzyme II DNA polymerase, and 1 μL DNA. Adjust with sterile distilled water to have a total volume of 20 μL for one reaction. Add one tube with sterile Milli-Q® water instead of DNA as negative control and one tube with DNA of an MC-producing laboratory strain as a positive control.
4. Use the following program for PCR cycling: initial denaturation at 95°C for 3 min, 35 cycles: 94°C for 30 sec, 58°C for 30 sec, 72°C for 60 sec, and a final extension step at 72°C for 10 min.
5. Run the PCR reactions on a 1.5% agarose gel in 0.5× TAE buffer (including EtBr or Midori Green) and include a size marker that covers the size of the expected amplification product, 250 bp (*Anabaena/Nodularia*), 247 bp (*Microcystis*), and 249 bp (*Planktothrix*), respectively.
6. Photograph the gel under UV transillumination.
7. Document and compare amplification product sizes of the DNA samples to that of the positive control sample and to the size marker.

SOP 6.3.4 Notes

Please note that the PCR protocol may be subject to changes depending on the PCR kit used and the recommendations of the manufacturer.

If analyzing several samples, it is more convenient to prepare a larger volume of mix (= PCR master mix) excluding DNA, aliquot this mix (19 μL) first into the PCR tubes, and after that add the DNA samples and controls to tubes. Some extra mix should be prepared for covering the pipetting losses.

SOP 6.3.5 References

Rantala, A., Rajaniemi-Wacklin, P., Lyra, C., *et al.* (2006) Detection of microcystin-producing cyanobacteria in Finnish lakes with genus-specific microcystin synthetase gene E (*mcy*E) PCR and associations with environmental factors, *Applied and Environmental Microbiology*, **72**, 6101–6110.

Vaitomaa, J., Rantala, A., Halinen, K., *et al.* (2003) Quantitative real-time PCR for determination of microcystin synthetase E copy numbers for *Microcystis* and *Anabaena* in lakes, *Applied and Environmental Microbiology*, **69**, 7289–7297.

SOP 6.4

PCR Detection of Anatoxin Biosynthesis Genes Combined with RFLP Differentiation of the Producing Genus

*Anne Rantala-Ylinen and Kaarina Sivonen**

Department of Food and Environmental Sciences, Division of Microbiology and Biotechnology, University of Helsinki, Helsinki, Finland

SOP 6.4.1 Introduction

This SOP describes a PCR-based detection technique for anatoxin (ANA) biosynthesis genes and the subdifferentiation by RFLP analysis. The method was initially published by Rantala-Ylinen *et al.* (2011). See Fig. 6.5 for an overview. Reference strains are given in Table S6.1 (Appendix).

SOP 6.4.2 Experimental

SOP 6.4.2.1 Materials

- DyNAzyme II DNA polymerase, 10× optimized DyNAzyme buffer (Thermo Fisher Scientific), and deoxynucleotides, e.g. dNTP Mix 10 mM each (Thermo Fisher Scientific).
- Primers (stock solutions 100 pmol μL^{-1}):
  - *ana*C-gen F (5′–TCTGGTATTCAGTCCCCTCTAT–3′)
  - *ana*C-gen R (5′–CCCAATAGCCTGTCATCAA–3′).
- Isolated nucleic acids of laboratory strains or field samples.
- Isolated nucleic acid of an ANA-producing strain.
- Sterile Milli-Q® water.
- 0.5× TAE buffer (stock 50× Tris-Acetic acid-EDTA, pH 8.0).
- 1× TBE buffer (Tris-borate EDTA, 89 mM TRIS Base, 89 mM boric acid, 2 mM EDTA, pH 8.0).
- Agarose (LE, analytical grade).

*Corresponding author: kaarina.sivonen@helsinki.fi

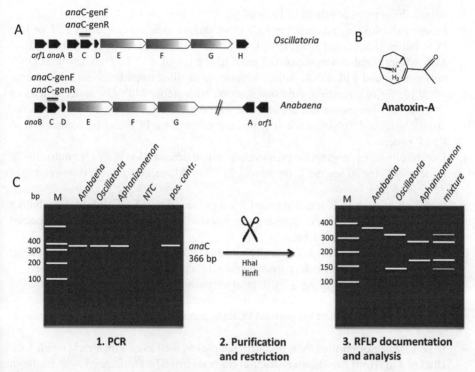

Figure 6.5 *Overview of PCR detection of anatoxin biosynthesis genes combined with RFLP differentiation of the producing genus. (A) Schematic representation of* ana *operons in Oscillatoria and* Anabaena; *the detected gene fragment is indicated with a black line. (B) Structure of anatoxin-a. (C) Standard workflow of procedure. NTC = non-template control.*

- TopVision agarose (Thermo Fisher Scientific).
- EtBr or Midori Green.
- PCR product purification kit.
- HhaI and HinfI FastDigest restriction enzymes and FastDigest Green Buffer (Thermo Fisher Scientific).

SOP 6.4.2.2   Equipment

- Thermocycler.
- Electrophoresis equipment.
- UV transilluminator with documentation.
- NanoDrop™ ND-1000 spectrophotometer.
- Adjustable mechanical pipettes (100–1000 µL, 20–200 µL, 1–10 µL).

SOP 6.4.3   Procedure

1. Adjust the amount of isolated DNA of laboratory strains to approximately 20 ng µL^{-1} and that of environmental samples to approximately 50 ng µL^{-1}.

2. Adjust the amount of primer to 10 pmol μL^{-1}.
3. Pipette the following mixture: For each DNA sample to be analyzed, mix 2 μL of 10×
   PCR buffer (final concn. 1×), 0.4 μL of dNTP mix (final concn. 200 μM), 1 μL of
   each primer (final concentration 0.5 pmol μL^{-1}), 0.5 μL (=1 U) DyNAzyme II DNA
   polymerase, and 1 μL DNA. Adjust with sterile distilled water to have a total volume
   of 20 μL for one reaction. Add one reaction with sterile Milli-Q® water instead of
   DNA as negative control and one reaction with DNA of an ANA-producing *Anabaena*,
   Oscillatoria, and *Aphanizomenon* as positive controls for PCR and for comparison of
   RFLP patterns.
4. Use the following program for PCR cycling: initial denaturation at 95°C (2 min), 30–35
   cycles: 94°C for 30 sec, 58°C for 30 sec, 72°C for 30 sec, and a final extension step at
   72°C for 5 min.
5. Run 5 μL of each PCR reaction on a 1.5% agarose gel in 0.5× TAE buffer (including
   EtBr or Midori Green) and include a size marker that covers the size of the expected
   amplification product, 366 bp.
6. Photograph the gel under UV transillumination. Presence of an amplification product
   indicates the sample contains potential anatoxin-a or homoanatoxin-a producers.
7. Purify PCR products using a PCR product purification kit, e.g. Amicon Ultra-0.5
   (Merck Millipore) following manufacturer's instructions.
8. Measure concentration of the purified PCR products with a NanoDrop™ spectrophotometer.
9. Digest 200 ng of purified PCR product (samples as well as control strains) with 1 μL
   HhaI or 1 μL HinfI FastDigest restriction enzymes in 0.67× FastDigest Green Buffer in
   a total volume of 30 μL. Incubate at 37°C for 15 min and inactivate at 65°C for 20 min.
10. Separate the restriction fragments in 3% TopVision agarose gel in 1× TBE buffer.
    Include a size marker, e.g. O'GeneRuler low-range DNA ladder (Thermo Fisher
    Scientific), and restriction fragments obtained with control strains.
11. Document and compare restriction patterns of the samples studied to those of control
    strains in order to identify the genus of the potential anatoxin producer.

SOP 6.4.4   Notes

Please note that the PCR protocol may be subject to changes depending on the PCR kit
used and the recommendations of the manufacturer.

For analyzing several samples, prepare a larger volume of mix excluding DNA (= PCR
master mix); see SOP 6.3.

SOP 6.4.5   Reference

Rantala-Ylinen, A., Kana, S.E.C., Wang, H., *et al.* (2011) Anatoxin-a synthetase gene
cluster of the cyanobacterium *Anabaena* sp. strain 37 and molecular methods to detect
potential producers, *Applied and Environmental Microbiology*, **77**, 7271–7278.

SOP 6.5

PCR Detection of the Saxitoxin Biosynthesis Genes, *sxt*A, *sxt*X, *sxt*H, *sxt*G, and *sxt*I

Andreas Ballot[1]* and Samuel Cirés[2]

[1]*Norwegian Institute for Water Research, Oslo, Norway*
[2]*Department of Biology, University of Madrid, Madrid, Spain*

SOP 6.5.1   Introduction

Several genes (*sxt*A, *sxt*G, *sxt*H, *sxt*I, and *sxt*X) encoding biosynthesis of saxitoxins have been tested for use as markers for detection of potential producers of saxitoxins (Ballot *et al.*, 2010; Casero *et al.*, 2014). However, all genes tested have also been detected in a variety of cyanobacterial strains not producing saxitoxins. No genetic markers have been tested yet which are solely found in producers of saxitoxins. The specificity of primers for the detection of the *sxt*A gene has been tested and confirmed the presence of *sxt*A in a variety of cyanobacterial strains producing and not producing saxitoxins: *Aphanizomenon gracile, Aph. flos-aquae, Anabaenopsis elenkinii, Dolichospermum crassum, D. planctonicum, D. flos-aquae, D. lemmermannii, D. mendotae, Chrysosporum ovalisporum, C. bergii, Cuspidothrix issatschenkoi, Cylindrospermum* sp., *Geitlerinema* sp., *Phormidium* sp., and *Scytonema* sp. (Ballot *et al.*, 2010; Smith *et al.*, 2011; Casero *et al.*, 2014, Borges *et al.*, 2015). *Sxt*G, *sxt*H, and *sxt*X genes have been tested and their presence is confirmed in saxitoxins-producing and non-producing strains of *Aph. gracile* and *Aph. flos-aquae* whereas *sxt*I has been detected in saxitoxins-producing *Aph. gracile* (Casero *et al.*, 2014) and in non-producing *C. raciborskii* strain CENA 303 (Hoff-Risseti *et al.*, 2013). *SxtI* is not present in non-producing *Aph. flos-aquae* strain NIVA-CYA 689, which contains the other four saxitoxin-biosynthesis genes (*sxt*A, *sxt*G, *sxt*H, *sxt*X) (Casero *et al.*, 2014).

This protocol refers to the non-simultaneous detection of fragments of the following genes *sxt*A (683 bp), *sxt*G (893 bp), *sxt*H (812 bp), *sxt*I (910 bp), and *sxt*X (656 bp). The potential use of these same primers for multiplex PCR would require additional optimization. See Fig. 6.6 for an overview. Reference strains are given in Table S6.1 (Appendix).

*Corresponding author: andreas.ballot@niva.no

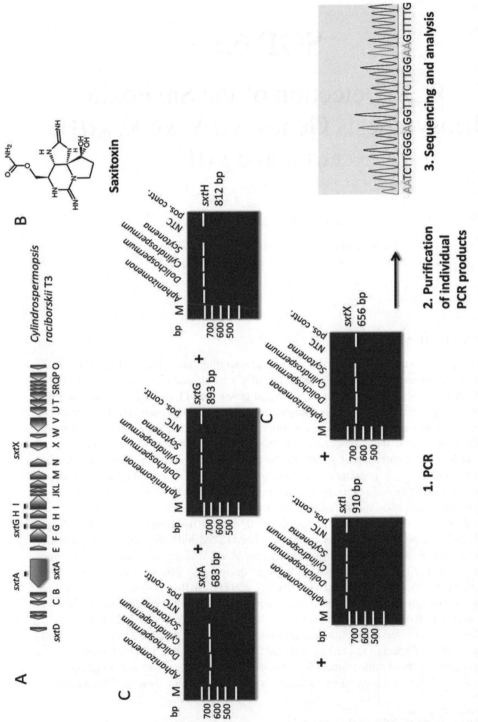

Figure 6.6 Overview of PCR detection of saxitoxin biosynthesis genes. (A) Schematic representation of sxt operon in Cylindrospermopsis raciborskii T3. The detected gene fragments are indicated with black lines. (B) Structure of saxitoxin. (C) Standard workflow of procedure. NTC non-template control.

SOP 6.5.2 Experimental

SOP 6.5.2.1 Materials

- PCR kit: e.g. Taq PCR core kit (Qiagen GmbH, Germany) including Taq DNA polymerase (5 U μL^{-1}), deoxynucleotide triphosphate mix (10 mM), 10× Qiagen PCR buffer, $MgCl_2$ solution.
- Primers (stock solutions 100 pmol μL^{-1}):
  - *sxt*A gene
  - *sxt*Af (5′–GCGTACATCCAAGCTGGACTCG–3′)
  - *sxt*Ar (5′–GTAGTCCAGCTAAGGCACTTGC–3′)
  - *sxt*G gene
  - *sxt*Gf (5′–AGGAATTCCCTATCCACCGGAG–3′)
  - *sxt*Gr (5′–CGGCGAACATCTAACGTTGCAC–3′)
  - *sxt*H gene
  - *sxt*Hf (5′–AAGACCACTGTCCCCACCGAGG–3′)
  - *sxt*Hr (5′–CTGTGCAGCGATCTGATGGCAC–3′)
  - *sxt*I gene
  - *sxt*If (5′–AGCGCTGCCGCTATGGTTGTCG–3′)
  - *sxt*Ir (5′–ACGCAATTGAGGGCGACACCAC–3′)
  - *sxt*X gene
  - *sxt*Xf (5′–GATGCAACCCATAAACTCGCAC–3′)
  - *sxt*Xr (5′–AAGGTACTCGTTTTCGTGGAGC–3′).
- Isolated genomic DNA of laboratory strains or field samples.
- Isolated nucleic acid of a STX-producing strain.
- Sterile Milli-Q® water.
- 0.5× TAE buffer (stock 50x Tris-Acetic acid-EDTA, pH 8.0).
- Agarose (Biotechnology grade).
- EtBr or Midori Green (0.2–0.5 $\mu g\ mL^{-1}$, usually about 1–3 μL of lab stock solution per 100 mL gel).

SOP 6.5.2.2 Equipment

- Thermocycler.
- Equipment for gel electrophoresis.
- UV transilluminator with gel documentation.
- Adjustable mechanical pipettes (100–1000 μL, 20–200 μL, 1–10 μL).

SOP 6.5.3 Procedure

1. Adjust the amount of isolated genomic DNA of laboratory strains to approximately 10–20 ng μL^{-1}. The amount of environmental DNA can be higher, since it contains DNA from several organisms.
2. Adjust the amount of primers to 10 pmol μL^{-1}.
3. Prepare the following mixture for the PCR reaction using a PCR kit, e.g. the Taq PCR core kit (Qiagen GmbH, Germany). For one PCR sample prepare mix (20 μL) which

contains 0.1 μL of Taq DNA polymerase (5 U μL^{-1}), 0.5 μL of deoxynucleotide triphosphate mix (10 mM), 2 μL of 10× Qiagen PCR buffer (final concentration 1×), 1 μL of one forward and one reverse primer (10 μM) for one of the selected genes, 14.4 μL sterile Milli-Q® water, and 1 μL of genomic DNA.

Additionally one reaction with sterile Milli-Q® water (negative control) and one reaction with DNA of a laboratory strain of a producer of saxitoxins (positive control) have to be prepared (see Table S6.1. Appendix, for potential positive controls). As pipetting can cause the loss of master mix, one or two extra reactions should be calculated for the master mix.

4. Use the following PCR program: initial denaturation step 94°C for 4 min, followed by 35 cycles of 94°C for 10 sec, 55°C for 20 sec, and 72°C for 1 min.
5. Test the PCR products on a 1.5% agarose gel (including EtBr or Midori Green) in 0.5× TAE buffer and include a DNA ladder that covers the size of the expected PCR amplification product, between 100 and 1000 bp.
6. Photograph and document the gel using UV transillumination (an example is given in Fig. 6.7).
7. Compare the PCR amplification product sizes to that of the positive control sample and the DNA ladder.

SOP 6.5.4   Notes

For PCR tests prepare a master mix of all ingredients minus genomic DNA.

A range of culture strains to be used as positive controls (e.g. *Aph. gracile* strains UAM529 and AB2008/31) can be found in Ballot *et al.* (2010) and Casero *et al.* (2014).

Please note that the PCR protocol may change depending on the PCR kit used and the manufacturer's protocol.

A safer alternative for the supposed EtBr is GelRed™ (Biotium, Hayward, Canada).

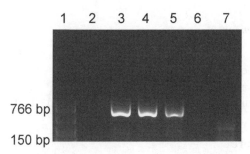

Figure 6.7 *Amplification of partial sxtA gene (683 bp) in culture samples of different cyanobacterial species (1.5% agarose gel stained with GelRed™). Lanes: (1) ladder; (2) negative control; (3) saxitoxin-producing* Aphanizomenon gracile *NIVA-CYA 666 (former AB2008/31); (4) saxitoxin-producing* Aphanizomenon gracile *NIVA-CYA 675 (former AB2008/47); (5) saxitoxin-producing* Aphanizomenon gracile *NIVA-CYA 677 (former AB2008/49); (6) non-producer* Cuspidothrix issatschenkoi *AB2008/12; (7) non-producer* Dolichospermum planctonicum *NIVA-CYA 649 (former AB2008/01).*

SOP 6.5.5   References

Ballot, A., Fastner J., and Wiedner, C. (2010) Paralytic shellfish poisoning toxin-producing cyanobacterium *Aphanizomenon gracile* in northeast Germany, *Applied and Environmental Microbiology*, **76** (4), 1173–1180.

Borges, H., Branco, L., Martins, M., *et al.* (2015) Cyanotoxin production and phylogeny of benthic cyanobacterial strains isolated from the northeast of Brazil, *Harmful Algae*, **43**, 46–57.

Casero, M.C., Ballot, A., Agha, R., *et al.* (2014) Characterization of saxitoxin production and release and phylogeny of *sxt* genes in paralytic shellfish poisoning toxin-producing *Aphanizomenon gracile*, *Harmful Algae*, **37**, 28–37.

Hoff-Risseti, C., Dörr, F.A., Schaker, P.D.C., *et al.* (2013) Cylindrospermopsin and saxitoxin synthetase genes in *Cylindrospermopsis raciborskii* strains from Brazilian freshwater, *PLOS ONE*, **8** (8), 1–14.

Smith, F.M.J., Wood, S.A., van Ginkel, R., *et al.* (2011) First report of saxitoxin production by a species of the freshwater benthic cyanobacterium, *Scytonema Agardh*, *Toxicon*, **57**, 566–573.

SOP 6.6

PCR Detection of the Cylindrospermopsin Biosynthesis Gene *cyr*J

Samuel Cirés[1]* and Andreas Ballot[2]

[1]*Department of Biology, University of Madrid, Madrid, Spain*
[2]*Norwegian Institute for Water Research, Oslo, Norway*

SOP 6.6.1   Introduction

The *cyr*J gene encodes a sulfotransferase involved in the synthesis of cylindrospermopsin by cyanobacteria. So far, *cyr*J has been detected only in cylindrospermopsin producing strains being absent in non-producing strains of *Anabaena*, *Aphanizomenon*, *Cylindrospermopsis*, and *Oscillatoria*, which nevertheless contained other *cyr* genes (e.g. *cyr*A, *cyr*B, *cyr*C)

*Corresponding author: samuel.cires@uam.es

(Mihali *et al.*, 2008; Mazmouz *et al.*, 2010; Ballot *et al.*, 2011). Therefore, among the 15 ORFs comprising the CYL-biosynthesis cluster, *cyr*J can be considered the only gene marker diagnostic for CYL production in cyanobacteria to date.

The present protocol describes the amplification of a *cyr*J fragment (584 bp) based on primers from Mihali *et al.*, (2008), and PCR conditions from Ballot *et al.*, (2011). See Fig. 6.8 for an overview. Reference strains are given in Table S6.1 (Appendix).

SOP 6.6.2   Experimental

SOP 6.6.2.1   Materials

- PCR kit: e.g. Taq PCR core kit (Qiagen GmbH, Germany) including Taq DNA polymerase (5 U μL^{-1}), deoxynucleotide triphosphate mix (10 mM), 10× Qiagen PCR buffer, $MgCl_2$ solution.
- Primers (stock solutions 100 pmol μL^{-1}):
  - cynsulF (5′–ACTTCTCTCCTTTCCCTATC–3′)
  - cylnamR (5′–GAGTGAAAATGCGTAGAACTTG–3′)
- Isolated genomic DNA of laboratory strains or field samples.
- Isolated nucleic acid of a CYL-producing strain.
- Sterile Milli-Q® water.
- 0.5× TAE buffer (stock 50× Tris-Acetic acid-EDTA, pH 8.0).
- Agarose (biotechnology grade).
- EtBr or Midori Green (0.2–0.5 μg mL^{-1}; usually about 1–3 μL of lab stock solution per 100 mL gel).

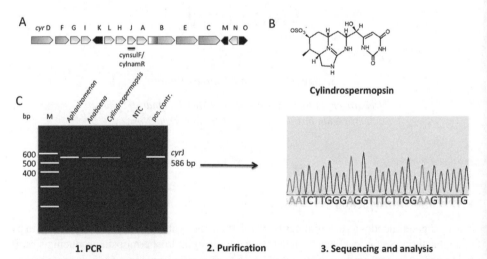

Figure 6.8 *Overview of PCR detection of cylindrospermopsin biosynthesis genes. (A) Schematic representation of* cyr *operon in* Cylindrospermopsis. *The detected gene fragment is indicated with a black line. (B) Structure of cylindrospermopsin. (C) Standard workflow of procedure. NTC = non-template control.*

SOP 6.6.2.2   Equipment

- Thermocycler.
- Equipment for gel electrophoresis.
- UV transilluminator with gel documentation.
- Adjustable mechanical pipettes (100–1000 µL, 20–200 µL, 1–10 µL).

SOP 6.6.3   Procedure

1. Adjust the amount of isolated genomic DNA of laboratory strains to approximately 10–20 ng µL^{-1}. Amount of environmental DNA can be higher, since it contains DNA from several organisms.
2. Adjust the amount of primers to 10 pmol µL^{-1}.
3. Prepare the following mixture for the PCR reaction:
   The PCR reactions were performed using the Taq PCR core kit (Qiagen GmbH, Germany). For one PCR sample prepare mix (20 µL) which contains 0.1 µL of Taq DNA polymerase (5 U µL^{-1}), 0.5 µL of deoxynucleotide triphosphate mix (10 mM), 2 µL of 10× Qiagen PCR buffer (final concentration 1×), 1 µL of each forward and reverse primer (10 µM), 14.4 µL sterile Milli-Q$^{®}$ water, and 1 µL of genomic DNA.
4. Additionally one reaction with sterile Milli-Q$^{®}$ water (negative control) and one reaction with DNA of a laboratory strain of a producer of cylindrospermopsin (positive control) have to be prepared (see Table S6.1, Appendix, for potential positive controls). As pipetting can cause the loss of master mix, one or two, extra reactions should be calculated for the master mix.
5. Use the following PCR program: initial denaturation step 95°C for 3 min, followed by 35 cycles of: 95°C for 45 sec, 50°C for 30 sec, and 72°C for 1 min, with a final elongation step of 72°C for 3 min.
6. Test the PCR products on a 1.5% agarose gel (including EtBr or Midori Green) in 0.5× TAE buffer (including EtBr) and include a DNA ladder that covers the size of the expected PCR amplification product, between 100 and 1000 bp.
7. Photograph and document the gel using UV transillumination (an example is given in Fig. 6.9).
8. Compare the PCR amplification product sizes to that of the positive control sample and the DNA ladder.

SOP 6.6.4   Notes

For several PCR tests prepare a master mix of all ingredients minus genomic DNA.

This protocol has been optimized for cyanobacterial cultures. Its use in field samples might in some cases require additional modifications not included here (e.g. addition of bovine serum albumin to the master mix and/or dilution of genomic DNA to overcome the effect of PCR inhibitors often present in natural samples).

A range of strains to be used as positive controls (e.g. *Cylindrospermopsis raciborskii* AWT 205, *Aph.* (*Chrysosporum*) *ovalisporum* strain ILC-164, *Aph.* (*Chrysosporum*) *ovalisporum* UAM 290, *Oscillatoria* PCC 7506) can be found in Mihali *et al.* (2008), Ballot *et al.* (2011), Cirés *et al.* (2014), and Mazmouz *et al.* (2010).

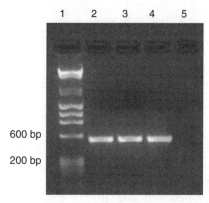

Figure 6.9 *Selective amplification of cyrJ gene in culture samples of different cyanobacterial genera (1.5% agarose gel stained with EtBr). Numbers in the left margin stand for length (base pairs) of fragments in the DNA ladder. Gel lanes: (1) ladder; (2) CYL-producing Aphanizomenon (Chrysosporum) ovalisporum UAM 290 (Cirés et al., 2014); (3) CYL-producing Anabaena lapponica Syke 966 (Spoof et al., 2006); (4) CYL-producing Oscillatoria sp. PCC 6506 (Mazmouz et al., 2013); (5) non-producer A. planctonica (Dolichospermum planctonicum) UAM 516 (Cirés et al., 2014).*

Please note that the PCR protocol may change depending on the PCR kit used and the manufacturer's protocol.

A safer alternative for the supposed EtBr is GelRed™ (Biotium, Hayward, Canada).

SOP 6.6.5   References

Ballot, A., Ramm, J., Rundberget, T., *et al.* (2011) Occurrence of non-cylindrospermopsin-producing *Aphanizomenon ovalisporum* and *Anabaena bergii* in Lake Kinneret (Israel), *Journal of Plankton Research*, **33**, 1736–1746.

Cirés, S., Wörmer, L., Ballot, A., *et al.* (2014) Phylogeography of cylindrospermopsin and paralytic shellfish toxin-producing Nostocales cyanobacteria from Mediterranean Europe (Spain), *Applied and Environmental Microbiology*, **80** (4), 1359–1370.

Mazmouz, R., Chapuis-Hugon, F., Mann, S., *et al.* (2010) Biosynthesis of cylindrospermopsin and 7-Epicylindrospermopsin in *Oscillatoria* sp. strain PCC 6506: Identification of the *cyr* gene cluster and toxin analysis, *Applied and Environmental Microbiology*, **76** (15), 4943–4949.

Mihali, T.K., Kellmann, R., Muenchhoff, J., *et al.* (2008) Characterization of the gene cluster responsible for cylindrospermopsin biosynthesis, *Applied and Environmental Microbiology*, **74**, 716–722.

Spoof, L., Berg, K.A., Rapala, J., *et al.* (2006) First observation of cylindrospermopsin in *Anabaena lapponica* isolated from the boreal environment (Finland), *Environmental Toxicology*, **21** (6), 552–560.

SOP 6.7

PCR from Single Filament of Toxigenic *Planktothrix*

Qin Chen[1,2], *Guntram Christiansen*[1,3], *and Rainer Kurmayer*[1]*

[1]*University of Innsbruck, Research Institute for Limnology, Mondsee, Austria*
[2]*College of Natural Resources and Environment, Northwest A & F University, Yangling, P. R. China*
[3]*Miti Biosystems GmbH, Max F. Perutz Laboratories, Vienna, Austria*

SOP 6.7.1   Introduction

This SOP describes how to amplify DNA fragments from rather low amounts (picograms) of template DNA originating from single filaments/colonies of toxigenic cyanobacteria (see also SOP 2.3 and SOP 5.4). Using advanced molecular toolkits such as engineered DNA polymerases makes it possible to perform dozens of PCR experiments amplifying large DNA fragments (>1000 bp) from one single specimen. One application is to observe the mutations occurring within gene clusters encoding cyanotoxin synthesis. See Fig. 6.10 for a flow diagram of the various steps. Reference strains are given in Table S6.1 (Appendix).

SOP 6.7.2   Experimental

SOP 6.7.2.1   Materials

- Reference DNA from strain *P. agardhii* strain NIVA-CYA126/8.
- Phire Hot Start II DNA polymerase (Thermo Fisher Scientific).
- Dream Taq polymerase (Thermo Fisher Scientific).
- dNTPs mixture (10 mM each).
- Sterile Milli-Q® water.
- PCR tubes or PCR (96 well) plate.
- Primers for "entry" PCR:
  - *psa*AB-IGS fwd 5′–GGGTGGTACTTGCCAAGTCTCT–3′;
    rev 5′–CGACGTGTTGTCGGGTCTT–3′ (PCR product size: 669 bp).

*Corresponding author: rainer.kurmayer@uibk.ac.at

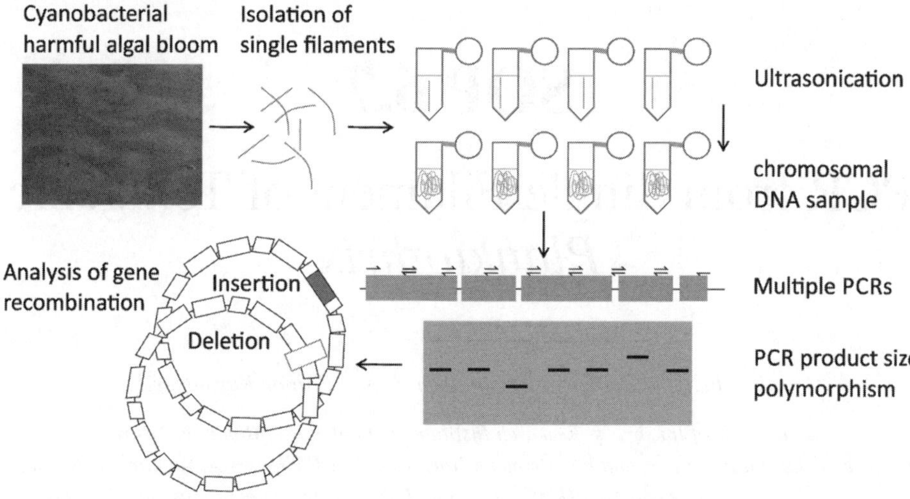

Figure 6.10   *Flow diagram showing steps of* Planktothrix sp. *filament isolation and PCR analysis to observe mutations inactivating toxin synthesis in cyanobacteria (see SOP 6.7).*

- Primers to detect partial deletion of the *mcy*D gene:
    - F*mcy*2+ 5′–AGCAGTTAATAATGATGGCG–3′;
    - F*mcy*2- 5′–GGCTGGGTTCAATAATATTAA–3′ (PCR product size 3628 bp).
    - For a complete primer list enabling the amplification of the entire *mcy* gene cluster without interruption see Table S6.2 in the Appendix.
    - Gel electrophoresis agarose (0.8%), TBE buffer (Sambrook and Russell, 2001).
    - Midori Green Advance (Nippon Genetics Europe GmbH, distributed through Biozym) for DNA staining.
    - DNA size ladder (appropriate DNA fragment size range).

SOP 6.7.2.2   Equipment

- Adjustable mechanical pipettes (1–10 μL, 2–20 μL, 20–200 μL).
- PCR cycler (96 well format).
- Gel electrophoresis and gel documentation.

SOP 6.7.3   Procedure

1. Perform one "entry" PCR in order to confirm the presence of DNA after sonification of each individual filament (SOP 5.4).
2. Pipette the following mixture: 1 μL of Dream Taq PCR buffer (10×), 0.4 μL of MgCl$_2$ (50 mM), 0.3 μL of dNTPs (10 mM each), 0.3 μL of each primer (10 pmol μL^{-1}), 0.05 μL of Dream Taq polymerase, 6.65 μL sterile Milli-Q® water, and 1.0 μL DNA template (10 μL reaction volume).

3. Run PCR cycle program using an initial denaturation step at 94°C for 3 min, followed by 35 cycles (denaturation at 94°C for 30 sec, annealing at 60°C for 30 sec, and elongation at 72°C for 30 sec), and a final elongation step at 72°C for 1 min.
4. Load PCR product onto the agarose gel (0.8%) and document positive samples (PCR product size 670 bp).
5. Perform the PCR to detect mutations in reaction mixtures of 10 μL, containing 2 μL of PCR dilution buffer (5×), 0.2 μL of dNTPs (Kapa Biosystems, Wilmington, MA, USA), 0.5 μL of each primer (10 pmol μL^{-1}), 0.2 μL of Phire Hot start II DNA polymerase, 5.6 μL sterile Milli-Q$^{®}$ water, and one μL DNA template.
6. Run PCR cycle program including an initial denaturation step at 98°C for 30 sec, followed by 40 cycles (denaturation at 98°C for 5 sec, annealing at 60°C for 20 sec, and elongation at 72°C for 70 sec), and final elongation (72°C, 1 min).
7. Load PCR product onto the gel and document mutations by size polymorphism when compared with the reference (Table S6.1, Appendix).

SOP 6.7.4   Notes

For a complete primer list enabling the amplification of the entire *mcy* gene cluster in Planktothrix sp. without interruption, see Table S6.2 (Appendix).

This technique will detect mutations by polymorphism in PCR product size. By using 16 primer pairs designed to yield fragments of 3.5 kb covering the whole *mcy* gene cluster (>50 kb) the mutations occurring within the entire microcystin synthetase can be analyzed.

Because it is impossible to resolve size differences of a few bp on the gel under standard agarose conditions, point mutations as well as shorter fragmented deletions/insertions (<100 bp) will not be detected.

For a general introduction to gel electrophoresis, see Sambrook and Russell (2001). As an alternative to EtBr, the nontoxic dye Midori Green is recommended.

SOP 6.7.5   References

Chen, Q., Christiansen, G., Deng, L., and Kurmayer, R. (2016) Emergence of nontoxic mutants as revealed by single filament analysis in bloom-forming cyanobacteria of the genus *Planktothrix*, *BMC Microbiology*, **16**, 1–12.

Sambrook, J. and Russell, D. (2001) *Molecular Cloning: A laboratory manual*, 3rd edn, Cold Spring Harbor Laboratory Press, Cold Spring Harbor, New York.

SOP 6.8

Analysis of Microcystin Biosynthesis Gene Subpopulation Variability in *Planktothrix*

*Rainer Kurmayer**

Research Institute for Limnology, University of Innsbruck, Mondsee, Austria

SOP 6.8.1   Introduction

The genetic variation within NRPS adenylation domains has been found most useful to characterize subpopulations genetically at a high resolution. Particularly adenylation domains involved in the activation of variable amino acids in position 2 of the microcystin molecule (the *mcy*BA1 region) have been found to be highly variable, for example when compared with intergenic spacer regions of other (housekeeping) genes the level of divergence is high (e.g. Kurmayer *et al.*, 2002).

This SOP describes how to amplify (larger) DNA fragments from NRPS adenylation *mcy*BA1 domains encoding the activation of the variable amino acid in position 2 of the microcystin peptide molecule in *Planktothrix* sp. PCR fragments are then analyzed by RFLP allowing to characterize (sub)population genetic diversity (e.g. Kurmayer and Gumpenberger, 2006).

SOP 6.8.2   Experimental

SOP 6.8.2.1   Materials

- Reference genomic DNA from strain *P. agardhii* NIVA-CYA126/8.
- Phusion polymerase (Thermo Fisher Scientific).
- dNTPs mixture (10 mM each).
- Sterile Milli-Q® water.
- PCR tubes or PCR (96 well) plate.

---

*Corresponding author: rainer.kurmayer@uibk.ac.at

- Primers:
  - ○ *mcy*BA1:
    - *mcy*BA1totfwd 5′–CACCTAGTTGAAGAACAAGTTCT–3′;
    - *mcy*BA1totrev 5′–AGACTTGTTTAATAGCAAAGGC–3′; (1693 bp) (Kurmayer and Gumpenberger, 2006).
- Gel electrophoresis agarose (0.8%), TBE-buffer (Sambrook and Russell, 2001).
- Midori Green Advance (Nippon Genetics Europe GmbH, distributed through Biozym) for DNA staining.
- DNA size ladder (appropriate DNA fragment size range).
- Restriction enzymes, AluI, TseI (New England Biolabs).

SOP 6.8.2.2   Equipment

- Adjustable mechanical pipettes (1–10 µL, 2–20 µL, 20–200 µL).
- PCR cycler (96 well format).
- Gel electrophoresis and gel documentation.

SOP 6.8.3   Procedure

1. For PCR analysis of isolated filaments/colonies (SOP 2.3) or isolated strains (SOP 3.1) prepare 1 µL of a sample (DNA) (SOP 5.1 and 5.4), which is subsequently incubated into reaction tubes for PCR.
2. *mcy*BA1 amplifications are performed in a volume of 20 µL, containing 4 µL of PCR dilution buffer (5× HF buffer containing 7.5 mM $MgCl_2$), 0.4 µL of dNTPs (10 mM each), 1.0 µL of each primer (10 pmol $µL^{-1}$), 0.2 µL of Phusion PCR polymerase, 12.4 µL sterile Milli-Q® water and 1 µL DNA template.
3. The PCR thermal cycling protocol includes an initial denaturation at 98°C for 3 min, followed by 35 cycles at 98°C for 30 sec, an annealing temperature of 60°C for 20 sec, and elongation at 72°C for 30 sec; final elongation.
4. Load PCR product (4 µL) onto the agarose gel (0.8%) and document positive samples (PCR product size 1.6 kbp).
5. PCR amplification product are digested using AluI and TseI according to the manufacturer's instructions. Digestions are performed in a volume of 15 µL, containing 1× restriction buffer (provided), 1 unit of the restriction enzyme (5 U $µL^{-1}$), and 7 µL of the PCR products and incubated at 37°C (AluI) and 65°C (TseI) for 1 h (or overnight).
6. For the electrophoresis of restriction fragments, 2% agarose is used. To facilitate restriction fragments separation gel electrophoresis is performed with low voltage (90 kV for a 15 cm wide gel) and for 2 h (pictures should be obtained every 30–60 min), see Fig. 6.11.

SOP 6.8.4   Notes

Both restriction enzymes are also available as fast digest (FD) enzymes to achieve digestion in 5–15 min under the specified conditions. Overnight digestion remains an option.

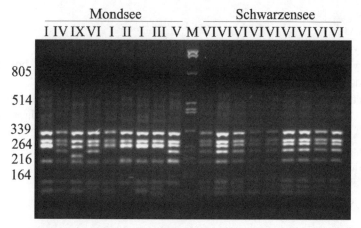

Figure 6.11 *Restriction type of mcyBA1-PCR products (1693 bp) from individual Plank-tothrix strains isolated from Lake Mondsee or Lake Schwarzensee, Salzkammergut area, Upper Austria. The roman numerals indicate different restriction profiles. M, λ PstI size marker. For a full description of results see Kurmayer and Gumpenberger(2006).*

As an alternative to single filament or strain isolation, DNA extracted from water samples (SOP 5.3) is used as a template and *mcy*BA1 PCR products obtained from samples are purified and cloned according to standard procedure (Sambrook and Russel, 2001), resulting in a so-called *mcy*BA1 clone library.

For PCR products obtained with Phusion polymerase (i.e. lacking A-overhangs generated by the Taq polymerase), a blunt end PCR cloning kit has to be used according to manufacturer's instructions. Following overnight growth of transformed colonies on the agar, the colonies carrying the insert fragment are picked randomly and re-amplified using the *mcy*BA1 primers and a Taq polymerase. These single *E. coli* clone PCR products are then digested following the conditions given above. By means of a classical sampling (rarefaction) curve, the restriction type diversity of a specific sample can be measured and quantified (e.g. Kurmayer and Gumpenberger, 2006).

SOP 6.8.5   References

Kurmayer, R. and Gumpenberger, M. (2006) Diversity of microcystin genotypes among populations of the filamentous cyanobacteria *Planktothrix rubescens* and *Planktothrix agardhii*, *Molecular Ecology*, **15** (12), 3849–3861.

Kurmayer, R., Dittmann, E., Fastner, J., and Chorus, I. (2002) Diversity of microcystin genes within a population of the toxic cyanobacterium *Microcystis* in Lake Wannsee (Berlin, Germany), *Microbial Ecology*, **43**, 107–118.

Sambrook, J. and Russell, D. (2001) *Molecular Cloning: A laboratory manual*, 3rd edn, Cold Spring Harbor Laboratory Press, Cold Spring Harbor, New York.

SOP 6.9

PCR Detection of Microcystin Biosynthesis Genes from Food Supplements

Vitor Ramos[1,2]*, *Cristiana Moreira*[1]*, and *Vitor Vasconcelos*[1,2]

[1]*Interdisciplinary Centre of Marine and Environmental Research (CIIMAR/CIMAR), University of Porto, Matosinhos, Portugal*
[2]*Faculty of Sciences, University of Porto, Porto, Portugal*

SOP 6.9.1   Introduction

Here we present a multiplex PCR-based protocol for the identification of *Microcystis* and simultaneous detection of toxigenic, microcystin-producing strains from different genera that may be present in "blue–green algae" (BGA) food supplements. This PCR technique was primarily developed by Saker *et al.* (2007) and consists of the use of three primer sets within a single PCR mixture to produce amplicons of varying sizes. This SOP is linked to SOP 2.4 and SOP 5.6. The SOPs are primarily intended to help out designing an effective plan for early warning prediction for food supplements quality and food safety assurance. See also Chapter 11 (Section 11.3.5) for additional rationale of these approaches. A flow diagram of the various steps is shown in Fig. 6.12. Reference strains are given in Table S6.1 (Appendix).

SOP 6.9.2   Experimental

SOP 6.9.2.1   Materials

- Isolated genomic DNA from food supplements samples (see SOP 5.6).
- DNA from a known microcystin producer strain (positive control; use the same isolate as that used to prepare the internal control for extraction efficiency, i.e. the spiked biomass in SOP 2.4).

*Corresponding author: vtr.rms@gmail.com

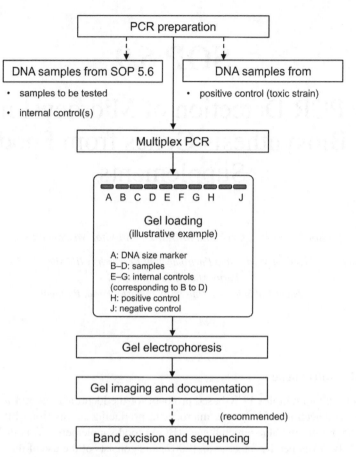

Figure 6.12 *Flow diagram outlining the procedural steps for the detection of microcystin synthetase genes in food supplements, by multiplex PCR (see SOP 6.9).*

- Water for molecular biology, e.g. sterile Milli-Q® water.
- Taq DNA polymerase and 10× (typically) PCR reaction buffer.
- $MgCl_2$ (25 mM).
- Deoxynucleotide triphosphates (dNTPs), 10 mM mix (2.5 mM each).
- Multiplex primer sets (see also Figure 6.13) (working solution at 10 µM, each primer):
  - Micr184F/Micr431R (220 bp) : Micr184F (5′–GCCGCRAGGTGAAAMCTAA–3′) and Micr431R (5′–AATCCAAARACCTTCCTCCC–3′), Neilan et al. (1997)
  - *mcy*A-Cd1F/*mcy*A-Cd1R (297 bp), (see Table 6.1) and
  - PCβF/PCαR (650 bp), (see Table 6.1)
- Filter tips.
- Eppendorf tubes and PCR tubes, DNase-free.
- PCR chamber (preferably).
- PCR-cooler and 0°C Benchtop cooler (cold blocks; preferable to an ice bath).
- Agarose.
- Molecular-weight size marker.

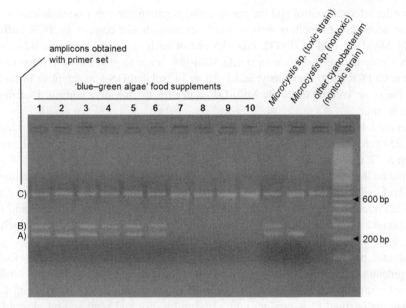

Figure 6.13 *Example of a 1.5% agarose gel run in 1× TAE buffer showing the PCR results obtained by a comparable approach described in this SOP. The primer set (A) was developed to amplify a* Microcystis-*specific fragment of the 16S rRNA gene with an expected size of 220 bp (Neilan et al. 1997); the primer set (B) was designed to amplify a 297 bp conserved fragment of mcyA, present in microcystin-producing species belonging to different genera (Hisbergues et al., 2003); and, the primer set (C) is for a 650 bp fragment of the phycocyanin operon, and it is used as positive control for the presence of cyanobacterial DNA (Neilan et al. 1995). Lanes 1–10 are samples from commercially obtained BGA food supplements of* Aphanizomenon flos-aquae *(1–6),* Nostoc flagelliforme *(7–8), and "Spirulina" (9–10). The last lane is a 100 bp DNA ladder. As can be observed, the three PCR products in lanes 1 to 6 very likely indicate that these samples were contaminated with a microcystin-producing* Microcystis *sp.*

- 1× TAE buffer.
- EtBr.

SOP 6.9.2.2   Equipment

- Thermocycler.
- Gel electrophoresis equipment and gel imaging system.
- UV transilluminator.
- Adjustable mechanical pipettes (0.5–10 μL, 2–20 μL, 20–200 μL, 100–1000 μL).

SOP 6.9.3   Procedure

1. Gently vortex and briefly centrifuge all solutions after thawing.
2. Prepare PCR reactions in a clean, UV irradiated and decontaminated environment. Preferably, make a master mix in an Eppendorf tube and then aliquot it to individual reaction tubes. Prepare (N + 1) × 20 μL reactions, being N the number of samples

plus the internal control and the positive and negative controls (the additional reaction is to account for pipetting errors). Each reaction should contain 1× PCR buffer, 2.5 mM MgCl$_2$, 1 mM of dNTPs mix, 0.5 µM of each of the primers, and 0.5 U of Taq DNA polymerase. Adjust with sterile Milli-Q® water to give a total volume of 20 µL for each PCR tube. Previously, add 1–10 ng of isolated DNA as template in each tube or the same volume of sterile Milli-Q® water for the negative control. Distribute the master mix.

3. Use the following PCR conditions: initial denaturation step at 95°C for 2 min followed by 35 cycles of 95°C for 90 sec, 56°C for 30 sec, 72°C for 50 sec, and a final extension step at 72°C for 7 min.

4. Separate the PCR products by electrophoresis on a 1.5% (w/v) agarose gel stained with EtBr. Load 5 µL of PCR product from each reaction and include a size marker, the spiked/internal control(s) and the positive and negative controls. Observe and photograph under UV transillumination. The result of a similar multiplex PCR approach using the same three primer sets here employed can be seen in Figure 6.13.

5. Optional, but recommended (see Section 11.3.3). Excise from gel and purify the DNA fragment(s) of interest (which typically is the band corresponding to the toxin synthetase gene) using the remaining volume of the PCR mixture on a new agarose gel. Ideally, clone the purified PCR product into a plasmid vector and send several plasmid DNA clones from a single sample for sequencing.

SOP 6.9.4   Notes

If possible, use a PCR chamber and cold blocks for tubes instead of an ice bath (the latter is more prone to contaminations).

The same procedure can be applied using primers developed for the detection and differentiation of other toxin biosynthesis genes (see Table 6.1). For instance, Saker *et al.* (2005) successfully used for the detection of hepatotoxin synthetase genes in food supplements the primer set HEPF/HEPR (Jungblut and Neilan, 2006), which beside a microcystin-related gene also amplifies a gene involved in the biosynthesis of nodularin.

Note that even in cases where the DNA is degraded or in the presence of contaminants (SOP 5.6), samples may still be suitable for downstream applications. PCR reactions can be expected to be successful if the smear extends above (i.e. includes fragments larger than) the expected amplicon size when visualizing the extracted total genomic DNA (0.8% agarose gel, w/v; see SOP 5.6) and provided that the DNA does not have any contaminants that act as inhibitors.

Contaminants (potential PCR inhibitors) can either be introduced during the isolation process or, on the contrary, be already present in the original sample. To circumvent this problem, additional steps can be undertaken to minimize their presence. For instance, try a PCR additive such as BSA (bovine serum albumin) or DMSO (dimethyl sulfoxide). Still, owing to the large number of aspects related to it (e.g. existence of different types of contaminants, which require different approaches), this is beyond the scope of this SOP.

Supplements may be biologically contaminated by one cyanobacterial species or by a mixture of species or genera. These cyanobacteria might not be toxic, but the sample purity is lost. Producers, quality inspectors, or researchers that aim to monitor/test the purity of the edible algae, in different stages of their life cycle, may want a more thorough characterization of the diversity present in the sample (i.e. achieve an identification at the species/genus level; see Chapter 4). Since samples are dominated by the edible algae, a partial or whole community analysis based on 16S rRNA gene sequencing needs to be followed (see Chapters 9 and 10 for different possibilities). In combination, a culture-dependent approach can also be used (see Chapter 3 and related SOPs), even though it has some limitations (e.g. viability and culturability of dried cyanobacteria might be difficult; resources and time needed).

SOP 6.9.5 References

Hisbergues, M., Christiansen, G., Rouhiainen, L., *et al.* (2003) PCR-based identification of microcystin-producing genotypes of different cyanobacterial genera, *Archives of Microbiology*, **180**, 402–410.

Jungblut, A.D. and Neilan, B.A. (2006) Molecular identification and evolution of the cyclic peptide hepatotoxins, microcystin and nodularin, synthetase genes in three orders of cyanobacteria, *Archives of Microbiology*, **185**, 107–114.

Neilan, B.A., Jacobs, D., and Goodman, A.E. (1995) Genetic diversity and phylogeny of toxic cyanobacteria determined by DNA polymorphisms within the phycocyanin locus, *Applied and Environmental Microbiology*, **61**, 3875–3883.

Neilan, B.A., Jacobs, D., Blackall, L.L., *et al.* (1997) rRNA sequences and evolutionary relationships among toxic and nontoxic cyanobacteria of the genus *Microcystis*, *International Journal of Systematic and Evolutionary Microbiology*, **47**, 693–697.

Saker, M.L., Jungblut, A.D., Neilan, B.A., *et al.* (2005) Detection of microcystin synthetase genes in health food supplements containing the freshwater cyanobacterium *Aphanizomenon flos-aquae*, *Toxicon*, **46**, 555–562.

Saker, M.L., Welker, M., and Vasconcelos, V.M. (2007) Multiplex PCR for the detection of toxigenic cyanobacteria in dietary supplements produced for human consumption, *Applied Microbiology and Biotechnology*, **73**, 1136–1142.

7

Quantitative PCR

Anne Rantala-Ylinen[1], Henna Savela[2], Kaarina Sivonen[1]*, and Rainer Kurmayer[3]*

[1] *Department of Food and Environmental Sciences, Division of Microbiology and Biotechnology, University of Helsinki, Helsinki, Finland*
[2] *Department of Biochemistry/Biotechnology, University of Turku, Turku, Finland*
[3] *Research Institute for Limnology, University of Innsbruck, Mondsee, Austria*

7.1 Introduction

Unlike conventional PCR (see Chapter 6), which shows a general presence of a toxin biosynthesis gene, qPCR allows determination of actual gene copy numbers present in the samples or of the proportions of potentially toxic cells in a population of cyanobacteria. In qPCR, amplification of the target sequence is followed real-time at each cycle via detection of fluorescent signal from dyes that are most often used as intercalating dyes (e.g. SYBR Green labeling). QPCR is based on the principle that the target sequence is doubled in each cycle and that the start of amplification reflects the amount of target sequence present: the more target sequence, the earlier the amplification exceeds a set threshold. The cycle when this happens is called a threshold cycle, C_T, or quantification cycle, C_q(-value). Detection in the exponential phase of PCR and use of fluorescent dyes makes qPCR sensitive and gives it a wide dynamic range and allows for detection of more than a twofold difference in DNA quantity between samples.

QPCR can be used to measure gene copy numbers and changes in levels of gene transcription. It is especially useful when studying sample series (e.g. seasonal dynamics of toxigenic cyanobacteria), effects of various environmental variables (e.g. nutrients or light conditions)

*Corresponding authors: kaarina.sivonen@helsinki.fi; rainer.kurmayer@uibk.ac.at

Molecular Tools for the Detection and Quantification of Toxigenic Cyanobacteria, First Edition.
Edited by Rainer Kurmayer, Kaarina Sivonen, Annick Wilmotte and Nico Salmaso.

on gene transcript levels, or following changes in transcription in different time points in growth experiments. Several assays for quantification of different microcystin biosynthesis genes (e.g. *mcy*A, *mcy*B, *mcy*D, *mcy*E) have already been established (examples listed in Table S7.1, Appendix), among which a few have been selected serving different purposes (see SOPs 7.3–7.7). However, only few assays have thus far been designed for the quantification of cylindrospermopsin biosynthesis genes (Campo *et al.*, 2013), saxitoxin biosynthesis genes (Savela *et al.*, 2015), and for anatoxin biosynthesis genes (Wang *et al.*, 2015). The most recent technique includes digital droplet PCR which quantitates a target DNA sequence based on PCR of a partitioned DNA sample. The number of PCR positive and PCR negative partitions is used to determine the absolute number of target DNA molecules (without a calibration curve). Experience using ddPCR is little available, but first comparisons with qPCR were achieved (Te *et al.*, 2015).

7.2 Primer/Probe Design

Primer design starts with the selection of the target gene, in this case a gene or genes responsible for toxin biosynthesis. Second, the specificity degree of the primers should be decided. They can be designed to amplify the target in only one producer genus/species or to amplify the same gene in several genera/species. Specificity of the primers is more crucial, if they are used to detect a single gene variant among diverse background DNA like in environmental samples than if they are used on a strain in culture experiments. Sequences of the target genes are available in public databases, such as the NCBI's GenBank. An alignment of the target gene sequences of different strains/species as well as similar gene sequences can be used to determine whether the gene is actually appropriate for primer design and then to identify suitable regions for primers. Specificity of the selected primer sequences should be checked *in silico* using (e.g. Primer-BLAST) and *in vivo* using reference strain material (Table S6.1, Appendix) in order to find out if they accidentally recognize other, non-target-gene regions or organisms. For the assays using SYBR Green or other intercalating dyes, designing of primers is enough, but for other chemistries, one or two probes are needed in addition to primers (Wong and Medrano, 2005). In the commonly used hydrolysis probe method, one probe is designed to attach to the amplified region and labeled on the 5′ end with a reporter dye and on the 3′ end with a quencher dye (see Bustin *et al.*, 2012).

Regardless of the detection chemistry, primers used in qPCR are designed to amplify a region as short as possible, typically 70–100 bp, of the target gene. Shorter amplicons are more efficiently amplified than longer ones and thus doubling of the amount of PCR product in each cycle is more probable to happen, resulting in reliable amplification (i.e. the amplification efficiency is 100%). Primers should be 15–25 bp long and have G+C content between 20–70%. Melting temperatures of the primers should be closely matched. Probe length is optimally under 30 bp and the melting temperature should be approximately 10°C higher than that of primers. Modified nucleotides such as locked nucleic acids (LNA) can be used to enhance specificity and stability of the binding of primers and probes. Minor groove binders (MGB) are attached to hydrolysis probes to stabilize binding and to increase their melting temperature (e.g. in AT-rich sequences). Many oligonucleotide synthesis companies offer to check the quality of the oligonucleotides before ordering (e.g. with regard to generation of primer dimers or stem-loop regions that would inhibit the binding of a probe to the target DNA). See Table 7.1 for a list of possible methodological phenomena and possible reasons.

Table 7.1 *Troubleshooting qPCR results*

Phenomenon	Possible Reason	Possible Solution
No amplification, not even in standard curve/positive control samples	PCR reaction not successful • Enzyme inactive • Missing or inappropriate primers/probes • Problem with the PCR mix	Repeat PCR reaction and carefully check inclusion of all reagents
PCR amplification in negative controls	• Contamination with template DNA • Fluorescent signal originates from primer dimers	• Check and re-prepare all reagents • Melting temperature of primer dimers is usually lower than that of correct amplicon
PCR amplification only in standard curve/positive control samples	PCR reaction successful, but: • No toxin genes present in the samples studied • PCR-inhibiting substances in the environmental samples studied	• Copy number of toxin gene under detection limit, reanalyze with less diluted template. • Dilute environmental DNA to lower concentration of inhibiting substance and repeat qPCR reaction.
PCR products of incorrect melting temperature (SYBR Green assays)	Unspecific PCR products or primer dimers	• Repeat PCR with higher annealing temperature • Design/select new primers for toxin biosynthesis genes • Optimize primer concentration used
Amplification curve not linear in the linear dynamic range	Efficiency of amplification reduced • Possibly owing to inhibiting substances present	• Dilute environmental DNA to lower concentration of inhibiting substance and repeat qPCR reaction. • Evaluate DNA extraction method to diminish inhibiting substances
High variation in C_q-values of replicate samples	• Unprecise pipetting of templates • Low assay efficiency causes random changes in amplification especially when target sequence is scarce	• Use larger volumes (>2 µL) to increase accuracy of pipetting and diminish proportional variation between replicates. • Optimize assay to increase efficiency and dynamic range of amplification

7.3   Optimization

Optimization is important for the proper performance of qPCR assays. It should be noted that, although using an assay described in the literature, its performance should be verified and optimized if needed, since differences in qPCR kits and PCR instruments used can affect the assay results. First things to be optimized are the annealing temperature of PCR, and primer and probe concentrations. The annealing temperature should be high enough to minimize formation of primer dimers and other unspecific products but low enough to allow efficient binding of primers. Optimal primer/probe concentrations are determined by testing several combinations of different concentrations of forward and reverse primers and the probe. The aim is to find primer/probe concentrations that show the highest signal magnitude and the minimum C_q-value. Specificity of the assay (i.e. amplification of only the target sequence) should be assessed by performing the qPCR using DNA from target and non-target cyanobacterial strains (Table S6.1, Appendix). The selection of strains depends on the assay (e.g. if designing a *Microcystis mcyE*-specific assay) test strains should include microcystin-producing and non-producing *Microcystis* strains as well as microcystin/nodularin-producing strains of other genera. However, sequencing of the amplicons is needed to make sure that the correct target sequence is amplified. Sensitivity of the qPCR assay (i.e. how the presence of competitive DNA affects the PCR performance) can be tested (e.g. by adding varying amounts of DNA from another strain(s) that have similar gene sequences to the target gene sequence). In each reaction, the amplification should start at the same C_q-value, if the competing DNA does not interfere with the amplification of the target.

An ideal qPCR assay is efficient and has a wide dynamic range (i.e. the target sequence amount is doubled in each cycle and the efficiency is not affected by the starting amount of target sequence). Thus, on a semi-logarithmic scale there is a linear relationship between the initial amount of DNA and C_q-values. Efficiency and dynamic range can be assessed by analyzing serially (e.g. tenfold) diluted samples and generating a standard curve. The slope of the standard curve reflects the efficiency of the assay. For a graph where C_q-value is on the y-axis and log (DNA amount in the template) on the x-axis, the slope of -3.322 equals 100% efficiency, while for a graph where $\log_{10}$ (DNA amount in the template) is on the y-axis and C_q-value on the x-axis, the slope value is -0.301. The dynamic range should cover the whole concentration scale of unknown samples.

7.4   Absolute Quantification

In DNA studies, absolute quantification of target sequence copy numbers is often used. This is achieved by comparing amplification (C_q-value) of the sample studied to amplification of standards, the quantity of which is known. C_q-values of standard dilutions are used to draw a standard curve, and the amount of target gene copies in the samples studied is calculated using the equation of the standard curve. Standard dilutions can be made from genomic DNA, a plasmid containing the target, or a PCR amplicon. Reliable quantification requires that the efficiency of amplification is the same both in the unknown samples and the standards (e.g. sample impurities often present in the environmental samples can inhibit polymerase function and lead to errors in copy number estimates). In addition, the C_q-values of unknown samples should be in the range of standard dilutions. If the C_q-values are too small, the samples should be diluted and if they are too high, a larger volume of the sample should be used.

7.5 Relative Quantification

In order to quantify (non)toxigenic subpopulations, primers specified to amplify both the total population of a species as well as part of the toxigenic subpopulation have been designed (e.g. Kurmayer and Kutzenberger, 2003). By using reference genes (e.g. phycocyanin (PC)-IGS (intergenic spacer region) or 16S rRNA), this approach also allows methodological artefacts (e.g. inhibition of PCR reaction due to unknown compounds co-isolated with DNA) or physiological variation (e.g. in the number of gene copies per cell) to be controlled for. Another possibility is to use the ΔC_q approach based on a multiplex qPCR targeting one gene involved in the toxin production (e.g. *mcy* gene) and one gene present in all the cells (e.g. encoding phycocyanin, or 16S rRNA). To start, standard curves are set up on several strains for the two genes (e.g. phycocyanin (PC-IGS) and *mcy*B) by relating the known quantity of DNA to the C_q-value. Then, the resulting ΔC_q between the two genes (e.g. $\Delta C_q = C_q$ for the PC gene -C_q for *mcy*B) is deduced from the regression equations. Finally, an equation established from the ΔC_q values allows estimating the percentage of *mcy*+ cells in the total population of cyanobacteria. The validation of this ΔC_q equation is based on the comparison of a theoretical ΔC_q equation with an experimental equation obtained on samples containing a mix of MC-producing strain and non-MC-producing strain (e.g. Briand *et al.*, 2009).

Further, relative quantification is used in gene expression studies, and the quantity of the target mRNA (cDNA) is expressed as an x-fold change compared to a calibrator sample that is considered a onefold expression. The calibrator sample is, for example, a time point zero sample in time course studies or an untreated sample. Quantities in unknown samples can be determined using either a standard curve or calculated using the C_q-values of unknown and calibrator samples, for example by comparative C_q-value method (Livak and Schmittgen, 2001).

7.6 Calibration of qPCR Results

In order to translate C_q-values to real numbers standard curves relating the DNA amount in the template to the C_q-value are required (1) to quantify the target genotype in equivalents of gene copies or cell numbers and (2) to determine the specificity and sensitivity of the qPCR assay. In general, standard curves based on pre-determined DNA amount in the template are established by relating the known DNA concentrations to the C_q-value of the diluted samples. Using single copy genes the number of gene (genome) copies per cell can be calculated using pre-established standard curves (Kurmayer and Kutzenberger, 2003). It is probably most accurate using a plasmid carrying the gene fragment of interest (*ca.* 100 bp) and relating the template DNA concentration (0.1 pg–1 ng) to the obtained C_q-value. Another possibility is PCR amplification of the qPCR amplification product (using a proof-reading DNA polymerase) and using the purified PCR product as a standard. Standard curves can also be prepared using genomic DNA of a toxin-producing strain (see SOP 7.3). Finally, using standard curves by relating cell concentrations to the C_q-value, it is possible

to quantify genotypes in cell numbers (e.g. cells of microcystin genotypes in a given volume of lake water).

7.7 General Conclusions

For a long time the qPCR technique has been the only quantitative technique available. Experiences using more recent techniques such as digital droplet PCR are not widely available. In qPCR-based techniques the semi-logarithmic standard curves cause limitations with regard to the accuracy in estimating genotype numbers and proportions, because minor deviations on a linear scale (C_q-values) are translated into larger deviations on a logarithmic scale (DNA amount in the template). As a rule of thumb, deviations of < 0.5 cycles from a specific C_q-value are due to experimental noise. This unspecific variation must be taken into account when translating the results on C_q-values to real (absolute and relative) numbers.

7.8 References

Briand, E., Escoffier, N., Straub, C., *et al.* (2009) Spatiotemporal changes in the genetic diversity of a bloom-forming *Microcystis aeruginosa* (cyanobacterium) population, *The ISME Journal*, **3**, 419–429.

Bustin, S.A., Zaccara, S., and Nolan, T. (2012) An introduction to the real-time polymerase chain reaction, in M. Filion (ed.), *Quantitative Real-Time PCR in Applied Microbiology*, Caister Academic Press, Norfolk, UK, **239**, 3–26.

Campo, E., Lezcano, M.Á., Agha, R., *et al.* (2013) First TaqMan assay to identify and quantify the cylindrospermopsin-producing cyanobacterium *Aphanizomenon ovalisporum* in water, *Advances in Microbiology*, **3**, 430–437.

Kurmayer, R. and Kutzenberger, T. (2003) Application of real-time PCR for quantification of microcystin genotypes in a population of the toxic cyanobacterium *Microcystis* sp., *Applied and Environmental Microbiology*, **69**, 6723–6830.

Livak, K.J. and Schmittgen, T.D. (2001) Analysis of relative gene expression data using real-time quantitative PCR and the 2(-Delta Delta C(T)) Method. *Methods (San Diego, Calif.)*, **25**, 402–8. doi: 10.1006/meth.2001.1262.

Savela, H., Spoof, L., Perälä, N., *et al.* (2015) Detection of cyanobacterial *sxt* genes and paralytic shellfish toxins in freshwater lakes and brackish waters on Åland Islands, Finland, *Harmful Algae*, **46**, 1–10.

Te, S.H., Chen, E.Y. and Gin, K.Y.-H. (2015) Comparison of quantitative PCR and droplet digital PCR multiplex assays for two genera of bloom-forming cyanobacteria, *Cylindrospermopsis* and *Microcystis*, *Applied and Environmental Microbiology*, **81**, 5203–5211.

Wang, S., Zhu, L., Li, Q., *et al.* (2015) Distribution and population dynamics of potential anatoxin-a-producing cyanobacteria in Lake Dianchi, China, *Harmful Algae*, **48**, 63–68.

Wong, M.L. and Medrano, J.F. (2005) Real-time PCR for mRNA quantitation, *BioTechniques* **39**, 75–85.

SOP 7.1

Optimization of qPCR Assays

*Rainer Kurmayer**

Research Institute for Limnology, University of Innsbruck, Mondsee, Austria

SOP 7.1.1   Introduction

This protocol describes how to (1) optimize reaction efficiency by adjusting primer (probe) concentrations, (2) determine the reaction efficiency, and (3) test the reproducibility of results obtained through qPCR experiments. Potential reference strains are given in Table S6.1 (Appendix).

SOP 7.1.2   Experimental

SOP 7.1.2.1   Materials

- qPCR master mix, e.g. TaqMan® Universal PCR Master Mix (Thermo Fisher Scientific) or iTaq™ Universal Probes Supermix (Bio-Rad).
- Primers and probes (stock solutions 100 µM).
- Sterile Milli-Q® water.

SOP 7.1.2.2   Equipment

- A qPCR instrument, e.g. ABI7300 instrument (Applied Biosystems).
- Centrifuge with a plate rotor/adaptor.
- Adjustable mechanical pipettes (100–1000 µL, 20–200 µL, 1–10 µL).

*Corresponding author: rainer.kurmayer@uibk.ac.at

SOP 7.1.3   Procedure

1. Prepare a standard DNA template dilution that leads to a C_q-value located in the linear dynamic range ($C_q \sim 20-25$).
2. Optimize primer concentrations, testing several combinations of different concentrations of forward and reverse primers (Table 7.2).
3. Prepare a PCR mix as indicated in SOP 7.3. For one 25 μL reaction, pipet 12.5 μL of 2× TaqMan® Universal PCR Master Mix or 2× iTaq Supermix (final concn. 1×), 5.5 μL of sterile Milli-Q® water, variable volume (μL) of both primers (final concn. 50, 300, 900 nM = fmol μL^{-1}) and 0.5 μL of probe (250 nM). Aliquot 20 μL of the PCR master mix to each of the PCR tubes/PCR plate well and add 5 μL of standard dilutions and negative control. Measurements are performed in triplicate.
4. Determine the minimum forward/reverse primer concentration to achieve the minimum C_q-value with high reproducibility. Using this concentration, determine the optimal probe concentration by testing 50–250 nM in 50 nM intervals and determine the minimum probe concentration to achieve the minimum C_q-value with high reproducibility.
5. The optimized forward/reverse primer and probe concentrations are used to establish standard dilution series in order to determine the amplification efficiency (Fig. 7.1).

Table 7.2   *Final forward/reverse primer concentrations (all in three replicates) to be compared to achieve the minimum C_q-value using standard dilutions*

Forward/Reverse Primer (nM)	50	300	900
50	50/50	50/300	50/900
300	300/50	300/300	300/900
900	900/50	900/300	900/900

(A)

Log$_{10}$ copies in template	mean C_q-value	ΔC_q
7.48	17.74 ± 0.07	
6.48	21.04 ± 0.07	3.30
5.48	24.55 ± 0.03	3.51
4.48	27.97 ± 0.1	3.42
3.48	31.39 ± 0.21	3.42
2.48	35.13 ± 0.82	3.74
1.48	37.22 ± 0.67	2.09

(B)

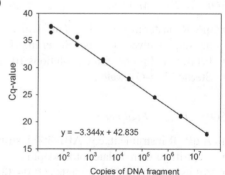

Figure 7.1   *(A) Example of amplification using a purified PC-IGS fragment (66 bp) from serially (tenfold) diluted samples (mean +/− standard deviation), and (B) calculation of a linear standard curve. The slope of the standard curve reflects the efficiency of the assay. The amplification efficiency (E) is calculated from* $E = (10^{-1/slope} - 1) \times 100$, *a slope of −3.322 equals 100% efficiency (Becker et al., 2000). Note that amplification efficiency decreases below 100 copies per template (data from Schober et al., 2007).*

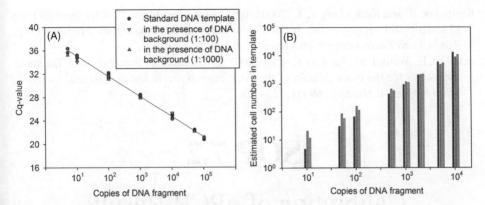

Figure 7.2 *Testing the influence of background DNA on the robustness of estimates on (A) C_q-values obtained through a DNA dilution series; (B) cell numbers estimated from the DNA dilution series (note: the DNA background was extracted from a field sample, 4th of May, 2000 and contained theoretically 34 cells of* Aphanizomenon *sp., 312 cells of* Limnothrix *sp., 116 cells of* Oscillatoria redeckei, *0.3 cells of* Microcystis *sp., and 55 cells of* Planktothrix agardhii *in the template) (from Kurmayer and Kutzenberger, 2003).*

6. The same dilution series is then tested both in the absence or presence of a background DNA. Typically, background is added at a dilution of 1:100 or 1:1000 and the measurements are compared to controls without background. The background is either prepared from environmental DNA free of the target (toxigenic) genotype or from DNA of most closely related strains that do not have the target genotype.

7. To quantify the influence of background effects, the number of cells or gene copies estimated in the presence of a natural background is divided by the cell number or gene copies estimated in the absence of a natural background. The more the ratio deviates from one, the stronger the background effects.

SOP 7.1.4 Notes

Quantification errors typically occur toward both ends of the calibration curve. Consequently, for field sample analysis measurements should be directed toward the central region of the standard curve (C_q-value ~20–30), which is found to be more resistant against background effects.

Owing to the linear-log calibration curves, the technique is inherently sensitive to variations in the slope induced by minor (uncontrollable) variations of the C_q-values. Thus deviation from theoretical values also has to be validated in logarithmic terms. For example, the deviation in cell number calculation from the theoretical value from the example given in Fig. 7.2 ranges between 87 and 198% (background 1:100) or 100–112% (background 1:1000). However, if analyzed and interpreted on logarithmic scale the cell number variation is accurately matched (see also Kurmayer and Kutzenberger, 2003).

SOP 7.1.5 References

Becker, S., Böger, P., Oehlmann, R., and Ernst, A. (2000) PCR bias in ecological analysis: A case study for quantitative Taq nuclease assays in analyses of microbial communities, *Applied and Environmental Microbiology*, **66**, 4945–4953.

Kurmayer, R. and Kutzenberger, T. (2003) Application of real-time PCR for quantification of microcystin genotypes in a population of the toxic cyanobacterium *Microcystis* sp., *Applied and Environmental Microbiology*, **69**, 6723–6730.

Schober, E., Werndl, M., Laakso, K. *et al.* (2007) Interlaboratory comparison of Taq nuclease assays for the quantification of the toxic cyanobacteria *Microcystis* sp., *Journal of Microbiological Methods*, **69** (1), 122–128.

SOP 7.2

Calibration of qPCR Results

*Rainer Kurmayer**

Research Institute for Limnology, University of Innsbruck, Mondsee, Austria

SOP 7.2.1   Introduction

Standard curves are required (1) to quantify the target genotype in absolute numbers (gene copies, cell numbers) and (2) to determine the specificity and sensitivity of a specific qPCR assay. For each gene locus a standard curve based on pre-determined DNA amount is established by relating the known DNA amount concentration to the C_q-value obtained from the diluted sample standard. The various steps of qPCR calibration using either DNA aliquots of cells or gene copies per template are shown in Fig. 7.3. Potential reference strains are given in Table S6.1 (Appendix).

SOP 7.2.2   Experimental

SOP 7.2.2.1   Materials

- qPCR master mix, e.g. TaqMan® Universal PCR Master Mix (Thermo Fisher Scientific) or iTaq™ Universal Probes Supermix (Bio-Rad).
- Primers and probes (stock solutions 100 μM).
- Sterile Milli-Q® water.

*Corresponding author: rainer.kurmayer@uibk.ac.at

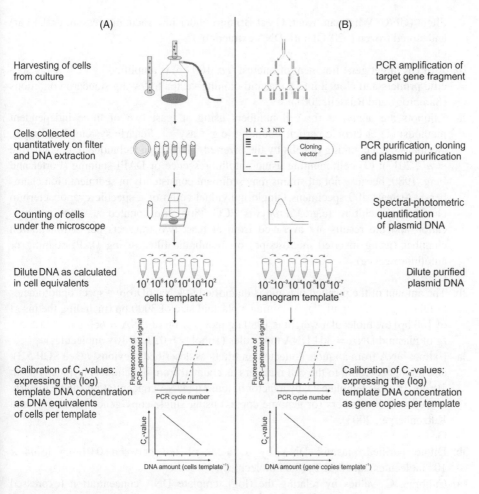

(A)

Harvesting of cells
from culture

Cells collected
quantitatively on filter
and DNA extraction

Counting of cells
under the microscope

Dilute DNA as calculated
in cell equivalents

$10^7 10^6 10^5 10^4 10^3 10^2$
cells template^{-1}

Calibration of C_q-values:
expressing the (log)
template DNA concentration
as DNA equivalents
of cells per template

Fluorescence of
PCR-generated signal

PCR cycle number

C_q-value

DNA amount (cells template^{-1})

(B)

PCR amplification of
target gene fragment

M 1 2 3 NTC

Cloning
vector

PCR purification, cloning
and plasmid purification

Spectral-photometric
quantification
of plasmid DNA

Dilute purified
plasmid DNA

$10^{-2} 10^{-3} 10^{-4} 10^{-5} 10^{-6} 10^{-7}$
nanogram template^{-1}

Calibration of C_q-values:
expressing the (log)
template DNA concentration
as gene copies per template

Fluorescence of
PCR-generated signal

PCR cycle number

C_q-value

DNA amount (gene copies template^{-1})

Figure 7.3 *Flow diagram showing the various steps of qPCR calibration (see SOP 7.2). (A) DNA can be quantified either as aliquots of cells per template or (B) as gene copies per template.*

SOP 7.2.2.2   Equipment

- A qPCR instrument, e.g. ABI7300 instrument (Applied Biosystems).
- Centrifuge with a plate rotor/adaptor.
- Adjustable mechanical pipettes (100–1000 μL, 20–200 μL, 1–10 μL).

SOP 7.2.3   Procedure

1a. A pure culture of a cyanobacterium species is grown aseptically in batch culture in a nutrient medium at 20°C and saturated light conditions and harvested during the logarithmic growth phase. Cells are harvested by filtering a known volume onto glass fiber

filters (GF/C, Whatman, Kent, Great Britain) under low vacuum pressure (<0.4 bar) and stored frozen (−20°C) until DNA extraction.
Or

1b. A qPCR target gene fragment of interest (*ca.* 100 bp) is amplified by PCR using specific primers and cloned into a plasmid-cloning vector following standard conditions (Sambrook and Russell, 2001).

2a. Aliquots are analyzed for cell numbers using at least two of three independent methods: (1) electronic particle counting (e.g. Casy® 1, Schärfe system, Reutlingen, Germany); (2) counting of cells by the inverted microscope technique (e.g. Catherine *et al.*, 2017); (3) cell counting using autofluorescence or DAPI staining (Porter and Feig, 1980; because not all strains may sediment consistently in sedimentation chambers). At least 400 specimens (single individual cells) of a specific cyanobacterium clonal strain culture (e.g. *Microcystis* PCC 7806) are counted at 400× magnification and the results are averaged from at least two transects per sedimentation chamber (using inverted microscope) or membrane filter (using DAPI staining or autofluorescence).
Or

2b. The amount of the purified plasmid is quantified in a NanoDrop™ spectrophotometer (1 $OD_{260\,nm}$ = 50 ng μL^{-1}). Assuming a plasmid size of 3000 bp (including the insert of 100 bp) the molecular weight is 1980 kg mol^{-1} (1 bp of DNA = 660 g mol^{-1}) and 1 fg of plasmid DNA = 304 DNA molecules (1 mol = 6.023 × 10^{23} molecules).

3a. Extract DNA from aliquots collected on filters as described previously (see SOP 5.3), and dilute according to the cell numbers as counted from the aliquots (typically from 10^2 to 10^7 cells template^{-1}), by a factor of ten. The DNA concentration can be linearly related to cell numbers (or genome copies) using single copy genes (Kurmayer and Kutzenberger, 2003).
Or

3b. Dilute purified plasmid DNA by a factor of ten between 0.01 ng (3.04 × 10^6 molecules) and 0.01 fg (3.04 molecules).

4a. Calibrate C_q-values by relating the (log) template DNA concentration (expressed as equivalents of cells/template) to the obtained C_q-value using a linear regression curve. The slope of the linear regression curve should be close to −3.33 equal to 100% amplification efficiency. Amplification efficiencies (E) are calculated from $E = (10^{-1/slope} - 1) \times 100$.
Or

4b. Calibrate C_q-values by relating the (log) template DNA concentration (expressed as DNA molecules = gene copies/template) to the obtained C_q-value using a linear regression curve. The slope of the linear regression curve should be close to −3.33 equal to 100% amplification efficiency. Amplification efficiencies (E) are calculated from $E = (10^{-1/slope} - 1) \times 100$.

5a. The estimated number of cells in the PCR template finally is related to the original number of cells in the DNA extract sample (considering the dilution factor and filtration volume and original cell density) resulting in units of cells ml^{-1}.
Or

5b. The estimated number of gene copies in the PCR template finally is related to the original number of gene copies in the DNA extract sample (considering the dilution factor and filtration volume) resulting in units of gene copies ml^{-1}.

SOP 7.2.4 Notes

The specificity and the robustness of each qPCR assay should be tested by measuring the same dilution series for calibration in the presence/absence of background DNA. The background is either prepared from environmental DNA or from DNA of most closely related strains that do not have the target gene. Typically, background is added at a dilution of 1:100 or 1:1000 and the measurements are compared to controls without background.

Frequently the counting of individual cells from filamentous cyanobacteria (e.g. *Planktothrix*) is rather difficult as cell separations are hardly visible. One possibility is to estimate the length of a filament within the boundaries of the grid and calculate the cellular biovolume (e.g. Lawton *et al.*, 1999).

SOP 7.2.5 References

Catherine, A., Maloufi, S., Congestri, R. *et al.* (2017) Cyanobacterial samples: Preservation, enumeration and biovolume measurements, in: J. Meriluoto, L. Spoof, and G.A. Codd (eds), *Handbook on Cyanobacterial Monitoring and Cyanotoxin Analysis*. John Wiley & Sons, Ltd, Chichester, UK.

Lawton, L., Marsalek, B., Padisak, J., and Chorus, I. (1999) Determination of cyanobacteria in the laboratory, in I. Chorus and J. Bartram (eds), *Toxic Cyanobacteria in Water. A guide to their public health consequences, monitoring and management*, WHO, E and FN Spon, London, 347–368.

Porter, K. and Feig, Y. (1980) The use of DAPI for identifying and counting aquatic microflora, *Limnology and Oceanography*, **25**, 943–948.

Sambrook, J. and Russell, D. (2001) Chapter 1, *Molecular Cloning: A laboratory manual*, 3rd edn, Cold Spring Harbor Laboratory Press, Cold Spring Harbor, New York, NY.

SOP 7.3

Quantification of Potentially Microcystin/Nodularin-Producing *Anabaena*, *Microcystis*, *Planktothrix*, and *Nodularia*

Anne Rantala-Ylinen[1], Kaarina Sivonen[1]*, and Rainer Kurmayer[2]*

[1]*Department of Food and Environmental Sciences, Division of Microbiology and Biotechnology, University of Helsinki, Helsinki, Finland*
[2]*Research Institute for Limnology, University of Innsbruck, Mondsee, Austria*

SOP 7.3.1 Introduction

The qPCR assays described below are intended for quantification of potential microcystin-producing *Anabaena*, *Microcystis*, and *Planktothrix* and nodularin-producing *Nodularia* in environmental water samples via determination of *mcy*E, *mcy*B, or *nda*F gene copy numbers in the extracted nucleic acid samples. Quantification is performed using an external standard curve prepared from genomic DNA of a microcystin-producing *Anabaena*, *Microcystis*, or *Planktothrix*, or a nodularin-producing *Nodularia* strain. The method for *Anabaena*- and *Microcystis-mcy*E detection was originally published by Sipari *et al.* (2010) and the protocol below was modified from Rantala-Ylinen *et al.* (2011). The protocol for *Planktothrix-mcy*B was originally published by Ostermaier and Kurmayer (2009), and the protocol for quantification of *Nodularia-nda*F by Koskenniemi *et al.* (2007). Potential reference strains are given in Table S6.1 (Appendix).

---

*Corresponding author: rainer.kurmayer@uibk.ac.at; kaarina.sivonen@helsinki.fi

SOP 7.3.2 Experimental

SOP 7.3.2.1 Materials

- qPCR master mix, e.g. TaqMan® Universal PCR Master Mix with Uracil N-glycosylase (UNG) (Thermo Fisher Scientific) or iTaq™ Universal Probes Supermix (Bio-Rad) or LightCycler FastStart DNA Master SYBR Green I (Roche).
- Primers and probes (stock solutions 100 μM).
- *Anabaena-mcyE* (TaqMan assay):
  - 611F (5′–CTAGAGTAGTCACTCACGTC–3′) and
  - 737R (5′–GGTTCTTGATAGTTAGATTGAGC–3′);
  - probe 672P (5′–FAM–CAAGTTCCCACAATTCTTGGATTAGCAGC–TAMRA–3′)
- *Microcystis-mcyE* (TaqMan assay):
  - 127F (5′–AAGCAAACTGCTCCCGGTATC–3′) and
  - 247R (5′–CAATGGGAGCATAACGAGTCAA–3′);
  - probe 186P (5′–FAM–CAATGGTTATCGAATTGACCCCGGAGAAAT–TAMRA–3′).
- *Planktothrix-mcyB* (TaqMan assay):
  - *mcy*BA1fwd (5′–ATTGCCGTTATCTCAAGCGAG–3′) and
  - *mcy*BA1rev (5′–TGCTGAAAAAACTGCTGCATTAA–3′);
  - TaqMan probe (5′– FAM–TTTTTGTGGAGGTGAAGCTCTTTCCTCTGA–TAMRA–3′).
- *Nodularia-nda*F (SYBR Green assay):
  - *nda*F8452 (5′–GTGATTGAATTTCTTGGTCG–3′) and
  - *nda*F8640 (5′–GGAAATTTCTATGTCTGACTCAG–3′).
- Isolated nucleic acids of laboratory strains and environmental samples.
- Sterile Milli-Q® water.
- For TaqMan assays, optical PCR reaction plates/tubes, and adhesive covers/caps compatible with the instrument used.
- For LightCycler assays, glass capillaries, plastic caps, a capping tool, and centrifuge adapters.

SOP 7.3.2.2 Equipment

- A qPCR instrument, e.g. the ABI7300 instrument (Applied Biosystems) or the LightCycler 2.0 carousel (Roche).
- Centrifuge with a plate rotor/adaptor.
- Adjustable mechanical pipettes (100–1000 μL, 20–200 μL, 1–10 μL).

SOP 7.3.3 Procedure

1. Prepare standard solutions by making a dilution series containing 10^1, 10^2, 10^3, 10^4, 10^5, 10^6 copies of *Anabaena-mcyE* or *Microcystis-mcyE* gene, *Planktothrix-mcyB*, or *Nodularia-nda*F gene. The gene copy numbers could be estimated following SOP 7.2. Solutions are prepared using genomic DNA of a microcystin-producing *Anabaena*, *Microcystis*, or *Planktothrix* or nodularin-producing *Nodularia* strain, respectively. Prepare enough dilution for at least three replicate qPCR reactions and for making the next dilution solution, e.g. 5 μL of the 10^5-dilution + 45 μL of sterile Milli-Q® water to get a 10^6-dilution. The standard dilution series should be prepared just

before performing the qPCR. To avoid DNA degradation and change in the standard concentration, do not freeze-thaw the dilutions.

2. Adjust the amount of primers and probes to 10 µM.

3. Dilute environmental DNA samples with sterile Milli-Q® water, e.g. minimum 1:100. If the C_q-values of the environmental samples are not within limits set by the standard curve, change the dilution factor accordingly.

4. Set up the qPCR instrument for an assay using absolute quantification according to the manufacturer's instruction. Fill in information on sample identity and quantity of standard samples. Select the correct detector according to the fluorescent dye of the probe.

5. Use the following program for PCR cycling with TaqMan® Universal PCR Master Mix: UNG incubation at 50°C, polymerase activation, and DNA denaturation at 95°C for 10 min, 40 cycles: 95°C for 15 sec, 62°C (60°C when using *Planktothrix mcy*BA1) for 60 sec. If using iTaq™ Universal Probes Supermix, no UNG incubation step is needed and the initial polymerase activation and DNA denaturation step is shorter: 3 min at 95°C. Set data collection to happen at annealing/extension step. Change the sample volume to 25 µL (default value in the ABI7300 instrument is 50 µL).

   For the SYBR Green assays with the LightCycler, use the following program: denaturation at 95°C for 10 min (heating rate 20°C sec^{-1}); amplification of 45 cycles: 95°C for 0 sec, 63°C for 5 sec, and 72°C for 8 sec (heating rate 20°C sec^{-1}), and melting curve analysis from 58°C to 95°C (heating rate of 0.1°C sec^{-1}). During amplification, fluorescence is measured at the end of each cycle at 72°C and during melting curve analysis continuously through channel F1 (530 nm).

6. Pipette the following mixture:

   For *mcy*-gene quantification (*Anabaena-mcy*E, *Microcystis-mcy*E, *Planktothrix-mcy*B). Prepare a PCR mix for the samples studied, standard dilutions, and negative control (sterile Milli-Q® water), all in three replicates. Prepare some extra mix to cover for pipetting losses. For one 25 µL reaction, pipet 12.5 µL of 2 × TaqMan® Universal PCR Master Mix or 2× iTaq Supermix (final concn. 1×), 0.3 µM of both primers (0.9 µM when using *Planktothrix mcy*BA1) and 0.2 µM probe (0.1 µM when using *Planktothrix mcy*BA1). Aliquot 20 µL of the mix to each of the PCR tubes/PCR plate well and add 5 µL of samples, standard dilutions and negative control.

   For the *Nodularia-nda*F assay, prepare similarly a PCR mix including 1 µL LightCycler FastStart DNA Master SYBR Green I mix, 3.5 mM MgCl$_2$, and 0.35 µM both primers in a 10 µL total volume. Aliquot 8 µL of the PCR mix into capillaries and add 2 µL of template DNA, standard dilutions, and negative control. Centrifuge briefly the solution to the bottom of capillaries using the specific adapters.

7. After the PCR run is completed, analyze the amplification plot according to the manufacturer's instructions for absolute quantification. The ABI7300 System SDS program as well as the LightCycler software 4.05 draws the standard curve and calculates the gene copy numbers in samples studied automatically if copy numbers of standards are given in the sample sheet. Alternatively, copy numbers can also be determined by drawing the standard curve in a spreadsheet using the C_q-values of standards, and then using the equation of the curve to calculate copy numbers that correspond to C_q-values of the samples studied. Inspect the amplification plot also to evaluate the amplification quality. Signs of successful PCR include amplification of the standards in a manner similar to

that observed in previous optimization steps, amplification curves look normal (i.e. linear but not sigmoidal at logarithmic scale), and amplification curves of the replicate sample are clustered. Check also that no amplification has occurred in negative control wells, which might indicate contamination. For the analysis of melting curves, see introduction above.

8. The qPCR assay determines only the gene copy number present in each PCR tube. The copy numbers present originally in the environmental samples can be estimated by taking into account the dilution factor (e.g. 1:100), volume of the extracted DNA (e.g. 100 µL), and volume of the water sample (e.g. 200 mL) used for DNA extraction. For example, if 10^4 copies of *Microcystis-mcyE* gene was detected (5 µL of 1:100 dilution), then 10^6 copies were in 5 µL of the original DNA, 20×10^6 copies ($= 10^6 \times 100$ µL/5 µL) were in the whole DNA extraction, and 10^5 copies ($= 20 \times 10^6$ copies/200 mL) were present in 1 mL of the original water sample. Since *mcyE* is a single-copy gene, it can be concluded that 10^5 potential microcystin-producing *Microcystis* cells were present in 1 mL of the water sample studied.

SOP 7.3.4   Notes

In general, all measurements are performed in triplicate (three independent PCRs for aliquots of the same DNA template).

Increased quantification errors typically occur toward both ends of the calibration curve. For field sample analysis cell quantification should be directed toward the central region of the standard curves, which are found to be most resistant against uncontrolled factors.

Chemistry and instruments used may affect the performance of the assays. Thus, proper performance should be tested and optimized if needed before sample analysis.

SOP 7.3.5   References

Koskenniemi, K., Lyra, C., Rajaniemi-Wacklin, P., J *et al.* (2007) Quantitative real-time PCR detection of toxic *Nodularia* cyanobacteria in the Baltic Sea, *Applied and Environmental Microbiology*, **73**, 2173–2179.

Ostermaier, V. and Kurmayer, R. (2009) Distribution and abundance of nontoxic mutants of cyanobacteria in lakes of the Alps, *Microbial Ecology*, **58** (2), 323–333, doi: 10.1007/s00248-009-9484-1.

Rantala-Ylinen, A., Sipari, H., and Sivonen, K. (2011) Molecular methods: Chip assay and quantitative real-time PCR: In detecting hepatotoxic cyanobacteria, *Microbial Toxins – Methods in Molecular Biology*, **739**, 73–86, doi: 10.1007/978-1-61779-102-4_7.

Sipari, H., Rantala-Ylinen, A., Jokela, J. *et al.* (2010) Development of a chip assay and quantitative PCR for detecting microcystin synthetase E gene expression, *Applied and Environmental Microbiology*, **76**, 3797–3805.

SOP 7.4

Relative Quantification of *Microcystis* or *Planktothrix mcy* Genotypes Using qPCR

*Rainer Kurmayer**

Research Institute for Limnology, University of Innsbruck, Mondsee, Austria

SOP 7.4.1   Introduction

By using reference genes (e.g. encoding phycocyanin-IGS, or 16S rRNA) in qPCR, this SOP allows (1) to control for methodological influences (e.g. instrumentation, or inhibition of PCR reaction due to unknown compounds co-isolated with DNA) or physiological variation (e.g. in the number of gene copies per cell), and (2) to quantify the proportion of a specific toxigenic subpopulation. By the latter technique important information on the environmental variability or stability on the quantitative presence of a specific gene indicative of toxin production can be obtained (e.g. Kurmayer and Kutzenberger, 2003). For the latter, standard curves relating cell concentrations to the C_q-value are required (1) to quantify the target genotype in equivalents of cell numbers and also (2) to determine the specificity and sensitivity of a specific qPCR (e.g. Taq nuclease) assay. The quantified genotype numbers can be compared with cell numbers counted in the microscope during field observational studies (see also SOP 7.2). Potential reference strains are given in Table S6.1 (Appendix).

SOP 7.4.2   Experimental

SOP 7.4.2.1   Materials

- qPCR master mix, e.g. TaqMan® Universal PCR Master Mix (Thermo Fisher Scientific).
- Primers and probes (stock solutions 100 µM) (Table 7.3).
- Sterile Milli-Q® water.

---

*Corresponding author:rainer.kurmayer@uibk.ac.at

Table 7.3 *Oligonucleotide primers used for PCR amplification of the mcyB gene indicative of microcystin synthesis and the phycocyanin (PC)-IGS region as a reference (from Kurmayer and Kutzenberger, 2003; Schober and Kurmayer, 2006; Kurmayer et al., 2011)*

Primer	Locus	Sequence (5'–3')	Direction	Annealing (°C)	Amplified product (bp)
Microcystis					
30F	mcyB	CCTACCGAGCGCTTGGG	F	60	78
108R		GAAATCCCTAAAGATTCCTGAGT	R		
Blau53T		F[1]-CACCAAAGAAACACCCGAATCTGAGAGG-T[2]	F		
188F	cpcBA	GCTACTTCGACCGCGCC	F	60	66
254R		TCCTACGGTTAATTGAGACTAGCC	R		
TaqmanMaPC		F[1]-CCGCTGCTGTCGCCTAGTCCCTG-T[2]	F		
Planktothrix					
mcyBAF	mcyB	ATTGCCGTTATCTCAAGCGAG	F	60	76
mcyBA1R		TGCTGAAAAACTGCTGCATTAA	R		
mcyBA1T		F[1]-TTTTTGTGGAGGTGAAGCTCTTTCCTCTGA-T[2]	F		
PIPC-IGSF	cpcBA	GAGCAGCACTGAAATCCAAG	F	60	72
PIPC-IGSR		GCTTTGGCTGCTTCTAAACC	R		
PIPC-IGST		F[1]-TTTGGCTTGACGGAAACGACCAA-T[2]	F		

[1]-F, FAM;
[2]-T, TAMRA.

SOP 7.4.2.2 Equipment

- A qPCR instrument, e.g. ABI7300 instrument (Applied Biosystems).
- Centrifuge with a plate rotor/adaptor.
- Instrument for electronic particle counting or microscopy for cell number estimation.
- Adjustable mechanical pipettes (100–1000 µL, 20–200 µL, 1–10 µL).

SOP 7.4.3 Procedure

1. Typically, for both genes (reference and target gene) the standard curves based on pre-determined cell concentrations are established by relating the known DNA concentrations (in cell equivalents) to the threshold cycle of the diluted samples.
2. A strain culture of *Microcystis* or *Planktothrix* is grown under sterile conditions in batch culture in a nutrient medium (see SOP 3.5) and harvested during the logarithmic growth phase. Cells are harvested by filtering onto glass fiber filters (GF/C, Whatman) under vacuum pressure and stored frozen (−20°C) until DNA extraction (see SOP 2.1).
3. Extract DNA from frozen filters as described in (SOP 5.3).
4. Aliquots are analyzed for cell numbers using at least two of three independent methods: (1) electronic particle counting (e.g. Casy, Schärfe system), (2) counting of cells by the inverted microscope technique, (3) cell counting using autofluorescence or DAPI staining (SOP 7.2).
5. From the DNA extract six dilutions ranging from $1{:}10^2$ to $1{:}10^7$ of template DNA (equivalent to counted cells) are prepared and analyzed for both genes in the absence or presence of a background (SOP 7.1). To compare background effects the number of cells estimated in the presence of a natural background is divided by the cell number estimated in the absence of a natural background and a ratio of cells with background/cells without background is calculated. The more the ratio deviates from one, the stronger the background effects. In addition to optimization of qPCR assays calibration, curves used for the reference and the target gene should be parallel in slope, implying comparable amplification efficiency across the linear dynamic range (SOP 7.1).
6. Calibration curves are not extrapolated beyond the highest dilution, which is defined arbitrarily as the limit of quantification. The limit of detection (LOD) is typically very much outside the calibration curve. However, close to the LOD the signal amplification is more unpredictable (e.g. unclear amplification curve). One possibility to define the LOD is that a clear qPCR signal (linear amplification curve at logarithmic scale) must occur in triplicate qPCRs obtained from a specific aliquoted DNA template. All signals below this threshold are set to zero. The qPCR signals between zero and the quantification limit are adjusted to the corresponding quantification threshold. Proportions of the subpopulation are then calculated by dividing estimated cell numbers by cell numbers of the corresponding total population (Fig. 7.4).
7. PCR is initiated after 10 min at 95°C to activate the hot start polymerase, followed by 50 cycles of a two-step PCR, consisting of a denaturation step at 95°C (15 sec) and subsequent annealing and elongation steps at 60°C respectively (1 min each). Each measurement is performed in triplicate.
   For *Microcystis*, reactions are performed with a volume of 25 µL, containing 12.5 µL of 2× TaqMan® Universal PCR Master Mix (Thermo Fisher Scientific), 300 nM

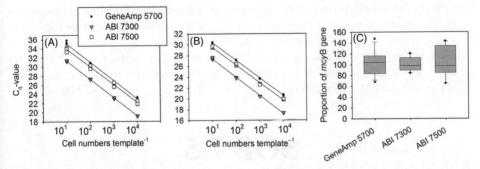

Figure 7.4 *Standard curves relating DNA amount (expressed in cell concentrations) to the threshold cycle for (A) target genotype mcyB, (B) reference genotype (PC-IGS), and (C) calculated mcyB gene proportions for three different qPCR systems. Despite methodological differences (sensitivity of different instruments) the range in proportion of mcyB genotype (estimated by cell equivalents template^{-1}) as calculated from calibration curves is comparable. Note that calculated proportions are partly exceeding 100% because of system-inherent inaccuracy due to application of semi-log calibration curves (see text above). Source: Data from Schober* et al. *(2007).*

(300 fmol μL^{-1}) of each primer, 100 nM of the TaqMan probe, 5 μL of template containing various amounts of genomic DNA, and filled up to 25 μL with sterile Milli-Q$^{®}$ water. For *mcy*B 900 nM of each primer and 250 nM of the TaqMan probe are used. Each measurement is done in triplicate (Schober *et al.*, 2007).

For *Planktothrix*, the 25 μL reaction mix consisted of 12.5 μL TaqMan$^{®}$ Universal PCR Master Mix (ABI), 5 μL of DNA template, and 50/300/100 or 900/900/100 nM of forward and reverse primers and TaqMan probes *cpc*BA and *mcy*BA1, each.

SOP 7.4.4 Notes

Because the noise within C_q-values induced by the semi-logarithmic calibration algorithm alone can mask differences in genotype proportion in populations, comparing and standardizing quantitative PCR results between laboratories prior to sample analysis is considered necessary. An inter-laboratory comparison of qPCR relative quantification between three instruments revealed (1) that all instruments were able to follow the variation in *mcy* genotype proportion both within mixtures of strains in the laboratory as well as in field samples and (2) the proportions of toxic genotypes were overestimated or underestimated by 0–72% and 0–50 %, each (Schober *et al.*, 2007).

SOP 7.4.5 References

Kurmayer, R. and Kutzenberger, T. (2003) Application of real-time PCR for quantification of microcystin genotypes in a population of the toxic cyanobacterium *Microcystis* sp., *Applied and Environmental Microbiology*, **69**, 6723–6730.

Kurmayer, R., Schober, E., Tonk, L., *et al.* (2011) Spatial divergence in the proportions of genes encoding toxic peptide synthesis among populations of the cyanobacterium *Planktothrix* in European lakes, *FEMS Microbiology Letters*, **317** (2), 127–137.

Schober, E. and Kurmayer, R. (2006) Evaluation of different DNA sampling techniques for the application of the real-time PCR method for the quantification of cyanobacteria in water, *Letters in Applied Microbiology*, **42**, 412–417.

Schober, E., Werndl, M., Laakso, K., *et al.* (2007) Interlaboratory comparison of Taq Nuclease Assays for the quantification of the toxic cyanobacteria *Microcystis* sp., *Journal of Microbiological Methods*, **69** (1), 122–128, doi: 10.1016/j.mimet.2006.12.007.

SOP 7.5

Quantification of Transcript Amounts of *mcy* Genes in *Planktothrix*

Guntram Christiansen[1,2] *and Rainer Kurmayer**

[1]*Research Institute for Limnology, University of Innsbruck, Mondsee, Austria*
[2]*Miti Biosystems GmbH, Max F. Perutz Laboratories, Vienna, Austria*

SOP 7.5.1 Introduction

This SOP describes how to quantify transcript amounts of *mcy* genes involved in microcystin synthesis in *Planktothrix* by means of qPCR using SYBR Green chemistry. The samples have been harvested for RNA extraction (see SOP 2.5), extracted for RNA (see SOP 5.7), and transcribed into cDNA (see SOP 5.8). The *mcy* transcript amount of *mcy* genes is related to two reference genes (phycocyanin and RNA polymerase). Protocols have been used by Christiansen *et al.* (2008) and Kurmayer *et al.* (2016). Potential reference strains are given in Table S6.1 (Appendix).

SOP 7.5.2 Experimental

SOP 7.5.2.1 Materials

- cDNA samples from SOP 5.8, 100 ng cDNA μL^{-1} and standards for each locus (as prepared in SOP 7.2).
- 96 PCR well plate (optical) and optical adhesive covers (for qPCR).

*Corresponding author: rainer.kurmayer@uibk.ac.at

Table 7.4 *Oligonucleotides for qPCR to determine transcript amounts of* mcy *genes* (mcyT – mcyJ) *in* Planktothrix *sp.*

Primer name	Sequence (5′–3′)	PCR product (bp)
mcy gene cluster (nine *mcy* genes)		
mcyTq+	AACGAAATATCGCGGCCC	
mcyTq−	CCAGGAGTTAGCAGTTGGGGT	106
mcyDq+	GGATGCTTTAATGCAGCAACG	
mcyDq−	TTCCCTGTATTCCAAGCTCCC	77
mcyEq+	ACCCTGACAAACCATTGATCA	
mcyEq-	GCGACTTATTCGCTGAAGATT	81
mcyGq+	GCACCAGGAGAACGTCGAGA	
mcyGq−	TTAAACAAATGCCTGACCCGG	88
mcyHq+	TGATTCCGAACTGGAAGGGA	
mcyHq-	GCTTTGATGCGGAAGTGCC	84
mcyAq+	ATGTTCGCACTCAGATGGCG	
mcyAq−	TGTCCCCCATCAAACTGTCC	75
mcyBq+	CAAAATTCATCGCCTAGCC	
mcyBq−	GGGTTTAATCAACAAGAGGC	87
mcyCq+	TGATTCGGGCTGAGGGTAACT	
mcyCq−	TTACGCCTCTGGTGTTCATCG	94
mcyJ+	CCCAATATTCCCCAATGGCT	
mcyJ−	TGCTTGCACCGCTTTAGCTC	79
Reference genes[1]		
rpo+	ACGGGTCAGACATCGTGGAA	
rpo−	GACGACGGACACGGGATTCT	70
PlPC+	GAGCAGCACTGAAATCCAAG	
PlPC−	GCTTTGGCTGCTTCTAAACC	72

[1] *rpo*, RNA polymerase, PLPC, phycocyanin gene *cpc*BA-IGS region (Schober and Kurmayer, 2006).

- RNase-free reaction tubes.
- Maxima SYBR Green qPCR master mix (Thermo Fisher Scientific).
- Sterile Milli-Q® water.
- Primer (10 µM) for cDNA locus amplification (Table 7.4).

SOP 7.5.2.2   Equipment

- Mechanical adjustable pipettes (1–10 µL, 2–20 µL, 20–200 µL), filter tips.
- qPCR cycler and quantitative data evaluation.

SOP 7.5.3   Procedure

1. Dilute the SYBR Green master mix (2×), for a 96 PCR well plate 619 µL of master mix are diluted with 446 µL PCR water (12.5 µL final PCR volume).
2. Aliquot PCR master mix (for each locus) and add corresponding primer pairs, 0.375 µL of forward and reverse primer each.

3. Pipette 11.5 µL in each well (triplicate measurements), add 1 µL of cDNA or 1 µL of PCR water (non-template control, NTC) into each well.
4. Close the PCR 96-well plate with the adhesive cover tightly. If there is visible liquid on the walls, the plate might be spun down shortly.
5. Run the qPCR cycle program including one step for DNA denaturation at 95°C (10 min) followed by 40 cycles of denaturation (95°C, 15 sec), annealing (60°C, 15 sec) and elongation (72°C, 20 sec), and the melting curve analysis consisting of a linear temperature rise from 60°C to 95°C (20 min).
6. After the qPCR cycle run is finished, check for eventual amplification signals of NTCs and inspect the amplification curves of all samples and the standards. If the NTCs are negative and the standards show the expected amplification response, the re-calculation of DNA amounts may proceed.
7. Determine the threshold cycle and re-calculate DNA amounts for each gene locus in the template using the external standard curves established for each gene locus independently.
8. For each gene locus under investigation, the transcript amount ratio as compared to the reference gene is calculated.

SOP 7.5.4   Notes

Because of the high sensitivity of qPCR and amplification of smallest amounts of DNA in the template a PCR hood (equipped with UV-light-irradiated circulation) is recommended and will exclude DNA template contamination.

It is recommended to sequence all gene loci to be analyzed for transcript amount to exclude single point mutations occurring within the primer binding regions and potentially interfering with qPCR.

The choice of appropriate multiple reference genes is important, which should ideally be independent of regulation under the experimental conditions (e.g. light availability might influence both the reference and the *mcy* genes under investigation).

The reference transcript amount of both the reference genes and the gene under investigation should be relatively close (e.g. in *Planktothrix*, *mcy* genes are transcribed at 1% of phycocyanin reference gene and 10% of the RNA polymerase *rpo* gene). If the transcript amount ratio reference gene/*mcy* gene is very high then random noise effects might interfere with transcript amount analysis.

SOP 7.5.5   References

Christiansen, G., Molitor, C., Philmus, B., and Kurmayer, R. (2008) Nontoxic strains of cyanobacteria are the result of major gene deletion events induced by a transposable element, *Molecular Biology and Evolution*, **25**, 1695–1704.

Kurmayer, R., Deng, L., and Entfellner, E. (2016) Role of toxic and bioactive secondary metabolites in colonization and bloom formation by filamentous cyanobacteria *Planktothrix*, *Harmful Algae*, **54**, 69–86.

Schober, E. and Kurmayer, R. (2006) Evaluation of different DNA sampling techniques for the application of the real-time PCR method for the quantification of cyanobacteria in water, *Letters in Applied Microbiology*, **42**, 412–417.

SOP 7.6

Quantification of Potentially Cylindrospermopsin-Producing *Chrysosporum ovalisporum*

Rehab El-Shehawy[1]* and Antonio Quesada[2]*

[1]*Institute IMDEA Water, Alcalá de Henares (Madrid), Spain*
[2]*Department of Biology, Autonomous University of Madrid, Madrid, Spain*

SOP 7.6.1   Introduction

The qPCR assay described below is a dual qPCR assay for the detection and quantification of the potentially cylindrospermopsin-producing *Aphanizomenon (Chrysosporum) ovalisporum* in a mixed DNA background (as obtained from field samples). A second aim is to distinguish between toxic *Cylindrospermopsis raciborskii* and toxic *A. ovalisporum*. The test is based on the quantitative determination of *cyr*J and *rpo*C genes in the extracted nucleic acid of the environmental samples. Quantification is performed using an external standard curve prepared from genomic DNA of a CYL-producing *A. ovalisporum*.

In order to detect CYL-producing *A. ovalisporum* in a mixed background and distinguish between toxic *C. raciborskii* and toxic *A. ovalisporum*, a dual qPCR assay is applied (in a sequential manner) by first running a qPCR using *cyr*J207F/R–*cyr*J207P (assay no. 1) to detect the presence of potentially CYL-producing *C. raciborskii* or *A. ovalisporum*, and then running a second qPCR using *rpo*C1148F/R–*rpo*C11830P (assay no. 2) to discriminate between CYL-producing *A. ovalisporum* (positive results will be obtained) or CYL-producing *C. raciborskii* (negative results or products below detection limit will be obtained). Potential reference strains are given in Table S6.1 (Appendix).

SOP 7.6.2   Experimental

SOP 7.6.2.1   Materials

- qPCR mastermix, e.g. QuantiFast HotStart TaqMan® PCR Master kit (Qiagen).

*Corresponding author: rehab@imdea.org; antonio.quesada@uam.es

Table 7.5   *Primers and probes used for quantitative cyrJ detection via qPCR*

Primers and probes	Sequence (5′–3′)	Reporter dye	Target gene
*cyrJ*207F	CCCCTACAACCTGACAAAGCTT		*cyrJ*
*cyrJ*207R	CCCGCCTGTCATAGATGCA		
*cyrJ*207P	AGCATTCTCCGCGGATCGTTCAGC	FAM	*rpoC*
*rpo*C1148F	CCGAAATGGACGGCTTGTT		
*rpo*C1111R	CAGTGACATTCCCAGTCTTTGG		
*rpo*C11830P	TGCGAGCGCATCTTTGGCCC	VIC	

- Primers and probes (stock solutions 100 µM) (Table 7.5).
- Isolated nucleic acids of laboratory strains and environmental samples (see SOP 5.3).
- Sterile Milli-Q® water.
- Optical PCR reaction plates/tubes and adhesive covers/caps compatible with the instrument used.

SOP 7.6.2.2   Equipment

- A qPCR instrument, e.g. ABI7300 instrument (Applied Biosystems).
- Centrifuge with a plate rotor/adaptor.
- Mechanical adjustable pipettes (1–10 µL, 2–20 µL, 20–200 µL), filter tips.

SOP 7.6.3   Procedure

1. Prepare standard solutions by making a dilution series containing 10^1, 10^2, 10^3, 10^4, and 10^5 copies of *cyr*J or *rpo*C genes. The gene copy numbers could be estimated following SOP 7.2.
   Prepare enough dilution for at least three replicate qPCR reactions. Always prepare the standard dilution series just before performing the qPCR. To avoid DNA degradation and change in the standard concentration, do not freeze-thaw the dilutions.
2. Adjust the amount of primers and probes to 10 µM.
3. Set up the qPCR instrument for an assay using absolute quantification according to the manufacturer's instruction. Do not forget to select the correct detector according to the fluorescent dye of the probe.
4. Use the PCR cycling program according to the qPCR TaqMan® manufacturer's instructions. Set annealing temperature at 56°C and 57°C for *cyr*J and *rpo*C, respectively. Change the sample volume to 25 µL (default value in the ABI7300 instrument is 50 µL).
5. After the PCR run is completed, analyze the amplification plot according to the manufacturer's instructions. Check the amplification plot to evaluate the amplification quality. Signs for successful PCR include amplification of the standards in a manner similar to that observed in previous optimization steps, amplification curves show a linear dynamic range (on log scale), and amplification curves of the replicate sample are clustered. Check also that no amplification has occurred in negative control wells, which might indicate contamination.

SOP 7.6.4   Notes

The rational behind using *cyr*J as a molecular marker to discriminate between potential CYL-producing and non-CYL-producing cyanobacterial genera is that non-CYL-producing genera lack *cyr*J (Ballot *et al.*, 2011).

Increased quantification errors typically occur toward both ends of the calibration curve. For field sample analysis, cell quantification should be directed toward the central region of the standard curves, which are found to be most resistant against uncontrolled factors.

Chemistry and instruments used may affect the performance of the assays. Thus, proper performance should be tested and optimized if needed before sample analysis.

SOP 7.6.5   References

Ballot, A., Ramm, J., Rundberget, T., *et al.* (2011) Occurrence of non-cylindrospermopsin producing *Aphanizomenon ovalisporum* and *Anabaena bergii* in Lake Kinneret (Israel), *Journal of Plankton Research*, **33** (11), 1736–1746.

Campo, E., Lezcano, M.Á., Agha, R., *et al.* (2013) First TaqMan assay to identify and quantify the cylindrospermopsin-producing cyanobacterium *Aphanizomenon ovalisporum* in water, *Advances in Microbiology*, **3**, 430–437.

SOP 7.7

qPCR Detection of the Paralytic Shellfish Toxin Biosynthesis Gene *sxt*B

Henna Savela*

Department of Biochemistry/Biotechnology, University of Turku, Turku, Finland

SOP 7.7.1   Introduction

This protocol describes the quantitative detection of one of the paralytic shellfish toxin biosynthesis genes, *sxt*B, in cyanobacteria (Fig. 7.5). The primer and probe design is based on *Anabaena circinalis* (*Dolichospermum circinale*), *Aphanizomenon flos-aquae*, *Cylindrospermopsis raciborskii*, *Lyngbya wollei*, and *Raphidiopsis brookii sxt*B sequences (Savela *et al.*, 2015). Potential reference strains are given in Table S6.1 (Appendix).

*Corresponding author: henna.savela@iki.fi

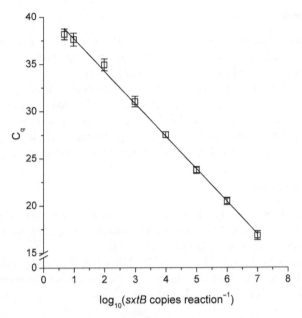

Figure 7.5 *Standard curve for the sxtB qPCR assay (y = 3.45x + 41.18, R^2 = 0.99, E = 94.9%), obtained using purified sxtB amplicons produced from A. circinalis CS-537/13 genomic DNA. Standard deviations of quantification cycles (C_q) are indicated. Source: Savela et al. (2015). Reproduced with permission of Elsevier.*

SOP 7.7.2   Experimental

SOP 7.7.2.1   Materials

- DyNAzyme II HotStart (Finnzymes/Thermo Fisher Scientific), or similar *Taq*-type DNA polymerase with $5' \rightarrow 3'$ exonuclease activity.
- 10× DyNAzyme II Hot Start Reaction Buffer.
- Deoxynucleotides, e.g. dNTP mix 10 mM each (Thermo Fisher Scientific).
- Primers (100 µM stock solutions):
  - *sxt*B_F2 (5′–TGTTGTGCTTGCTGCTCTATCAG–3′)
  - *sxt*B_R2 (5′–CAGCGTTTTCAGCGTAYCGAC–3′).
- Detection probe (100 µM stock solution).
- (5′–aminoC6-CAATCAAAGTTATGCTCCCTATACGA-Phos–3′). The original protocol describes the labelling of the detection probe at the amino C6 moiety with a fluorescent Tb(III)-chelate (Nurmi *et al.*, 2002; Savela *et al.*, 2015). Similar chelates are available, e.g. from Kaivogen Oy, Finland. Alternatively, another type of fluorescent label, such as FAM, can be used, but note that the protocol and reagent concentrations may require optimization.
- Quencher probe (100 µM stock solution).
- (5′–GGGAGCATAACTTTGATTG–BHQ®1–3′). The quencher probe is labelled at the 3′ end with Black Hole Quencher 1. Alternatively, use a quencher moiety with suitable spectral overlap with the fluorescent label of choice.
- Sterile Milli-Q® water or molecular biology grade water.

- 96-well PCR plates and optical quality caps compatible with the instrumentation.
- DNA from laboratory cultured cyanobacteria, or field samples.

SOP 7.7.2.2   Equipment

- A qPCR instrument, or a separate PCR instrument (e.g. Bio-Rad C1000) and a fluorescence reader (e.g. PerkinElmer Victor X4 2030 Multilabel Reader).
- Centrifuge with a plate rotor/adaptor.
- Mechanical adjustable pipettes (1–10 μL, 2–20 μL, 20–200 μL), filter tips.

SOP 7.7.3   Procedure

1. Prepare qPCR standards containing 5^0, 10^1, 10^2, 10^3, 10^4, 10^5, 10^6, and 10^7 *sxt*B copies per 4 μL. Standards can be prepared by using genomic DNA extracted from a paralytic shellfish toxin producing cyanobacterial strain. Alternatively, an *sxt*B amplification product produced by using the primers listed above can be used. Quant-iT PicoGreen dsDNA Assay Kit (Thermo Fisher Scientific) is recommended for quantification of all DNA prior to use as qPCR standards. See also SOP 7.2 for calibration of gene copy numbers.
2. Prepare 10 μM working solutions of both primers, and 1 μM and 100 nM working solutions of the quencher probe and detection probe, respectively.
3. Dilute samples as instructed in SOP 7.2.
4. Calculate the number of reactions: multiply the number of samples to be analyzed including the negative control (sterile Milli-Q® water or molecular biology grade water) and qPCR standards by three (all samples are to be run in triplicate). Prepare a few extra reactions to compensate for pipetting loss (multiplying the number of reactions by 1.1 is recommended).
5. Pipette the reaction mixture. For one 20 μL reaction, combine 2 μL of 10× DyNAzyme II Hot Start Reaction Buffer (final conc. 1×) 0.4 μL dNTP mix (final conc. 0.2 mM), 0.8 μL of each primer (final conc. 0.4 μM), 0.6 μL of detection probe (final conc. 3 nM), 0.48 μL of quencher probe (final conc. 24 nM), 0.1 μL (final conc. 0.2 U/reaction) of DyNAzyme II Hot Start DNA polymerase, and 4 μL of template: negative control, qPCR standard or unknown sample. Adjust to 20 μL with sterile Milli-Q® water.
6. Thermal cycling and fluorescence data acquisition: program the following thermal cycling sequence: 95°C for 5 min, then 40 cycles of 95°C for 30 sec, 62°C for 1 min. Starting at cycle 8 and ending at cycle 40, set a data acquisition step after every second cycle: after annealing and extension at 62°C, incubate at 35°C for 15 sec, then measure. A total of 17 measurement steps will be carried out. Control that the sample volume is set to 20 μL instead of the instrument default. If using a separate fluorometer, select the appropriate measuring protocol in advance, and pause the cycling sequence and perform the measurement after each 15 sec incubation at 35 °C.
7. Data analysis can be carried out using the manufacturer's software (if running a commercial qPCR instrument) or by using a spreadsheet. Quantification cycle (C_q) values for the standards and samples can be obtained from the qPCR instrument or, if using a fluorescence reader, from the raw data by calculating signal-to-background ratios: for all reactions separately, compare each individual measurement value to the starting signal level of the same reaction. Plot the signal-to-background ratios against PCR cycles; C_q-values can be determined from the resulting amplification plots. See SOP 7.1 for further instructions.

SOP 7.7.4 Notes

The required instrumentation will depend on the choice of label for the detection probe. Prompt fluorophores including, among others, FAM and TAMRA are compatible with most commercial qPCR instruments. Lanthanide labels (*e.g.* the Tb(III) chelate used in the original assay) require that the fluorescence reader is capable of time-resolved fluorometry. These types of instrument are manufactured e.g. by PerkinElmer. Always check that your fluorescent probes are compatible with the instruments available for your use.

SOP 7.7.5 References

Nurmi, J., Wikman, T., Karp, M., and Lövgren, T. (2002) High-performance real-time quantitative RT-PCR using lanthanide probes and a dual-temperature hybridization assay, *Analytical chemistry*, **74**, 3525–3532.

Savela, H., Spoof, L., Perälä, N., *et al.* (2015) Detection of cyanobacterial *sxt* genes and paralytic shellfish toxins in freshwater lakes and brackish waters on Åland Islands, Finland, *Harmful Algae*, **46**, 1–10.

SOP 7.8

Application of the Minimum Information for Publication of Quantitative Real-Time PCR Experiments (MIQE) Guidelines to Quantitative Analysis of Toxic Cyanobacteria

*Henna Savela**

Department of Biochemistry/Biotechnology, University of Turku, Turku, Finland

SOP 7.8.1 Introduction

This standard operating procedure outlines how the MIQE guidelines (Bustin *et al.*, 2009) can serve the design and execution of quantitative experiments on toxin-producing

*Corresponding author:henna.savela@iki.fi

cyanobacteria. This SOP is an adaptation of the original guidelines, and is applicable to quantification of DNA targets. To ensure the comparability and reproducibility of qPCR results and experiments, a set minimum level of information needs to be provided. Description of sampling schemes and sample preparation is particularly important when reporting on quantitative experiments on toxin-producing cyanobacteria. Without these details, it can be difficult to assess the impact and significance of the results. To accompany this SOP, a detailed checklist, based on the original by Bustin *et al.* (2009), has been prepared to aid and guide both in experiment design and manuscript preparation (Table 7.6).

SOP 7.8.2   Sampling

1. Describe the sample(s) and study site(s). If using cultured cyanobacteria, include species names, strain designations, and sources. If analyzing environmental samples, note the date of sampling and describe the collection technique in detail. In some cases it may be relevant to note the time of day as well. Assess the growth stage of cyanobacterial cultures, or bloom growth phase in the field. Record the volume and/or mass of each sample, cell density if available, and the number of biological replicates.
2. Describe the method used to harvest the cells: filtration, centrifugation, other? Describe the means used to avoid cross-contamination between samples, if any. It is recommended to use disposable sampling materials if possible. Otherwise, take care to adequately clean the equipment, e.g. filtration device, between samples, and document the method to minimize the chance for cross contamination.
3. Describe sample storage conditions and duration.

SOP 7.8.3   Sample Preparation and DNA Extraction

1. Provide details on the method used to extract DNA: the name and manufacturer of any reagent kit used, identity of additional reagents, as well as details on laboratory instrumentation. Provide references to published literature where appropriate. All modification to existing methods should be clearly explained for reproducibility, and all new methods should be described particularly well.
2. Assess the quality and yield of the resulting extracted DNA to ensure reliable quantification results. Spectrophotometric analysis (absorbance ratio A260/A280) is recommended for estimation of DNA purity. If using fluorescent dye detection to measure DNA concentration (e.g. PicoGreen) the method and instrumentation should be described in detail and/or references provided. Test all samples for amplification inhibition. This can be carried out by preparing and analyzing dilution series, or by spiking subsamples with a known quantity of target DNA.
3. Store DNA adequately: at least at –20°C. Avoid repeated freeze–thaw cycles. Describe DNA storage conditions (temperature, solvent, and volume) and duration.

SOP 7.8.4   Target Information and Oligonucleotide Design

1. The targeted sequence should be readily identifiable by making gene names and sequence database accession numbers available. If oligonucleotide design is based on

Table 7.6 *The minimum information for publication of quantitative real-time PCR experiments (MIQE) guidelines applied to quantitative analysis of toxin biosynthesis genes in cyanobacteria*

Item to check	Description	Importance[1]	SOP no.[2]
Target information			
	Gene of interest: gene symbol and accession number(s)	E	
	Cyanobacterium/cyanobacteria of interest (taxon)	E	
Sampling			
	Date and origin of sample	E	2.6
	Description of the study site and/or laboratory-grown strains	E	2.6
	Acknowledgment and contact details of person(s) who conducted the sampling	D	
	Sample collection method (plankton net, water samples, other)	E	2.1/2.2
	Sample type (depth-integrated, surface, other)	E	2.1
	Number of biological replicates	E	
	Identification of cyanobacteria in field samples	E	4.1
	Growth phase (laboratory culture and/or field samples)	E	2.1/3.5
Sample preparation			
	Description of cell harvesting method	E	2.1
	Description of method used to avoid cross-contamination	D	
	Volume and/or mass of sample(s) processed	E	2.6
	Cell density and/or total count (description of microscopic counting method)	D	2.6
	Sample storage conditions (temperature, duration, solvent, if any)	E	
DNA extraction			
	Complete DNA extraction and purification protocol, identity of kit(s)	E	5.3
	Complete description of modifications to existing methods	E	
	Details of DNase/RNase treatment	E[3]	
	DNA/RNA contamination assessment	E[3]	
	Details on nucleic acid quantification (method, instrument)	E	
	DNA purity and yield	D	5.6
	DNA storage conditions (temperature, duration, solvent, volume)	E	
	Inhibition testing using dilution series, by spiking, or similar	E	

Table 7.6 *(Continued)*

Item to check	Description	Importance[1]	SOP no.[2]
qPCR target and oligonucleotides			
	Primer sequences	E	
	Probe sequence(s), provide if applicable	E	
	Oligonucleotide design software identity and version	D	
	Details on any oligonucleotide modifications	E	
	Oligonucleotide manufacturer information and purification method	D	
	Amplicon length	E	
	Amplicon location and secondary structure analysis	D	
	Description of the *in silico* specificity screen	E	
	Information on any possibly interfering homologous genes/sequences	D	
	Sequence alignment	D	
qPCR protocol			
	Type of qPCR assay (intercalating dye, hydrolysis probe, *etc.*)	E	
	Reaction volume	E	
	DNA/template amount	E	
	Primer, probe, divalent salt (Mg^{2+}), and dNTP concentrations	E	7.1
	Polymerase, buffer/kit identity, manufacturer, and concentration	E	
	Chemical composition of the buffer	D	
	Concentrations and identities on any additives (dyes, DMSO and similar)	E	
	Identity and manufacturer of qPCR plastics (plates, caps)	D	
	Complete thermocycling protocol	E	7.1
	Method of reaction setup (manual/robotic)	D	
	Description of qPCR standards	E	7.2
	Instrument and manufacturer	E	
Assay validation			
	Evidence of optimization	D	7.1
	qPCR assay specificity	E	7.2, 7.3, 7.4
	C_q values for NTC reactions (if using dyes such as SYBR Green I)	E	
	Calibration curve(s): slope, y-intercept, coefficient of variation (R^2)	E	7.2
	Amplification efficiency, E	E	7.1, 7.2
	Confidence intervals for E, or standard error	D	

(Continued)

Table 7.6 *(Continued)*

Item to check	Description	Importance[1]	SOP no.[2]
	Linear dynamic range	E	7.1, 7.2
	C_q variation at LOD	E	7.2, 7.4
	Confidence intervals throughout quantification range	D	
	Evidence for LOD, qPCR standards	E	7.2, 7.4
	Evidence for limit of detection, cyanobacterial samples and in presence/absence of background DNA	D	7.2, 7.4
	For multiplex assays, LOD and E of each assay	E	
Data analysis			
	Software used for data analysis; qPCR and statistical (identity and version)	E	
	Statistical methods used to assess results significance	E	
	Methods used to determine C_q-values and outliers	E	
	Concordance of biological replicates	D	
	Number and stage of technical replicates	E	
	Repeatability (intra-assay variation)	E	
	Reproducibility (inter-assay variation, coefficient of variation)	D	
	Power analysis	D	
	C_q or raw data submission with real-time PCR data markup language (RDML)	D	

Source: Adapted from Bustin *et al.* (2009) with permission of the American Association for Clinical Chemistry
[1] Importance of the item to check: E, essential; D, desirable.
[2] For more information on the item to check see the indicated SOP in this book.
[3] Essential for qPCR assays to determine gene transcript amounts.

new sequences, these should be made available via submission either directly with the manuscript or to a database (e.g. GenBank).

2. To reach optimal amplification efficiency, short (up to 200 bp, optimally 70–100 bp) amplicons are preferred in qPCR assays. Reporting the amplicon length enables later verification of assay specificity.

3. All oligonucleotide sequences should ideally be provided; primer sequences without exception, and probe sequences if applicable. It is not recommended to use commercial probes with undisclosed sequences (Bustin *et al.*, 2009).

SOP 7.8.5 qPCR Protocol

1. Describe the qPCR reaction conditions in complete detail, including all reagent (essential such as polymerase, primers, and dNTP, as well as additives such as

DMSO), concentrations, reaction volume, template amount, thermocycling parameters, instrumentation, and software.
2. Describe the type of qPCR standard material used: amplicon, plasmid, or cell equivalents of DNA as well as how it was produced, purified, and quantified.

SOP 7.8.6 qPCR Validation

1. Confirm and provide evidence for primer specificity by running agarose gels and/or melting curve analysis after the qPCR run.
2. Prepare a standard curve covering a minimum concentration range of six orders of magnitude and provide its slope, y-intercept, coefficient of determination (R^2), and PCR efficiency (E). The efficiency can be calculated from the slope by using the equation $E = 10^{(-1/slope)} - 1 \times 100$.
3. Determine the limit of detection (LOD) using a dilution series of the target DNA. The minimum requirement is that a dilution series of standard material (amplicon, plasmid, etc.) should be analyzed, if possible with and without background DNA (environmental, or from related, non-target cultured cyanobacteria). Additionally, the LOD can be determined using sample DNA. At the LOD, probability of amplification should be 95%.

SOP 7.8.7 Data Analysis

1. Describe the qPCR and statistical software and methods used to analyze the results. Provide details on the method used to determine C_q-value cutoff and outliers, and report the results for NTCs as well as the stage and number of technical replicates.
2. Describe the repeatability and reproducibility of results in statistical terms.

SOP 7.8.8 Reference

Bustin, S., Benes, V., Garson, J., *et al.* (2009) The MIQE guidelines: Minimum Information for Publication of Quantitative Real-Time PCR Experiments, *Clinical Chemistry*, **55** (4), 611–622.

8

DNA (Diagnostic) and cDNA Microarray

Anne Rantala-Ylinen[1], Kaarina Sivonen[1]*, and Annick Wilmotte[2]

[1]*Department of Food and Environmental Sciences, Division of Microbiology and Biotechnology, University of Helsinki, Helsinki, Finland*
[2]*InBios – Center for Protein Engineering, University of Liège, Liège, Belgium*

8.1 DNA (Diagnostic) Microarray

8.1.1 Introduction

Diagnostic microarrays have been used to study cyanobacterial diversity in environmental samples. They usually include sequences (probes) for only one or a few genes allowing a number of genera/species to be detected. In contrast, cDNA microarrays are used for screening genome-wide changes in gene expression and include probes for all or most genes in one strain (Burja *et al.*, 2003). The purpose of this subchapter is to introduce the principle of a diagnostic microarray (= DNA chip) that uses a ligation detection reaction (LDR) and universal microarray (Gerry *et al.*, 1999) to simultaneously detect and identify all potential microcystin and nodularin producers present in a sample (Rantala *et al.*, 2008; Sipari *et al.*, 2010). This is especially useful when monitoring environmental samples that can contain many cyanobacterial genera, including toxin-producing strains.

*Corresponding author: kaarina.sivonen@helsinki.fi

Molecular Tools for the Detection and Quantification of Toxigenic Cyanobacteria, First Edition.
Edited by Rainer Kurmayer, Kaarina Sivonen, Annick Wilmotte and Nico Salmaso.
© 2017 John Wiley & Sons Ltd. Published 2017 by John Wiley & Sons Ltd.

8.1.2   Methodological Principles

8.1.2.1   *Amplification of mcyE/ndaF Genes*

Prior to LDR and hybridization, it is necessary to amplify by PCR the *mcy*E/*nda*F genes present in the environmental samples. This is accomplished by using the so-called general primers that are able to recognize and amplify the *mcy*E/*nda*F genes of all the main producer genera (Rantala *et al.*, 2004; 2008), (see also SOP 6.2). Conditions of PCR amplification can affect the outcome of a DNA-chip assay dramatically. PCR conditions (e.g. annealing temperatures) should be optimized to allow the unbiased amplification of all the gene variants, even those present in a low percentage (Rantala, 2007). Preferential amplification of some gene variant(s) can leave others totally undetected, thus distorting the results. In addition, for successful amplification, the DNA extracted from an environmental sample needs to be representative of the whole cyanobacterial community and be free of impurities inhibiting the PCR polymerase (see Chapters 5 and 6).

8.1.2.2   *Ligation Detection Reaction (LDR)*

The LDR step is a basis for identification of the different *mcy*E/*nda*F genes present in the samples. First, the PCR products are recognized by two probes. The "discriminating probe" is designed to include at the 3′ end 1–2 nucleotides that are unique to the target gene sequence and is labeled with a fluorescent dye (Cy3) at the 5′ end. The "common" probe is phosphorylated at the 5′ end and has a complementary ZipCode (cZipCode) sequence attached to its 3′ end. The ligation occurs only if both the "discriminating" and "common" probes bind to the target PCR product and if the nucleotides at the probe junction perfectly match and base pair with the target. As a result of ligation, a single molecule is formed with a Cy3 label at the 5′ end and a cZipCode sequence at the 3′ end. During the following LDR cycles, new pairs of "discriminating" and "common" probes are attached to the same PCR product, leading to amplification of the signal from the different *mcy*E/*nda*F genes present.

8.1.2.3   *Hybridization*

The LDR reactions are hybridized onto universal microarrays (Gerry *et al.*, 1999) via the cZipCode sequences that recognize the ZipCodes (24 bp oligonucleotides) printed on glass slides. The array is called "universal" since the same ZipCodes/cZipCodes can be used with any set of probes attached to them. ZipCode sequences have no counterparts in the target sequences, and they have been designed to prevent cross-hybridization between different ZipCodes, reducing background hybridization signal levels. They have similar T_m values, allowing each ZipCode and cZipCode pair to hybridize with the same efficiency in the assay conditions. In this application, each ZipCode corresponding to a certain *mcy*E/*nda*F gene variant is spotted in four replicates on the arrays. In addition, each array contains eight replicate spots of the hybridization control ZipCode and six replicate spots of the LDR control ZipCode (Rantala *et al.*, 2008). Although the LDR reaction contains all probes and cZipCodes, only those probe pairs that annealed to the target and were ligated during the LDR can carry the fluorescent Cy3 dye to the spots. In contrast, Cy3-labelled "discriminating" probes that were not ligated to the corresponding "common" probes are washed away in the process. Thus, fluorescence is detected by laser scanning only from the spots that correspond to the *mcy*E/*nda*F gene variants originally amplified. These signals allow identifying the microcystin- and nodularin-producing cyanobacterial genera present in the sample.

8.1.3 General Conclusions

The DNA-chip assay is able to simultaneously detect all main microcystin- and nodularin-producing cyanobacterial genera. In addition, eight environmental samples can be analyzed in one assay (high-throughput assay). The specificity of the LDR probes enables the differentiation of the highly similar sequence variants of different toxin producers. The main drawbacks of the method are the expensive equipment needed and the fact that the assay is not commercialized; each component must be custom-made and thus cannot be used without prior optimizations in each laboratory.

8.1.4 References

Burja, A.M., Dhamwichukorn, S., and Wright, P.C. (2003) Cyanobacterial postgenomic research and systems biology, *Trends in Biotechnology*, **21**, 504–411.

Gerry, N.P., Witowski, N.E., Day, J., *et al.* (1999) Universal DNA microarray method for multiplex detection of low abundance point mutations, *Journal of Molecular Biology*, **292**, 251–262.

Rantala, A. (2007) Evolution and detection of cyanobacterial hepatotoxin synthetase genes, PhD thesis, Edita Prima Oy, Helsinki, Finland, http://urn.fi/URN:ISBN:978-952-10-4369-7, accessed 23 February 2017.

Rantala, A., Fewer, D.P., Hisbergues, M., *et al.* (2004) Phylogenetic evidence for the early evolution of microcystin synthesis, *Proceedings of the National Academy of Sciences USA*, **101**, 568–573.

Rantala, A., Rizzi, E., Castiglioni, B., *et al.* (2008) Identification of hepatotoxin-producing cyanobacteria by DNA-chip, *Environmental Microbiology*, **10**, 653–664.

Sipari, H., Rantala-Ylinen, A., Jokela, J., *et al.* (2010) Development of a chip assay and quantitative PCR for detecting microcystin synthetase E gene expression, *Applied and Environmental Microbiology*, **76**, 3797–3805.

8.2 cDNA Microarray for Cyanobacteria

*Hans C.P. Matthijs and J. Merijn Schuurmans**

*Department of Aquatic Microbiology, Institute for Biodiversity and Ecosystem Dynamics,
University of Amsterdam, Amsterdam, The Netherlands*

8.2.1 Introduction

Each living cell contains thousands of genes which are expressed at different moments in time and at different rates. Additionally, expression is for a large part regulated. An assembly of all expressed genes is called a "transcriptome." A transcriptome is studied by isolating native RNA from cyanobacterial cells and binding its derived transcribed cDNA to a pre-designed microarray, which is a flat glass slide with a printed matrix of precisely defined bound nucleotides (nt) at thousands of fixed positions/spots. Each spot contains a unique nucleotide. The DNA sequences bound to those spots are designed on the basis of a previously fully analyzed genome. Using array designer programs, short (minimally 18 nt) or long (60 nt) probes are designed that provide unique recognition of a single gene fragment in the cDNA pool. The probes should comply with preset requirements regarding hybridization temperature and stringency, and should not form hairpin duplexes. As a luxury alternative to specifically selected probes, a tiling array may be designed in which the complete genome is divided into 60 nt units that are bound to (a much larger number of) array matrix spots. In the tiling array design, probes are made against both strands of the genomic DNA, through which knowledge on involvement of antisense RNA has been achieved.

8.2.2 Principles of Microarray Use

After the design, several companies can perform custom-made array printing. An array is the size of a microscope cover slide and may contain up to 144k spots. In our practice, we used either 16k or 60k spot arrays that we acquired from the Agilent company for *Synechocystis* PCC 6803 and *Microcystis* PCC 7806, respectively. A total of eight arrays are printed on one slide, each with a size of about 1 × 1 cm, which is very similar to a microscope cover slide. For proper statistics, at least three and up to five biological replicates are used, each loaded on a different array. Arrays that are not used can be protected by covering for future use. Each array can be used once. According to the Agilent specifications, loading consists of a mixture of cDNA derived from control cells and colored with one dye, and of an equally abundant cDNA specimen derived from treated cells and colored with a second dye. Two pools of cDNA are thus hybridized to each array. One pool serves as a control and is prepared from cells that were grown in a reference condition. The other pool is from cells that were grown in different conditions, which are often referred to as "treatments." The relative abundance of gene transcripts defines how much of each is bound to the nucleotides in the spots on the array. Each spot contains hundreds of equal nucleotides and binding is a matter of chance for the probe and its target to meet. Technical replicates

*Corresponding author: J.M.Schuurmans@uva.nl; jmschuurmans@gmail.com

are not really needed, and the formerly applied dye-swaps may also be omitted, depending on the design of the study. The cDNA needs to fulfill certain requirements concerning its purity and quantity. After labeling with fluorescent dye, one or two replicate samples are loaded over the array for overnight hybridization. After washes, the relative amounts of dye one and dye two retrieved from each spot in an array reader device serve as an indication of the relative transcription of genes in control and treated cells. Raw data need extensive bioinformatics analysis: if three arrays of biological replicates are run, their relative cDNA quality, binding efficiency to the array, and dye incorporation may differ. This requires numerical normalization before ratios can be derived and statistically evaluated between the biological replicates. Finally, the lists of signal data per spot are connected to the information about the gene represented in each spot. Usually, multiple nucleotides are designed for a single gene, and during data analysis, all signals for a single gene will theoretically report an equal up or down regulation. Experience tells us which oligonucleotides in the spots, despite the efforts spent in array design, may give wrong estimations, be it by non-specificity or a lower uniformity in hybridization efficiency. Finally, the changes in relative gene transcription will show which genes are downregulated (repressed) relative to the control or upregulated (induced). Virtually all genes have some kind of regulation, and thresholds need to be defined over which genes are considered significantly regulated. If upregulation is defined as > 1.8 times, a larger number of regulated genes will be obtained than when stringency is defined as 2.0 or more. Next to these intuitively settled thresholds (with 0.5 times as a typical upper limit for downregulation), or expressed as log2 variants +1 or −1, the final selection of data is based on statistical evaluation of the data quality where p values smaller than 0.05 are considered acceptable. The regulated genes can be arranged to a preset ordering as designed in CyanoBase (http://genome.microbedb .jp/cyanobase/Synechocystis), which then permits the overview of paths of metabolism that are actually regulated between control and treatment, using the Kegg database (http://www .genome.jp/kegg/pathway.html) information for in-depth interpretation.

8.2.3 Considerations for Experimental Design

Study of a certain metabolic trait requires a very precise definition of culture conditions. For example, if a nutrient is depleted in a treatment, the culture will grow less densely than the control. As a consequence, the control will show a light limitation as compared to the treatment culture. Hence, to compensate for such secondary effects, the nutrient-limited culture must be exposed to a light regime that mimics the consequences of the increase of the cell density in the control culture. Another important consideration is that storage of nutrients as macromolecules (e.g. glycogen, polyhydroxyalkanolates, or lipids for C, cyanophycine for N, or polyphosphate for P) may require a prolonged incubation time before signaling and responsiveness to changes in nutrient supply can be observed. Fast signals are usually seen after 1 h; slow signals may need a waiting time of 6 h. Additionally, fast-responsive and slower-responsive genes exist. Hence, measurement of a change at a single time point may not be sufficient, and a time series is recommended. Costs are a seriously limiting factor, though prices of arrays have gone down sharply over the last 15 years, including chemicals and payment for machines and tools. The real primary costs of an array may still be €140 at the time of writing, labor for the bioinformatics analysis not included.

8.2.4 Microarray: Practical Approach

Culture samples (40–50 mL) were collected, immediately cooled on ice to stop any further gene transcription, and centrifuged for 10 min at 4000 × g and 4°C in a pre-cooled centrifuge. The pellets were re-suspended in 1 mL TRIzol® (Life Technologies, Thermo Fisher Scientific Inc., Grand Island, NY, USA), transferred to screwcap tubes, and flash frozen in liquid nitrogen. The tubes were stored at –80°C until extraction. RNA was extracted using beads (0.5 mm Bashing Beads; Zymo Research, Orange, CA, USA) to facilitate cell disruption and subsequent chloroform phase separation according to supplier's instructions. After the phase separation step, the DirectZol RNA MiniPrep kit (Zymo Research) was used for RNA purification, including the optional DNase I treatment. The purity and quality of the extracted RNA was assessed using a NanoDrop™ ND-1000 spectrophotometer (NanoDrop™ Technologies) and Bioanalyzer assays (Agilent 2100 bioanalyzer, Agilent Technologies, USA). Single-stranded cDNA preparation, labelling, and hybridization onto the microarray slides were performed as described in Eisenhut *et al.* (2007). The microarray slides were scanned with an Agilent Microarray Scanner (model G2505B) with default settings. Microarray analysis can be performed using commercial *Synechocystis* cDNA microarray version 2.0 (TaKaRa Biotech), which covers 3076 of 3264 ORFs of the whole genome, a custom-designed array for *Microcystis aeruginosa* PCC 7806, or any other custom design.

8.2.5 Microarray: Data Analysis

Signal intensities for probes were obtained from the scanned microarray image using Agilent Technologies' Feature Extraction software version 10.5.1.1 (Protocol: GE1_105_Dec08). The array data were analyzed with the R package Limma version 3.18.13 (Smyth, 2005; www.bioconductor.org; www.R-project.org). After background correction ("minimal method"), within-array normalization was applied using global loess normalization. Between-array normalization was applied using A-quantile normalization (Bolstad *et al.*, 2003). Normalized values for replicated probes on the arrays were averaged. Differential expression was statistically evaluated by means of the Limma package based on a linear model with the signal intensities from the control arrays as reference. For each point, p values were calculated from the moderated t statistic, and were adjusted for multiple hypothesis testing by controlling the false discovery rate (Storey and Tibshirani, 2003). Log_2 gene expression values of $< –0.9$, or > 0.9 (similar to Nodop *et al.*, 2008 and Schwarz *et al.*, 2011), and p values < 0.05 were considered differentially expressed. Visual representation was done using hierarchical clustering and heatmaps with the h clust and heatmap.2 functions of the R gplots package version 2.13.0.

8.2.6 References

Bolstad, B.M., Irizarry, R.A., Astrand, M., and Speed, T.P. (2003) A comparison of normalization methods for high density oligonucleotide array data based on variance and bias, *Bioinformatics*, **19**, 185–193.

Eisenhut, M., Von Wobeser, E.A., Jonas, L., *et al.* (2007) Long-term response toward inorganic carbon limitation in wild type and glycolate turnover mutants of the cyanobacterium *Synechocystis* sp. strain PCC 6803, *Plant Physiology*, **144**, 1946–1959.

Nodop, A., Pietsch, D., Hocker, R., *et al.* (2008) Transcript profiling reveals new insights into the acclimation of the mesophilic fresh-water cyanobacterium *Synechococcus elongatus* PCC 7942 to iron starvation, *Plant Physiology*, **147**, 747–763.

Schwarz, D., Nodop, A., Hüge, J., *et al.* (2011) Metabolic and transcriptomic phenotyping of inorganic carbon acclimation in the Cyanobacterium *Synechococcus elongatus* PCC 7942, *Plant Physiology*, **155**, 1640–1655.

Smyth, G.K., Michaud, J., and Scott, H.S. (2005) Use of within-array replicate spots for assessing differential expression in microarray experiments, *Bioinformatics*, **21**, 2067–2075.

Storey, J.D. and Tibshirani, R. (2003) Statistical methods for identifying differentially expressed genes in DNA microarrays, *Methods in Molecular Biology*, **224**, 149–157.

SOP 8.1

DNA-Chip Detection of Potential Microcystin and Nodularin Producing Cyanobacteria in Environmental Water Samples

*Anne Rantala-Ylinen and Kaarina Sivonen**

Department of Food and Environmental Sciences, Division of Microbiology and Biotechnology, University of Helsinki, Helsinki, Finland

SOP 8.1.1   Introduction

This SOP describes a DNA-chip method for simultaneous detection and identification of microcystin (MC) and nodularin (NOD) biosynthesis genes specific for the genera *Anabaena*, *Microcystis*, *Planktothrix*, *Nostoc*, and *Nodularia*. The method includes a PCR step, in which *mcy*E/*nda*F gene variants present in the samples are amplified, and an LDR step, in which probes recognize their target sequences among the amplified PCR products. In the last step, ligated probe pairs are hybridized onto a universal microarray through the attached ZipCode sequences. The specificity and sensitivity of the method was tested with several cyanobacterial strains and applied to lake and Baltic Sea water samples (Rantala, 2007; Rantala *et al.*, 2008). The protocol below was modified from Rantala-Ylinen *et al.* (2011).

---

*Corresponding author: kaarina.sivonen@helsinki.fi

SOP 8.1.2   Experimental

SOP 8.1.2.1   Materials

SOP 8.1.2.1.1   PCR Step

- SUPER TAQ plus polymerase (5 U μL^{-1}) and 10× SUPER TAQ plus buffer (HT Biotechnology Ltd), and deoxynucleotides, e.g., dNTP Mix 10 mM each (Thermo Fisher Scientific).
- Primers (stock solutions 100 pmol μL^{-1})
  - *mcy*E–F2 (5′–GAAATTTGTGTAGAAGGTGC–3′) or *mcy*E–F2b (5′–TGAAATTTG TGTAGAAGGTG–3′)
  - *mcy*E–R4 (5′–AATTCTAAAGCCCAAAGACG–3′).
- Isolated nucleic acids of water samples and laboratory strains.
- Sterile Milli-Q® water.
- 0.5 × TAE buffer (stock 50× Tris-Acetic acid-EDTA, pH 8.0).
- Agarose (LE, analytical grade).
- Ethidium bromide solution (10 mg mL^{-1}) or Midori Green.
- PCR product purification kit, e.g. E.Z.N.A. Cycle-Pure kit (Omega Bio-tek, Inc.).
- Bovine serum albumin (BSA) acetylated (10 mg mL^{-1}).

SOP 8.1.2.1.2   LDR Step

A so-called Oligomix is prepared by mixing discriminating probes (250 fmol μL^{-1}) with a fluorescent Cy3 label at 5′ ends and common probes (250 fmol μL^{-1}) phosphorylated at 5′ ends and with a cZipCode sequences attached to the 3′ ends (Rantala *et al.*, 2008). Exposure to light should be avoided in all the protocol steps to prevent loss of fluorescence.

- LDR control oligonucleotide (25 fmol μL^{-1}):
  5′–AGCCGCGAACACCACGATCGACCGGCGCGCGCAGCTGCAGCTTGCT CATG–3′. Store as aliquots at –20°C.
- *Pfu* DNA ligase (4 U μL^{-1}) and 10× *Pfu* DNA ligase buffer (Stratagene, Agilent Technologies).
- Sterile Milli-Q® water.

SOP 8.1.2.1.3   Hybridization Step

Custom-made microarray slides with ZipCodes that correspond to the probe pairs and cZipCodes of OligoMix (Castiglioni *et al.*, 2004; Sipari *et al.*, 2010). Protect slides from light and dust during storage. Use gloves and forceps, when handling slides. Avoid touching/scratching the printed area.

- Saline-sodium citrate (SSC) buffer (20× concn.).
- Sodium dodecyl sulfate, SDS (10% solution).
- Bovine serum albumin ≥ 96% (BSA).

- Prehybridization solution: 5× SSC, 1% BSA, 50 mL. Store at 4°C if prepared on a previous day.
- Washing solution I: 1× SSC, 0.1% SDS.
- Washing solution II: 0.1× SSC.
- Sterile Milli-Q® water.
- Salmon testes DNA (10 mg mL^{-1}).
- Hybridization control (10 fmol µL^{-1}): 5′–Cy3-GTTACCGCTGGTGCTGCCGCCGTA –3′. Store as aliquots at –20°C.

SOP 8.1.2.2   Equipment

- Thermocycler.
- Electrophoresis equipment.
- UV transilluminator with documentation system.
- Spectrophotometer.
- Hybridization chamber system; the protocol below uses a custom-made chamber and press-to-seal silicone isolator (1.0 × 9 mm; Schleicher & Schuell BioScience, Dassel, Germany) that enables the simultaneous hybridization of eight samples.
- Temperature-controlled water bath.
- Laser scanner, e.g. GenePix Autoloader 4200AL (Axon Instruments, Inc).
- Software for analyzing laser scans, e.g. GenePix Pro Microarray Acquisition and Analysis Software for GenePix Microarray Scanners (Axon Instruments, Inc.).

SOP 8.1.3   Procedure

SOP 8.1.3.1   PCR Step

1. Adjust the amount of isolated DNA of environmental samples to 30–40 ng µL^{-1} and that of laboratory strains (controls) to 10–20 ng µL^{-1}.
2. Adjust the amount of primers to 10 pmol µL^{-1}
3. Pipette the following mixture:
   a. For each sample to be analyzed, mix 2 µL of 10× SUPER TAQ plus buffer (final concn. 1×), 0.5 µL of dNTP mix (final concn. 250 µM), 1 µL of each primer (final concn. 0.5 pmol µL^{-1}), 0.2 µL (=1 U) SUPER TAQ plus polymerase, and 1 µL DNA. Adjust with sterile Milli-Q® water to total volume of 20 µL. Prepare enough PCR mix (excluding DNA) for eight environmental DNA samples (two replicates of each), a positive (DNA of MC or NOD producing strain), and a negative (sterile Milli-Q® water) control sample as well as some extra volume to cover for pipetting losses. Aliquot 19 µL of this PCR mix into the PCR tubes. Then, add the DNA samples and controls to the tubes.
4. Use the following program for PCR cycling: initial denaturation at 95°C for 3 min, 30 cycles: 94°C for 30 sec, 53°C for 30 sec, 68°C for 60 sec, and a final extension step at 68°C for 10 min.
5. Combine the two replicated PCR reactions and run 5 µL of each reaction on a 1.5% agarose gel in 0.5× TAE buffer (incl. EtBr) and include a size marker that covers the

size of the expected amplification product, *ca.* 810 bp, e.g. GeneRuler 100 bp Plus DNA ladder (Thermo Fisher Scientific).

6. Photograph the gel under UV transillumination.
7. Document and compare amplification product sizes of the environmental water samples to that of the positive controls and to the size marker.
8. Purify the rest (*ca.* 35 μL) of the combined eight PCR reactions with E.Z.N.A. Cycle-Pure Kit according to manufacturer's instructions. Elute DNA with 50 μL sterile water.
9. Measure DNA concentration with a spectrophotometer.

SOP 8.1.3.2   LDR Step

1. For LDR, 25 fmol ($=0.025$ pmol) of PCR product is needed. This corresponds to 13.4 ng of PCR product with a length of 810 bp ($=0.025$ pmol $\times$ 810 bp $\times$ mass of one base pair ($=660$ pg pmol^{-1}) $\times$ 1 ng/103 pg). Calculate how many μL of PCR product are needed by dividing 13.4 ng by the DNA concentration of PCR product determined in previous step of the protocol. If the volume needed is < 2 μL, dilute the PCR product prior to preparing the LDR reaction to increase the accuracy of pipetting.
2. For one LDR reaction, add 2 μL 10× Pfu ligase buffer, 1 μL of OligoMix, 1 μL of LDR control oligo, and X μL ($=13.4$ ng) of PCR product. Adjust with sterile Milli-Q$^{®}$ water to have a total volume of 19 μL. Keep the tubes on ice.
3. Heat the reactions for 2 min at 94°C (use a PCR machine), spin, and place the tubes back on ice.
4. Add 1 μL of Pfu DNA ligase to each reaction. Spin again.
5. Use the following program for LDR cycling: 94°C for 2 min, 30 cycles: 90°C for 30 sec and 63°C for 4 min, and a final denaturation 94°C for 2 min, and cooling 4°C for 10 min. Store at –20°C if not used immediately for hybridization.

SOP 8.1.3.3   Hybridization Step

1. Warm the water bath to 42°C and let the prehybridization solution warm in the bath, e.g. in a 50 mL tube. After the temperature is reached, set the microarray glass into the tube and incubate for 1 h in the dark.
2. Wash the glass with sterile Milli-Q$^{®}$ water five times for 30 sec. Dry by centrifugation at 200 $\times$ g for 5 min. The slide can be centrifuged in the tube.
3. Increase the temperature of the water bath to 65°C and set up the hybridization chamber system.
4. Prepare the hybridization mix for eight samples plus one extra to cover for pipetting losses:
   a. For nine reactions, mix 243 μL of sterile Milli-Q$^{®}$ water, 146.7 μL 20× SSC (final concn. 5× SSC), 6.3 μL salmon testes DNA (final concentration 0.1 mg mL^{-1}) and 9 μL hybridization control and mix by vortexing.
5. Add 45 μL of the hybridization mix to the LDR reactions (20 μL) prepared in the LDR step of the protocol (see above). Incubate at 94°C for 2 min in a PCR machine, and chill on ice. Spin the tubes. Transfer 65 μL of these mixes to individual hybridization

chambers carefully to avoid the formation of air bubbles that would interfere with the hybridization of cZipCodes to ZipCodes.

6. Place the hybridization chamber in a plastic box. Create a water-saturated environment, e.g. by placing wet tissues in the bottom of the box. This prevents hybridization mixes from evaporating from the chambers. Submerge the box in water (65°C) and incubate for two hours in the dark. Warm up 50 mL of washing solution I in the water bath during hybridization. Seal the tube carefully.

7. Open the hybridization chamber in washing solution I (room temperature) to avoid drying the hybridization mix on the hot glass slide, which would cause high background signal levels.

8. Wash the slide in 50 mL tubes in the following sequence:
   a. washing solution I (that was pre-warmed) in the 65°C water bath for 15 min
   b. washing solution II at room temperature for 5 min
   c. Milli-Q® water at room temperature three times for 5 min
   d. in every washing step, mix a couple of times by inverting the tube.

9. Dry the slide by centrifugation at 200 × g for 5 min. Store at room temperature in the dark. The slide should be scanned immediately or on the next day after hybridization.

SOP 8.1.3.4   Signal Detection and Image Analysis

1. Depending on the microarray scanner in use, follow the manufacturer's instruction. On the settings, select the wavelength 532 nm that is used for Cy3 dye. Set other parameters, e.g. PMT gain and laser power in such a way that the signal level is not oversaturated (e.g. signal value over 65535 in Axon laser scanner).

2. Scan all the eight arrays and save the images in a format compatible with the image analysis program used (16-bit TIFF format in GenePix Pro program).

3. Open the scanned image and import the file (GAL file in GenePix Pro program) that contains data on the arrayed spots (= features), e.g. size (diameter), position in the array, identity of the ZipCode printed in each position.

4. Following the manufacturer's instructions, perform the image analysis to get at least information on the mean/median signal intensity and signal-to-noise ratio (SNR) ratios of the spots, and mean/median signal intensity of the background.

5. Save the results including a JPEG figure of the scanned image.

6. Export the results to a spreadsheet to evaluate the signals from the spots. A particular *mcyE/ndaF* gene is considered present in the sample, if three conditions are fulfilled by the signals of the corresponding spots (four replicate spots in one array):
   a. SNR of a spot is greater than or equal to three
   b. over 70% of the spot's signal intensity is two standard deviations higher than the background signal intensity (column % > B532+2SD in the results sheet)
   c. at least two of the replicated spots fulfill the first and second conditions.

7. To compare signals of a certain gene in different hybridizations or arrays, the signals first need to be normalized. To do this, the average gene-specific (four replicates) signal intensity is presented as a percentage of average signal intensity of either hybridization (eight replicates) or LDR (six replicates) control spots.

SOP 8.1.4 Notes

The DNA chip technology that is proposed here builds upon earlier experience from the fabrication of a DNA chip targeting the 16S rRNA genes of cyanobacteria (Castiglioni *et al.*, 2004). During the PCR steps, there is a potential bias in *mcy* genotypes' abundances and more rare genotypes may possibly be underrepresented. The combination of several independent PCR products obtained from one sample may help to reduce the potential bias and improve the detection of the less abundant *mcy* genotypes.

SOP 8.1.5 References

Castiglioni, B., Rizzi, E., Frosini, A., *et al.* (2004) Development of a universal microarray based on the ligation detection reaction and 16S rRNA gene polymorphism to target diversity of cyanobacteria, *Applied and Environmental Microbiology*, **70**, 7161–7172.

Rantala, A. (2007) Evolution and detection of cyanobacterial hepatotoxin synthetase genes, PhD thesis, Edita Prima Oy, Helsinki, Finland. Available at: http://urn.fi/URN:ISBN: 978-952-10-4369-7.

Rantala, A., Rizzi, E., Castiglioni, B., *et al.* (2008) Identification of hepatotoxin-producing cyanobacteria by DNA-chip, *Environmental Microbiology*, **10**, 653–664.

Rantala-Ylinen, A., Sipari, H., and Sivonen, K. (2011) Molecular methods: Chip assay and quantitative real-time PCR: In Detecting Hepatotoxic Cyanobacteria, *Methods in Molecular Biology*, **739**, 73–86, doi: 10.1007/978-1-61779-102-4_7.

Sipari, H., Rantala-Ylinen, A., Jokela, J., *et al.* (2010) Development of a chip assay and quantitative PCR for detecting microcystin synthetase E gene expression, *Applied and Environmental Microbiology*, **76**, 3797–3805.

SOP 8.2

cDNA Microarrays for Cyanobacteria

J. Merijn Schuurmans* and Hans C.P. Matthijs

Department of Aquatic Microbiology, Institute for Biodiversity and Ecosystem Dynamics,
University of Amsterdam, Amsterdam, The Netherlands

SOP 8.2.1   Introduction

This SOP describes an Agilent-based cDNA oligonucleotide microarray method to quantify global mRNA transcript expression in cyanobacteria. For *Synechocystis* PCC 6803 and *Microcystis aeruginosa* PCC 7806, Agilent microarray chips are available. The method includes sample collection and preparation of pure RNA for subsequent cDNA synthesis and hybridization on Agilent chips. Additionally, the array analysis is discussed using the Limma software package in R. A flow diagram of the entire procedure is shown in Fig. 8.1.

SOP 8.2.2   Experimental

This procedure describes sample collection, preparation, array hybridization, and array analysis for RNA global expression pattern analysis of cyanobacteria.

SOP 8.2.2.1   Materials for Sample Collection, Preparation, and Hybridization

- Sterile 50 mL tubes.
- RNase-free Eppendorf tubes.
- RNase-free screwcap tubes.
- Gloves (nitrile; don't use latex).
- Liquid nitrogen (N_2 (l)).
- Bucket with ice.
- 0°C freezer capacity.
- TRIzol® reagent.
- Chloroform (without isoamyl alcohol).
- 95–100% ethanol.
- Zymo Direct-Zol™ RNA miniprep kit (Zymo Research).

---

*Corresponding author: J.M.Schuurmans@uva.nl; jmschuurmans@gmail.com

- ZR BashingBead™ Lysis Tubes (2.0 mm) (Zymo Research).
- RNase-free water (fresh Milli-Q®).
- Random hexamers (Invitrogen).
- Superscript II Reverse transcriptase and buffer (Invitrogen).
- Dithiothreitol (0.1M DTT).
- dNTP mix (Invitrogen).
- Cy3 and Cy5 fluorescent DNA labelling dyes (1 mM, Amersham).
- NaOH (1M) and HCl (1M).
- QIAquick™ PCR purification kit (Qiagen).

SOP 8.2.2.2   Equipment for Sample Collection, Preparation, and Hybridization

- Heat block.
- Pipettes and RNase-free filter tips.
- Tube and Eppendorf centrifuges capable of cooling.
- Beadbeater.
- Vortex machine.
- NanoDrop™ (ND-1000) spectrophotometer.
- –80°C freezer capacity.
- Fume hood.

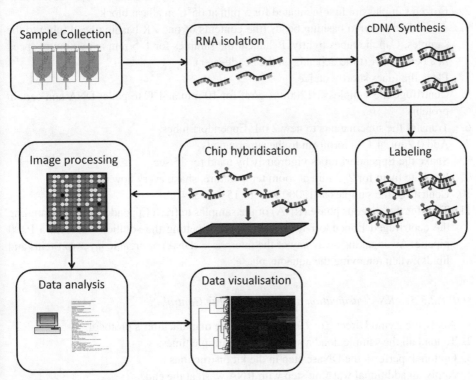

Figure 8.1 *Workflow for two color DNA microarray experiments. (See color plate section for the color representation of this figure.)*

- 2100 bioanalyzer (Agilent).
- Microarray scanner G2505B (Agilent).
- Hybridization chamber (Agilent)

SOP 8.2.3   Procedure

SOP 8.2.3.1   Sample Collection

1. Collect culture samples (10–50 mL of a culture containing approximately 1×10^7 cells mL^{-1}, adjust the sample volume depending on the available amount of cells), and put them on ice.
2. Centrifuge the samples for 5 min at $4000 \times$ g and 4°C in a pre-cooled centrifuge.
3. Remove the supernatant without disturbing the cell pellet.
4. Re-suspend the cell pellets in 1 mL TRIzol® (Life Technologies, Grand Island, NY, USA) immediately after centrifugation in a fume hood.
5. Transfer the TRIzol® mixtures to RNase-free Eppendorf tubes or screwcap vials.
6. The mixtures can be flash frozen in liquid nitrogen and stored at –80°C or immediately used for the following steps.

SOP 8.2.3.2   Cell Lysis and Phase Separation

1. When proceeding immediately with the RNA isolation, continue to step 3 of this list.
2. Frozen samples are first incubated for 5 min at 65°C in a heat block.
3. Add 0.5 mm Zymo bashing beads (the contents of one ZR bashing bead tube) to the samples. Cap the tubes tightly. Beadbeat the samples for 1–5 min at maximum speed in a beadbeater (time depends on the beadbeater).
4. Place the tubes shortly on ice.
5. Centrifuge the samples at $12000 \times$ g/rcf for 10 min at 4°C to pellet DNA and certain proteins.
6. Transfer the supernatants to new 2 mL Eppendorf tubes.
7. Add 0.2 mL 4°C chloroform to the supernatants.
8. Shake the Eppendorf tubes vigorously by hand for 15 sec.
9. Incubate them for 2–3 min at room temperature, shake every now and then.
10. Centrifuge the samples at $12000 \times$ g for 15 min at 4°C.
11. Transfer the aqueous phase (RNA) of the samples to fresh Eppendorf tubes by angling the centrifuged Eppendorf tubes at 45° and pipetting the solution out with a P200 pipette. Avoid taking away any of the interphase (DNA) or organic layer (proteins and lipids) when removing the aqueous phase.

SOP 8.2.3.3   RNA Purification and Initial Quality Control

1. Apply the Zymo Direct-Zol™ kit following the manufacturer's protocol.
2. To load all the sample, load a spin column multiple times.
3. Optional: perform the DNase step in the kit instructions.
4. Apply an additional washing step with RNA wash at the end.
5. Perform an extra spin in an empty collection tube to ensure the membrane is dry.

6. Elute the column with 70 μL and then 30 μL DNase/RNase-free Milli-Q® water (two elution steps).
7. Measure the concentration and purity of the RNA samples with a NanoDrop™ spectrophotometer.
8. Keep the samples on ice during the measurements.
9. Store the RNA samples at −80°C or continue to the next step.

SOP 8.2.3.4   RNA Quality Check

1. Load the RNA samples into a bioanalyzer (Agilent) using the manufacturer's protocol.
2. Samples of a concentration of more than 50 ng μL^{-1} and showing no signs of degradation can be used for cDNA synthesis.

SOP 8.2.3.5   cDNA Synthesis and Labeling

1. Mix 10 μg of total RNA with 0.5 μg random hexamers in 15 μL total volume.
2. Incubate 10 min at 70°C.
3. Chill for 10 min at 4°C (on ice).
4. Add 6 μL 5× reverse transcriptase buffer, 3 μL DTT (0.1 M), 1.4 μL sterile Milli-Q® water and 0.6 μL dNTP mix (25 mM) and mix gently.
5. Add 2 μL Cy3 or Cy5 (1 mM) dye.
6. Add 2 μL Superscript II reverse transcriptase.
7. Mix carefully and incubate at room temperature (RT) for 10 min.
8. To start cDNA synthesis, transfer the samples to 42°C for 110 min.
9. Hydrolyze RNA by adding 1.5 μL NaOH (1 M) and putting the sample at 70°C for 10 min.
10. Neutralize the alkali by adding 1.5 μL of HCl (1 M).
11. Apply the QIAquick™ PCR purification kit according to the manufacturer's instructions.
12. Check the concentration of Cy-labelled cDNA on the NanoDrop™ spectrophotometer.
13. The labelled cDNA is now ready for hybridization.

SOP 8.2.3.6   Agilent Array Hybridization

1. Denature equal amounts of Cy3 and Cy5 labelled cDNA at 98°C for 3 min.
2. Mix with 125 μL hybridization buffer and 25 μL control target solution.
3. Apply the hybridization solution to the microarray surface.
4. Hybridize the solution to the array in an Agilent hybridization oven for 17 h at 60°C.
5. Wash the slides according to the Agilent cDNA microarray kit protocol.
6. After washing, the slides can be scanned with an Agilent microarray scanner using default settings.
7. Spot fluorescence intensity can be extracted using Feature Extraction software 8.1.1.1.
8. Mean signal intensity of each spot was used for further analysis.

SOP 8.2.3.7   Array Analysis

1. For array analysis the Limma package from Bioconductor in the R statistics environment is used.
2. Create a text file with the experiment description with the sample, file containing data and the dye scheme, for example:

Sample Filename

Cy5 Cy3

cult_1 US22502421_254750310003_S01_GE2_107_Sep09_1_1.txt t_0_75 t_2

cult_2 US22502421_254750310003_S01_GE2_107_Sep09_1_2.txt t_4   t_8

cult_1 US22502421_254750310003_S01_GE2_107_Sep09_1_3.txt t_0   t_4

---

3. Attach spot types using a SpotType file:

SpotType	**ControlType**	**color**	**cex**
Gene	*	black	0.2
PosContr	1	red	0.2
NegContr	-1	blue	0.2

---

4. Remove the controls from the data.
5. Perform a background correction.
6. Normalize the data within arrays.
7. Normalize the data between arrays.
8. Fit a linear model according to your experimental design.
9. Calculate log expression values using a contrast matrix and calculate empirical Bayesian statistics for the dataset.
10. Save the calculated values into a text file for further analysis or plotting.

SOP 8.2.3.8   An Example R-Script with Use of Limma

```
#R script for 2 channel array analysis Microcystis
##################################################

# Can be done in Rstudio windows
library(limma)
##################################################
# Complete = with self/self hyb
targetsFile <- "Exp_descr.txt"
targets <- readTargets(file=targetsFile, sep="\t")
show(targets)
# Import raw data:
RG <- read.maimages(targets$Filename, source="agilent",
quote="")
```

```
dim(RG)
show(RG)
summary(RG$R)
# SpotTypes attachement:
spottypes <- readSpotTypes(file = "SpotTypeFile.txt")
show(spottypes)
RG$genes$Status <- controlStatus(spottypes, RG)
# REMOVING CONTROLs:
isGene <- RG$genes$Status == "Gene"
dim(RG[isGene,])
RG.raw <- RG[isGene,]
dim(RG.raw)
### Background Correction
### and Within Array Normalisation
RGb <- backgroundCorrect(RG.raw,method="normexp", offset=20)
MAb <- normalizeWithinArrays(RGb, method="loess")
RG_Norm <- RGb
MA_Norm <- MAb
### Between Array Normalisation
MA_scaled <- normalizeBetweenArrays(MA_Norm,
method="Aquantile")

### Linear Model
design <- modelMatrix(targets, ref="reference")
fit_simple <- lmFit(MA_scaled, design)
contrast.matrix <- makeContrasts(t_1, t_2, t_4, t_8, t_24,
levels=design)
fit_simple2 <- contrasts.fit(fit_simple, contrast.matrix)
fit_simple2 <- eBayes(fit_simple2)
results <- decideTests(fit_simple2, method = "separate",
adjust.method="BH", p. value=0.001)
write.fit(fit_simple2, file = "simple_fit.txt", digits=4,
adjust="fdr", sep="\t")
###End
```

SOP 8.2.4   Notes

SOP 8.2.4.1   Sample Collection

For a typical batch culture (OD_{750} ~0.2 or ±1×10^7 cells · mL^{-1}) 50 mL is needed for sufficient material.

Samples can also be stored for at least one month, after addition of TRIzol® and freezing with liquid nitrogen (N_2 (l)), at −80°C.

SOP 8.2.4.2   Cell Lysis and Phase Separation

Lysis with beads and TRIzol® is more effective than TRIzol® alone.

Actual effective beadbeat time depends on the beadbeater used.

In high-fat-content samples, a layer of fat collects above the supernatant. Remove and discard this fatty layer.

The 4°C spins are essential for phase separation. Room temperature spins may result in variable phase separation thus resulting in variable RNA yields. Use straight chloroform in the above step; no isoamyl alcohol is needed.

SOP 8.2.4.3   RNA Purification and Initial Quality Control

Minimum results needed for Micro-array analysis:

- RNA concentration, larger than 50 ng μL^{-1}.
- RNA quality (A_{260}/A_{280}) $\geq$ 1.8.
- Ideally the RNA quality (A_{260}/A_{230}) $\geq$ 1.8; however, this is not essential.

The NanoDrop™ cannot distinguish between RNA and DNA. The NanoDrop™ cannot be used to detect degraded RNA; however, degraded RNA can cause an increased absorbance at 260 nm.

- If A_{280} is >0.5, the reading is out of the linear range; dilute the sample.
- Partially dissolved RNA samples have an A_{260}/A_{280} ratio < 1.6.
- A A_{260}/A_{280} ratio of $\geq$ 1.8 is generally accepted as "pure" for RNA. If the ratio is lower, it may indicate the presence of protein, phenol, or other contaminants that absorb strongly at or near 280 nm.
- Expected A_{260}/A_{230} values are commonly in the range of $\geq$ 1.8. If the ratio is appreciably lower than expected, it may indicate the presence of contaminants which absorb at 230 nm.
- Guanidine isothiocyanate (component of TRIzol®) has a very strong effect on the OD A_{260}/A_{230} ratio, which at the 0.5% contamination level drops below 0.5.

Phenol (component of TRIzol®) has a very strong effect on the quantification of RNA. Note that, at a 0.5% contamination level, the measured concentration is over three times as high as the actual value of 50 ng μL^{-1}. This strong disturbing effect of phenol on the quantification of RNA is due to the contaminating absorption peak at 270 nm. At very low RNA concentrations below 10 ng μL^{-1} this contaminating peak is often even confused for RNA. If the peak is at 270 nm it arose from a contamination.

No spin column RNA purification step was included, to include small RNAs. The components of total RNA include: ribosomal RNA (rRNA) ~90% of total RNA, transfer RNA (tRNA) ~5% of total RNA, small nuclear RNAs ~1% of total RNA, and messenger RNA ~1–3% of total RNA.

SOP 8.2.4.4   RNA Quality Check

Samples with a lower concentration that are not degraded can be used, but will need to be concentrated.

SOP 8.2.4.5   cDNA Synthesis and Labeling

Alternatively, an RNaseH treatment can be applied.

SOP 8.2.4.6   Agilent Array Hybridization

Instead of washing solution 3 in the Agilent protocol, acetonitrile can be used to avoid precipitate formation on the slide.

SOP 8.2.4.7   Array Analysis

To check on the quality of the array data and to visualize the normalization steps, plots can be made (boxplots; MAplots) in between steps. In most cases a standard offset to 20 can be used (normexp; offset = 20). Generally "Loess" normalization is used.

There are many different ways available, and each has its advantages and disadvantages. Generally speaking, "aquantile" is used. See also the Limma guide and documentation (http://www.bioconductor.org/packages/release/bioc/vignettes/limma/inst/doc/usersguide .pdf or Bolstad *et al.* (2003).

The generated file can be used for further processing into clustering, toptables, heatmaps, or other ways of presenting or visualizing the dataset. Note that replicate spots are not averaged in this analysis. This can be done after normalization (using the avereps function in the Limma package in R) or after data interpretation. Note that *p* values should never be averaged. A possibility to merge *p* values is to perform Fisher's method using the metap package in R.

SOP 8.2.5   Reference

Bolstad, B.M., Irizarry, R.A., Astrand, M., and Speed, T.P. (2003) A comparison of normalization methods for high density oligonucleotide array data based on variance and bias, *Bioinformatics*, **19**, 185–193.

9

Analysis of Toxigenic Cyanobacterial Communities through Denaturing Gradient Gel Electrophoresis

Iwona Jasser[1]*, Aleksandra Bukowska[1], Jean-Francois Humbert[2], Kaisa Haukka[3], and David P. Fewer[3]

[1]*Department of Microbial Ecology and Environmental Biotechnology, Faculty of Biology, University of Warsaw, Warsaw, Poland*
[2]*Institute of Ecology and Environmental Sciences, UPMC, Paris, France*
[3]*Department of Food and Environmental Sciences, Division of Microbiology and Biotechnology, University of Helsinki, Helsinki, Finland*

9.1 Introduction

Denaturing gradient gel electrophoresis (DGGE) is a powerful method that allows the detection of single base-pair changes in a DNA sequence and has been widely applied to study the molecular ecology and diversity of microorganisms (Boutte *et al.*, 2008; Fernandez-Carazo *et al*,. 2011; Zhan and Sun, 2012). DNA fragments of the same length but with different nucleotide sequences are separated (Muyzer and Smalla, 1998; Muyzer, 1999). The separation of DNA strands is based on differences in melting behaviors of double-stranded DNA in a polyacrylamide gel with a linear gradient of denaturants in DGGE (Myers *et al.*, 1987).

*Corresponding author: jasser.iwona@biol.uw.edu.pl

Molecular Tools for the Detection and Quantification of Toxigenic Cyanobacteria, First Edition.
Edited by Rainer Kurmayer, Kaarina Sivonen, Annick Wilmotte and Nico Salmaso.
© 2017 John Wiley & Sons Ltd. Published 2017 by John Wiley & Sons Ltd.

A GC-rich region is attached to the studied fragment through PCR in order to prevent complete dissociation of DNA strands. This GC-clamp has a high melting temperature, and does not dissociate in the concentration of denaturants used, resulting in a Y-shaped partially double-stranded tertiary structure that retards migration through the polyacrylamide gel. The GC-clamp is 40 bp in length, and is attached to either the 5′ or the 3′ end of the fragment by PCR amplification (Sheffield *et al.*, 1989).

DNA sequences are composed of melting domains, which melt at the same temperature or concentration of the denaturant. When PCR fragments are separated in a polyacrylamide gel containing a gradient of denaturants, the position where fragment migration halts is determined by the melting domain with the lowest temperature (Myers *et al.*, 1987). This forms the basis for separating PCR products from one another in a polyacrylamide gel with a gradient of denaturants and allows discrimination of single base-pair differences.

9.2   Main Applications of the Method

The main application of DGGE in microbial ecology is the study of the genetic diversity of microbes in their own environment without the need to isolate and culture these organisms (Muyzer, 1999). The method provides a means of gaining rapid and inexpensive insights into the genetic diversity of microbial communities. In microbial ecology the DGGE fingerprinting method typically relies on the isolation and amplification of genes encoding the 16S rRNA genes and subsequent physical separation of these fragments to provide a snapshot of the genetic diversity of the microbial community. It can be used to study the diversity of the overall cyanobacterial community, diversity within the taxon, or to distinguish between different organisms with specific features in an environmental sample. DGGE is also useful in the detection of unknown genotypes (e.g. in less studied habitats). In addition, since the results of DGGE are semi-quantitative, it can provide information about the structure of microbial communities.

9.3   Possible Applications

The DGGE method has been used to study the genetic diversity of cyanobacteria (Janse *et al.*, 2003; Taton *et al.*, 2003; Becraft *et al.*, 2011; Jasser *et al.*, 2013) as well as toxigenic cyanobacteria (Boutte *et al.*, 2008; Kim *et al.*, 2010). The DGGE method has also been used to distinguish cyanobacteria taxa possessing the genes responsible for the production of toxins by cyanobacteria directly, as well as determining the diversity of toxin biosynthetic genes within a community. A number of DGGE-based methods have been developed, which target the microcystin biosynthetic pathway directly to provide insights into the genetic diversity of microcystin-producing cyanobacteria (Table 9.1). These methods are useful to monitor toxigenic cyanobacteria in natural environments (Fewer *et al.*, 2009; Ye *et al.*, 2009; Yen *et al.*, 2012; Bukowska *et al.*, 2014). No DGGE methods have been developed yet for anatoxin, saxitoxin, or cylindrospermopsin pathways, but as more sequence information emerges for the biosynthetic pathways responsible for these toxins it will become feasible. The orientation of the clamp, the priming sites, and the selection of amplified fragment must be optimized, either empirically or using theoretical predictions and calculations of the melting behavior (Myers *et al.*, 1987).

Table 9.1 Loci used in DGGE analysis

Locus	Forward primer	Reverse primer	Target	Reference
16S	GGGGAAT(T/C)TT CCGCAATGGG	GACTACTGGGGTA TCTAATCCCATT and GACTACAGGGG TATCTAATCCCTTT	All genera	Nübel et al. 1997
ITS	G(T/C)CACGCC CGAAGTC(G/A) TTAC	CTAACCACCTGA GCTAAT	All genera	Janse et al. 2003
cpcBA	AACCTATGTAG CTTTAGGAGTACC	CTTAAGAAACGA CCTTGAGAATC	All genera	Kim et al. 2010
mcyE	GTTAACTTTTTT GGGCATAGT CCTGA	TTGCAGAAATTC TTTAGGCTGTAAATC	All genera	Fewer et al. 2009
mcyJ	TAGCTAAAGCAG GGTTATCG	TCTTACTATTAACC CGCAGC	Microcystis	Kim et al. 2010
mcyA	AAAATTAAAAG CCGTATCAAA	AAAAGTGTTTTATTA GCGGCTCAT	All genera*	Hisbergues et al. 2003 Ye et al. 2009 Bukowska et al. 2014
GC clamp	CGCCCGCCGCG CCCCGCGCCCG GCCCGCCGCCC CCGCCCC			Sheffield et al. 1989

*They are considered universal, but in our experience they preferentially amplify *Microcystis* and *Planktothrix*.

9.4 DGGE Procedure

1. PCR amplification.
2. Pouring DGGE gels with appropriate gradient of denaturant.
3. Separation of PCR products on the DGGE gel for set amount of time.
4. Post-staining of DGGE gel.
5. Documentation of band patterns for clustering analyses.
6. Optional excision, re-amplification by PCR, purification, and cycle-sequencing of bands.

The recognition of diversity of studied cyanobacteria can be realized in two ways: (1) through clustering analyses or (2) sequencing of separated bands (Fig. 9.1). Based on the position of the bands in the gel and their intensity, image analysis software and clustering analyses allow for the quantification of the diversity of the communities. DGGE fingerprints also facilitate observation of the changes in subsequent populations in time, without the need for their identification. On the other hand, well-separated and visible

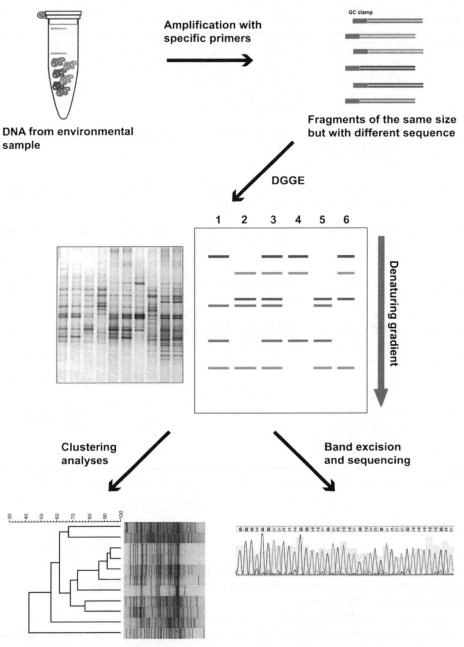

Figure 9.1 *Scheme showing the principle of DGGE analysis with two main applications: clustering analyses and sequencing of separated bands. (See color plate section for the color representation of this figure.)*

bands on a DGGE profile can be excised from the gel and sequenced after re-amplification. This second approach enables verification of the taxa, which DNA has been earlier amplified with specific primers. The DGGE procedure is described step by step in SOP 9.1.

9.5 General Conclusions Including Pros and Cons of the Method

The DGGE method relies on amplification of DNA through PCR and therefore has the potential to detect even minute amounts of DNA (see Chapter 6). The method is relatively easy to perform and inexpensive and allows insights into the genetic diversity and structure of multiple microbial communities and comparison of communities in time and space. It can be considered the first step followed by next-generation sequencing (NGS) in studying new environments or a main approach when the funds for sequencing are limited. However, the method is also time-consuming if sequence information is required and the method requires special equipment. Caution must be exercised using the method as artefacts can be introduced through PCR. There can also be biases introduced in the study of toxic cyanobacteria using toxin biosynthetic genes. Many of the methods developed in the literature are either intentionally or inadvertently specific to one type of toxic cyanobacteria, but the danger is researchers possibly being led to believe that there are no toxic cyanobacteria present when, in fact, there are (false negatives). DGGE conditions need to be optimized for each new application in order to obtain satisfactory results.

9.6 Optimization of the Method and Troubleshooting

A DNA fragment selected for separation should have an appropriate level of variability. Sequences, which can be separated using the DGGE technique, must differ by at least 1% (Muyzer *et al.*, 1993). The length of a separated DNA fragment usually does not exceed 600 bp, though in some cases larger fragments were also separated (Vainio and Hantula, 2000).

The concentrations of acrylamide most often used for environmental applications are 6–8%. However, sometimes, even gels up to 12% were employed (Green *et al.*, 2004). Usually, the smaller the DNA fragment, the higher the concentration of acrylamide used. A proper gradient range is essential for good separation. Concentration is the lowest at the top of the gel denaturants and the highest at the bottom. An overly wide gradient may cause incomplete separation of the different fragments, so that the diversity of the whole sample is not fully revealed. On the other hand, if the gradient is too narrow, the DNA can migrate out of the gel, especially the small fragments, which do not stop completely during electrophoresis. It is recommended to check individually what an appropriate gradient for each fragment would be. Beginning with the wider gradient, it should be gradually narrowed until obtaining a satisfactory result.

The next step, which needs optimization, is the runtime and voltage. The ranges used are 3–17 h and 50–250 V. Normally, the shorter the time, the higher the voltage applied. However, the longer electrophoresed gels seem to be of better quality. In some cases, higher voltage is used in the initial phase of DGGE followed by separation in the lower voltage. The buffer temperature in which the electrophoresis is performed is another very important factor. It must be maintained on a constant level during the entire duration of the DGGE; in most cases this is 60°C.

The amount of PCR product loaded on the gel depends on the length of the DNA fragment and its sequence heterogeneity. It is important to load the similar amount of amplicon on each lane. Usually, it is 300–500 ng per lane for environmental samples, while the amount of DNA needed for PCR reaction is 0.2–2 ng μL^{-1}. PCR cycling conditions must be optimized individually for each primer pair. However, it is helpful to apply the progressive reduction of annealing temperature in order to reduce non-specific annealing of the primers. A reference DNA sample with known migration pattern on each gel is essential for image analysis programs when comparing several DGGE digitalized profiles. It is very useful to create a standard of mixed variously migrating fragments, or to use the same sample as a standard. It is recommended for there to be a standard on at least 2–3 lines on each gel.

9.7 References

Becraft, D., Cohan, F.M., Kühl, M., *et al.* (2011) Fine-scale distribution patterns of *Synechococcus* ecological diversity in microbial mats of Mushroom Spring, Yellowstone National Park, *Applied Environmental Microbiology*, **77**, 7689–7697.

Boutte, C., Mankiewicz-Boczek, J., Komarkova, J., *et al.* (2008) Diversity of planktonic cyanobacteria and microcystin occurrence in Polish water bodies investigated using a polyphasic approach, *Aquatic Microbial Ecology*, **51**, 223–236.

Bukowska, A., Bielczynska, A., Karnkowska-Ishikawa, A., *et al.* (2014) Molecular (PCR-DGGE) versus morphological approach: analysis of taxonomic composition of potentially toxic cyanobacteria in freshwater lakes, *Aquatic Biosystems*, **10**, 2–11.

Fernandez-Carazo, R., Hodgson, D.A., Convey, P., and Wilmotte, A. (2011) Low cyanobacterial diversity in biotopes of the Transantarctic Mountains and Shackleton Range (80–82°S), Antarctica, *FEMS Microbiology Ecology*, **77**, 503–517.

Fewer, D.P., Köykkä, M., Halinen, K., *et al.* (2009) Culture-independent evidence for the persistent presence and genetic diversity of microcystin-producing *Anabaena* (Cyanobacteria) in the Gulf of Finland, *Environmental Microbiology*, **11**, 855–866.

Green, S.J., Freeman, S., Hadar, Y., and Minz, D. (2004) Molecular tools for isolate and community studies of Pyrenomycete fungi, *Mycologia*, **96**, 439–451.

Hisbergues, M., Christiansen, G., Rouhiainen, L., *et al.* (2003) PCR-based identification of microcystin-producing genotypes of different cyanobacterial genera, *Archives of Microbiology*, **180**, 402–410.

Janse, I., Meima, M., Kardinaal, W.E., and Zwart, G. (2003) High-resolution differentiation of cyanobacteria by using rRNA-internal transcribed spacer denaturing gradient gel electrophoresis, *Applied Environmental Microbiology*, **69**, 6634–6643.

Jasser, I., Królicka, A., Jakubiec, K., and Chróst, R.J. (2013) Seasonal and spatial diversity of picocyanobacteria community in the Great Mazurian Lakes derived from DGGE analyses of ITS region of rDNA and cpcBA-IGS markers, *Journal of Microbiology and Biotechnology*, **23**, 739–749.

Kim, S.G., Joung, S.H., A. , *et al.* (2010) Annual variation of *Microcystis* genotypes and their potential toxicity in water and sediment from a eutrophic reservoir, *FEMS Microbiology Ecology*, **74**, 93–102.

Muyzer, G. (1999) DGGE/TGGE method for identifying genes from natural ecosystems, *Current Opinion in Microbiology*, **2**, 317–322.

Muyzer, G. and Smalla, K. (1998) Application of denaturing gradient gel electrophoresis (DGGE) and temperature gradient gel electrophoresis (TGGE) in microbial ecology, *Antonie Van Leeuwenhoek*, **73**, 127–141.

Muyzer, G., De Waal, E., and Uitterlinden, A.G. (1993) Profiling of complex microbial populations by denaturing gradient gel electrophoresis analysis of polymerase chain reaction-amplified genes coding for 16S rRNA, *Applied Environmental Microbiology*, **59**, 695–700.

Myers, R.M., Maniatis, T., and Lerman, L.S. (1987) Detection and localization of single base changes by denaturing gradient gel electrophoresis, *Methods in Enzymology*, **155**, 501–527.

Nübel, U., Garcia-Pichel, F., and Muyzer, G. (1997) PCR primers to amplify 16S rRNA genes from cyanobacteria, *Applied Environmental Microbiology*, **63**, 3327–3332.

Sheffield, V.C., Cox, D.R., Lerman, L.S., and Myers, R.M. (1989) Attachment of a 40-base-pair G+C rich sequence (GC-clamp) to genomic DNA fragments by the polymerase chain reaction results in improved detection of single-base changes, *Proceedings of the National Academy of Sciences USA*, **86**, 232–236.

Taton, A., Grubisic, S., Brambilla, E., *et al.* (2003) Cyanobacterial diversity in natural and artificial microbial mats of Lake Fryxell (McMurdo Dry Valleys, Antarctica): A morphological and molecular approach, *Applied Environmental Microbiology*, **69**, 5157–5169.

Vainio, E.J. and Hantula, J. (2000) Direct analysis of wood-inhabiting fungi using denaturing gradient gel electrophoresis of amplified ribosomal DNA, *Mycological Research*, **104**, 927–936.

Ye, W., Liu, X., Tan, J., *et al.* (2009) Diversity and dynamics of microcystin-producing cyanobacteria in China's third largest lake: Lake Taihu, *Harmful Algae*, **8**, 637–644.

Yen, H.K., Lin, T.F., and Tseng, I.C. (2012) Detection and quantification of major toxigenic *Microcystis* genotypes in Moo-Tan reservoir and associated water treatment plant, *Journal of Environmental Monitoring*, **14**, 687–966.

Zhan, J. and Sun, Q. (2012) Diversity of free-living nitrogen-fixing microorganisms in the rhizosphere and non-rhizosphere of pioneer plants growing on wastelands of copper mine tailings, *Microbiological Research*, **167**, 157–165.

SOP 9.1

DGGE-*mcy*A Conditions

Aleksandra Bukowska* and Iwona Jasser

Department of Microbial Ecology and Environmental Biotechnology, Faculty of Biology,
University of Warsaw, Poland

SOP 9.1.1 Introduction

This SOP presents a detailed description of the DGGE procedure used to study occurrence and diversity of potentially microcystin-producing cyanobacteria. The analysis uses the *mcy*A gene to identify potentially toxic cyanobacteria in mixed assemblages of cyanobacteria from lakes of various trophic status. Principles and applications of DGGE analysis are presented in Chapter 9.

SOP 9.1.2 Experimental

SOP 9.1.2.1 Materials

- PCR reagents (see Chapter 6).
- Template DNA.
- Agarose electrophoresis reagents (see Chapter 6).
- Acrylamide/bis-Acrylamide 37.5:1, 40% solution, e.g. A7168, Sigma-Aldrich.
- TAE buffer 50×, e.g. 50× TAE buffer, CL86.1, Roth or buffer made individually of Tris, acetic acid, EDTA and deionized water. Dilute with deionized water to 1× working solution.
- Formamide 99.5%, e.g. 47671, Sigma-Aldrich.
- Deionized water.

*Corresponding author: a.bukowska@biol.uw.edu.pl

- Urea 99.5%, e.g. 51456, Fluka.
- Ammonium persulfate (APS) 98%, e.g. A3678, Sigma-Aldrich.
- Tetramathylethylenediamine (TEMED) 99%, e.g. T9281, Sigma-Aldrich.
- 96% ethanol.
- Loading dye, e.g. 6× orange loading dye solution, R0631, Life Technologies (Thermo Fisher Scientific).
- Reference DNA, mix of variously migrating on DGGE DNA fragments, prepare individually.
- SYBR Green I nucleic acid gel stain, 10000×, e.g. S9430, Sigma-Aldrich

SOP 9.1.2.2   Equipment

- PCR equipment (see Chapter 6).
- Agarose gel electrophoresis equipment (see Chapter 6).
- Laboratory balance, e.g. CPA224S-0CE, Sartorius.
- DGGE machine with equipment, e.g. DCode Universal Mutation Detection System (Bio-Rad), Fig. 9.2.
- Power supply, e.g. PowerPack Basis, Bio-Rad.
- Gradient maker, e.g. GM-100, CBS scientific.
- Magnetic stir plate and small stir bar.
- Needle and syringe.
- Transilluminator with camera, e.g. Benchtop 2UV Transilluminator, UVP and EOS 600D, Canon.
- Scalpel.

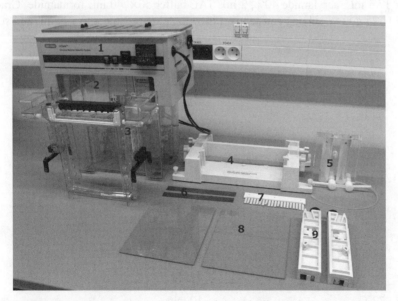

Figure 9.2 *DGGE machine with equipment. 1, lid; 2, buffer tank; 3, inner core; 4, casting stand; 5, gradient maker; 6, spacers; 7, comb, 8, glass plates; 9, clamps.*

SOP 9.1.3   Procedure

SOP 9.1.3.1   PCR of mcyA Gene

- Primer set (Hisbergues *et al.*, 2003; Ye *et al.*, 2009; Bukowska *et al.*, 2014):
  - *mcyA*-Cd1F (5′–AAAATTAAAAGCCGTATCAAA–3′)
  - *mcyA*-Cd1R (5′–AAAAGTGTTTTATTAGCGGCTCAT–3′).
- On the 5′ end of the forward primer the 40-nucleotide GC clamp (5′–CGCCCGCCGCGC CCCGCGCCCGGCCCGCCGCCCCCGCCCC–3′) is added.
- The DNA fragment amplified with this primer pair is about 300 bp. PCR mixture contains approximately 20 ng of DNA isolated from an environmental sample.
- For PCR procedure, see Chapter 6.
- See Table 9.2.

SOP 9.1.3.2   DGGE Electrophoresis

Denaturing gradient gel electrophoresis is carried out on 1-mm-thick vertical gels, containing 7% polyacrylamide (at an acrylamide/bis-acrylamide ratio of 37.5:1). Linear gradient of denaturants in gel increase from 20 to 60%.

SOP 9.1.3.2.1   Reagent Preparation

- 7% acrylamide solution (100 mL). Mix 17.5 mL acrylamide 40%, 2 mL TAE buffer 50×, and fill with deionized water to 100 mL.
- 7% acrylamide solution with 100% concentration of denaturants.
- Mix 17.5 mL, acrylamide 40%, 2 mL TAE buffer 50×, 40 mL formamide. Gradually, add 41.25 g urea, stirring constantly. Fill with deionized water to 100 mL and stir until

Table 9.2   *PCR cycling program*

Temperature	Duration	Cycles
94°C	4 min	
94°C	1 min	
59°C–47°C (decreasing)	1 min	20×
by 0.6°C every cycle		
72°C	1 min	
94°C	30 sec	
47°C	40 sec	10×
72°C	40 sec	
72°C	10 min	
4°C	∞	

urea is completely dissolved. Heating the mixture can help to dissolve urea. Protect the acrylamide solutions from light and store in 4°C no longer than 60 days.
* 10% APS
* Weigh 1 g APS and pour into the volumetric flask. Fill with deionized water to 10 mL. Stir until APS is completely dissolved. Split into 0.5 mL portions and freeze. Do not use thawed APS again.

SOP 9.1.3.2.2   Gel Preparation (Use Powder-Free Gloves and Paper Towels)

1. Wash the glass plates, spacers, and comb with detergent and wipe them with 96% ethanol.
2. Put the spacers on the sides of the back plate (larger) and place the front plate on this. Put the special alignment card included in the DGGE system between the glass plates and attach the clamps on each side. Place the sandwich assembly in the casting stand.
3. Prepare gel solutions with 20% and 60% denaturant concentrations, as indicated in Table 9.3.
4. Add the APS and TEMED *immediately* before pouring solutions into the gradient maker. Mix well.
5. Put the gradient maker on the stir plate and place on a platform (around 20 cm). Put the stir bar into the right chamber of the gradient maker, which is intended for the high-denaturant solution (chamber closer to the rubber hose). Set the casting stand containing the glass plate sandwich underneath. Place the end of the hose with the needle between the two glass plates.
6. Pour gel solutions into appropriate chambers in the gradient maker (low-denaturing to the left and high-denaturing to the right). Begin stirring and make sure that the low-denaturing solution leaks into the right chamber (different density of liquids is visible) and that the space between the glass plates is filling up. Get rid of air bubbles in the hose using a syringe, if the solution is not leaking.
7. After completely filling the space between the plates, take out the hose and pour deionized water into the chambers in order to flush away acrylamide and prevent its polymerization in the gradient maker, hose, and needle.
8. Put the comb between the glass plates.
9. Repeat the procedure and prepare second gel in a new sandwich assembly.
10. Allow gels to polymerize for at least 2 h.

Table 9.3 *Preparation of gel solutions*

20% denaturing gel	60% denaturing gel
3.2 mL, 7% acrylamide with 100% denaturants	9.6 mL, 7% acrylamide with 100% denaturants
12.8 mL, 7% acrylamide without denaturants	6.4 mL, 7% acrylamide without denaturants
12 μL TEMED	12 μL TEMED
120 μL, 10% APS	120 μL 10% APS

SOP 9.1.3.2.3   Electrophoresis

1. Prepare 7 L 1× TAE buffer (140 mL 50× TAE and 6860 mL deionized water) and pour into the DGGE tank. Place the lid on the tank, switch on the heater and mixing, and heat the buffer to about 65°C (runtime temperature is 60°C but during the sample loading the buffer temperature will decrease).
2. Take out the combs from polymerized gels. Attach the gel sandwiches to the inner core and place the core in the tank. Flush the wells with 1× TAE buffer using 1 mL pipette.
3. Mix the amplified DNA (300–500 ng) with 10 μL of loading dye. Load the samples into the wells using 200 μL capillary tips. Put reference DNA sample on at least 2–3 lanes on each gel. Do not load the samples on side lanes; load them with 30 μL of loading dye.
4. Place the lid on the tank, set the heating to 60°C, and turn on the mixing. If the temperature is below 60°C, wait until the buffer reaches the final temperature.
5. Turn on the recirculating pump and wait until the upper chamber is filled with the buffer.
6. Connect the power and run electrophoresis for 5 min at 200 V. Afterwards set 50 V and run DGGE for 16 h (960 min).
7. After electrophoresis, disconnect the power supply, turn off the heater, mixing, and pump, and wait 30 sec.
8. Remove the cover and take out the inner core from the tank. Remove the gel sandwich assembly from the core. Carefully take off the frontal glass plate. Let the gel remain on the back plate.
9. Mix 14 μL SYBR Green and 200 μL deionized water in the plastic container. Place the back plate with the gel in the container and shake until the gel detaches from the glass. Remove the plate from the container carefully, paying attention not to damage the gel. Stain in a dark place for 15 min.
10. Move the gel carefully onto the transilluminator and photograph it. Digitalized DGGE profiles can be used for clustering analyses.
11. Repeat the procedure with the second gel.

SOP 9.1.3.2.4   Excising the Bands and Reamplification

1. Select the bands, which you want to excise. Print the gel image and mark selected bands.
2. Excise the bands with sterile scalpel, trying to cut them out with the least amount of surrounding polyacrylamide. Expose gel to UV light for as short a time as possible. After each band sterilize scalpel with 96% ethanol and fire.
3. Place excised bands in sterile 1.5 mL Eppendorf-type tubes and add 40 μL sterile deionized water. Incubate for 24 h in 4°C.
4. Use the eluent as a template to re-amplification with the same primer set. Perform the PCR reaction according to the same procedure as described in Section 9.1.3.1.
5. If amplification was successful, run 15 μL of re-amplified band on DGGE according to the same procedure (see Section 9.1.3.2.3). If only one band, on the same position as original band is visible, the rest of the re-amplified PCR product can be used for sequencing.

SOP 9.1.4 Notes

Use powder-free gloves and paper towels.

Acrylamide is a neurotoxicant: always wear protective clothes and gloves.

Protect your skin and eyes from UV radiation.

SOP 9.1.5 References

Bukowska, A., Bielczynska, A., Karnkowska-Ishikawa, A., *et al.* (2014) Molecular (PCR-DGGE) versus morphological approach: analysis of taxonomic composition of potentially toxic cyanobacteria in freshwater lakes, *Aquatic Biosystems*, **10**, 2–11.

Hisbergues, M., Christiansen, G., Rouhiainen, L., *et al.* (2003) PCR-based identification of microcystin-producing genotypes of different cyanobacterial genera, *Archives of Microbiology*, **180**, 402–410.

Ye, W., Liu, X., Tan, J., *et al.* (2009) Diversity and dynamics of microcystin—producing cyanobacteria in China's third largest lake, Lake Taihu, *Harmful Algae*, **8**, 637–644.

10

Monitoring of Toxigenic Cyanobacteria Using Next-Generation Sequencing Techniques

Li Deng[1,2]*, Maxime Sweetlove[3]*, Stephan Blank[1], Dagmar Obbels[3], Elie Verleyen[3], Wim Vyverman[3], and Rainer Kurmayer[1]

[1] *Research Institute for Limnology, University of Innsbruck, Mondsee, Austria*
[2] *Institute of Virology, Helmholtz Zentrum Munich, Munich, Germany*
[3] *Laboratory of Protistology and Aquatic Ecology, Department of Biology, Ghent University, Ghent, Belgium*

10.1 Introduction

One of the major questions in toxic cyanobacterial research is "Who is there?" This question can be answered using various tools, including the long-lasting gold standard: sequence 16S ribosomal RNA (rRNA) gene amplicons generated by polymerase chain reaction (PCR) amplifying from genomic DNA to generate a community profile. Traditionally, 16S rRNA gene amplicon sequencing was performed by cloning and Sanger sequencing (capillary electrophoresis) of PCR amplicons. Because of intensive labor and costs, most studies using these techniques typically analyze fewer than 100 clones per sample. Next-generation

*Corresponding author: li.deng@helmholtz-muenchen.de; Maxime.Sweetlove@UGent.be

Molecular Tools for the Detection and Quantification of Toxigenic Cyanobacteria, First Edition.
Edited by Rainer Kurmayer, Kaarina Sivonen, Annick Wilmotte and Nico Salmaso.
© 2017 John Wiley & Sons Ltd. Published 2017 by John Wiley & Sons Ltd.

sequencing (NGS) brought two major advances: massive parallelization of the sequencing reactions and clonal separation of templates without the need to insert gene fragments in a host. This has simplified and increased the sequencing depth of 16S rRNA gene amplicons tremendously, and thus is now increasingly used in environmental microbiology studies for biodiversity assessment (i.e. metabarcoding), including monitoring toxic cyanobacteria in various environments (Tringe and Hugenholtz, 2008). See Fig. 10.1 for a standard workflow.

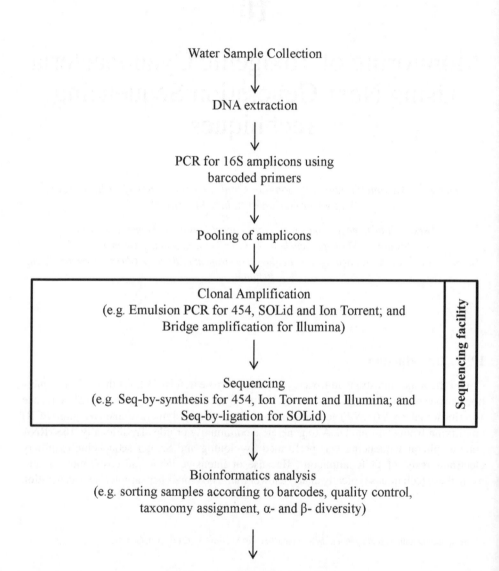

Figure 10.1 *Standard workflow for the NGS of 16S rRNA amplicons for monitoring toxic cyanobacteria (see SOP 5.3 for quantitative DNA extraction from environmental samples).*

It is generally agreed that for 16S rRNA gene sequencing studies the number of reads per sample is not paramount, as ~1,000 reads can generate the same patterns as multi-million reads datasets (Kuczynski *et al.*, 2010). Consequently, pooling multiple PCR amplicons in one sequencing run is cost efficient, given the large amounts of sequencing data the NGS platforms, even the benchtop sequencer, can generate (35 Mbp ~ 100 Gbp). Thus, barcodes which are unique, sample-specific oligonucleotides are introduced into the end of the primer for that purpose. Resulting sequences can be sorted according to the barcodes using bioinformatics. Furthermore, since for each sequencing run a large number of sequences is generated, bioinformatics-assisted exploration is needed to handle those large datasets and to extract the relevant sample-specific information. In general, those steps include (1) quality control of sequences, (2) assigning sequences to the original samples according to specific barcodes, (3) assigning sequences to operational taxonomic units (OTUs), (4) characterizing community composition using alpha- and beta-diversity estimates, and (5) taxonomic assignment of OTUs by means of a taxonomic database.

In this chapter, we describe an approach for sequencing 16S rRNA gene amplicons. The DNA extracts as obtained in Chapter 5 (see SOP 5.1) are used and the 16S rRNA gene is amplified using primers binding to conserved regions next to variable regions such as V3 or V6 (e.g. Andersson, Lindberg *et al.*, 2008). In contrast to (q)PCR (Chapters 6 and 7) the primer can be modified in order to contain the adaptor (specific to a sequencing platform) and barcode sequences (see below). The amplicons are purified, quality checked, quantified, and finally pooled at an equimolar ratio. A subsample of this pool is sent for sequencing following the instructions of the sequencing facility.

Water managers and authorities frequently have the task of surveying hundreds of kilometers of shoreline or a large number of water bodies. In studies where a large number of samples need to be analyzed for the presence of toxigenic cyanobacteria, high-throughput NGS provides a cost- and time-effective alternative to microscopic counting or traditional molecular characterization of community structure.

The method for sequencing 16S rRNA gene amplicons is a valuable and rapid complement to the conventional, systematic, and large-scale/long-term monitoring of cyanobacteria. Coupled with other specific markers and various filter membrane pore sizes, it can be further applied to phytoplankton community composition assessment in environmental samples. The approach provides a community profile of (toxigenic) cyanobacteria and other prokaryotes and allows a much more comprehensive view on microbial community composition compared to earlier technology to be obtained. Furthermore, this NGS method can be used to monitor the immigration presence and changes in the relative abundance of toxigenic genera into water bodies as a function of changing environmental conditions (eutrophication or global temperature rise). Using different primer sets, particular subgroups of the (cyanobacterial) community can be targeted.

10.2   Specific Procedures

10.2.1   16S rRNA Gene Amplicon Library Preparation

DNA extracted from environmental samples (see SOP 5.1) is used to amplify 16S rRNA using primers binding to conserved regions next to variable regions within the 16S rRNA. Resulting amplicons were amplified from corresponding hypervariable regions. The choice of primer sets largely depends on the downstream sequencing platforms (i.e. the length of amplicon sequences should be close to the read length of the various sequencing platforms; e.g. Roche 454, 500–1000 bp; Illumina, 100–700 bp; Ion Torrent, 100–300 bp).

The design of the primer sets is slightly different between sequencing platforms as well, but often contains an adaptor and barcode sequence besides the particular 16S rRNA primer sequences (e.g. Nübel *et al.*, 1997; Andersson *et al.*, 2008; Huse *et al.*, 2008), and see reviews for this specific topic (Baker *et al.*, 2003; Schloss, 2010; Klindworth *et al.*, 2013). The barcode sequences are often suggested in the sequencing platform manual (e.g. the 10-mer multiple identifiers (MIDs) from Roche 454). However, the choice of various barcode oligosequences also depends on the specific 16S rRNA primer sequences according to the following general rules: (1) barcode should not begin or end with the same nucleotide as the adaptor sequence, (2) barcode should not begin or end with the same nucleotide as 16S PCR amplification primer, (3) homopolymers of more than two nucleotides are not allowed, (4) barcodes must differ from one another by at least two nucleotides, and (5) annealing temperature should be the same among barcodes used in the same PCR run. It is also possible to generate PCR amplicons using 16S rRNA primers only and ligate barcodes afterwards. The advantage of using a fusion primer is that there is no need for an extra ligation step. The disadvantage, however, is that PCR amplification is done with large fusion primers which can form hairpin loops or cause non-specific amplification.

In order to enable the detection of rare taxa and to avoid errors, a rather low number of PCR cycles is used (*ca.* 20 cycles). To control for random effects during PCR, several replicate PCRs are performed using aliquots of the extracted DNA as a template. Resulting amplicons are combined for purification.

10.2.2 Amplicon Purification, Quantification and Pooling

Combined amplicons can be purified using gel purification as described in SOP 10.1.

In case of processing a large number of samples, gel purification can be replaced by using beads with a stringent size cutoff (e.g. AMPure beads).

Pooling all the amplicons in an equimolar ratio is a crucial step. In order to retrieve a similar number of reads for each of the samples, a precise DNA quantification method should be employed (e.g. PicoGreen quantification (or Qubit) (see SOP 10.1) and Agilent Bioanalyzer). Please note that even by doing the most accurate quantification and pooling there is still some unpredictable variability in the number of reads per sample.

10.2.3 Sequencing

There is a large difference in the output between the sequencing platforms in the marker, for example ~35 Mbp (454 GS Junior), ~2 Gbp (Ion Torrent PGM), ~10 Gbp (Ion Torrent Proton), ~15 Gbp (Illumina MiSeq), and ~180 Gbp (Illumina HiSeq X).

The methods (e.g. Emulsion PCR, bridge amplification, and sequencing) for the benchtop sequencer is often straightforward and is therefore not included in the SOP. Please note that, in general, NGS platforms are challenged by samples that have a low genetic diversity. This limitation was particularly problematic for the Illumina platforms (prior to the release of the

Real Time Analysis software v. 1.17.28 and MiSeq Control Software v. 2.2.0). A commonly used solution to solve this problem is to spike commercially prepared libraries of the PhiX phage genome to increase the amount of random control DNA. Illumina recommends that between 40 and 50% of the DNA should be PhiX.

10.2.4 Bioinformatic Exploration of Sequencing Results

High-quality reads are filtered by requiring: (1) an exact match to the barcode and primer sequences at the start of the sequence, (2) the absence of ambiguous nucleotides, (3) a minimum sequence length, and (4) an average quality score higher than 20.

Qualified reads can be processed using sequencing pipelines, for example MOTHUR (Schloss *et al.*, 2009; SOP 10.2), QIIME (Caporaso *et al.*, 2010), MICCA (Albanese *et al.*, 2015), and many forthcoming ones. Using QIIME as an example, this pipeline employs UCLUST to classify reads into OTUs according to sequence similarity, selecting the most abundant sequence from each OTU as a representative sequence for that OTU. PyNAST and UCLUST are employed for alignment and pairwise alignment with a minimum percent identity. Chimeric sequences are removed and taxonomy is assigned to bacterial OTUs by using the Basic Local Alignment Search Tool (BLAST) for each representative sequence against the continuously updated Ribosomal Database Project (RDP) (Cole *et al.*, 2009) and Greengenes (DeSantis *et al.*, 2006). Using the RDP database, the taxonomic composition can be resolved for domains, phyla, classes, orders, families, and genera.

In order to assign OTUs to species of cyanobacteria, high-quality sequences of cyanobacterial strains available from RDP, Greengenes, and other databases can be downloaded and manually cured to generate an in-house cyanobacterial database. The obtained OTUs from all environmental samples can then be blasted against this strain-specific database.

10.2.5 General Conclusions Including Pros and Cons of the Method

Identifying and measuring the abundance of toxigenic cyanobacteria by microscopy requires expert knowledge and longer analysis time. Alternatively, NGS of 16S RNA amplicons is straightforward, rapid, and has a high-throughput. However, caution must be exercised using this method as artifacts can be introduced through PCR (e.g. this approach suffers from biases related to incomplete specificity, incomplete coverage and chimeric errors (Sipos *et al.*, 2010; Haas *et al.*, 2011), which makes this approach semi-quantitative). Fortunately, artifacts can be reduced by carefully designed methods (e.g. Schloss *et al.*, 2011; Ibarbalz *et al.*, 2014). In addition, the taxonomic resolution of the method is still rather coarse, owing to the relatively short sequence lengths. Thus, in SOP 10.1 a rather longer PCR amplicon is amplified (~800 bp).

10.2.6 References

Albanese, D., Fontana, P., De Filippo, C., *et al.* (2015) MICCA: A complete and accurate software for taxonomic profiling of metagenomic data, *Scientific Reports*, **5**, Article number: 9743, doi: 10.1038/srep09743.

Andersson, A.F., Lindberg, M., Jakobsson, H., *et al.* (2008) Comparative analysis of human gut microbiota by barcoded pyrosequencing, *PLOS ONE*, **3** (7).

Baker, G.C., Smith, J.J., and Cowan, D.A. *et al.* (2003) Review and re-analysis of domain-specific 16S primers, *Journal of Microbiological Methods*, **55** (3) 541–555.

Caporaso, J.G., Kuczynski, J., Stombaugh, J., *et al.* (2010) QIIME allows analysis of high-throughput community sequencing data, *Nature Methods*, **7** (5) 335–336.

Cole, J.R., Wang, Q., Cardenas, E., *et al.* (2009) The Ribosomal Database Project: Improved alignments and new tools for rRNA analysis, *Nucleic Acids Research*, **37** (Database issue), D141–D145.

DeSantis, T.Z., Hugenholtz, P., Larsen, N., *et al.* (2006) Greengenes: A chimera-checked 16S rRNA gene database and workbench compatible with ARB, *Applied and Environmental Microbiology*, **72** (7) 5069–5072.

Haas, B.J., Gevers, D., Earl, A.M., *et al.* (2011) Chimeric 16S rRNA sequence formation and detection in Sanger and 454-pyrosequenced PCR amplicons, *Genome Research*, **21** (3), 494–504.

Huse, S.M., Dethlefsen, L., Huber, J.A., *et al.* (2008) Exploring microbial diversity and taxonomy using SSU rRNA hypervariable tag sequencing, *PLoS Genetics*, **4** (12).

Ibarbalz, F.M., Pérez, M.V., Figuerola, E.L., and Erijman, L. (2014) The bias associated with amplicon sequencing does not affect the quantitative assessment of bacterial community dynamics, *PlOS ONE*, **9** (6), e99722.

Klindworth, A., Pruesse, E., Schweer, T., *et al.* (2013) Evaluation of general 16S ribosomal RNA gene PCR primers for classical and next-generation sequencing-based diversity studies, *Nucleic Acids Research*, **41** (1).

Kuczynski, J., Liu, Z., Lozupone, C., *et al.* (2010) Microbial community resemblance methods differ in their ability to detect biologically relevant patterns, *Nature Methods*, **7** (10), 813–819.

Nübel, U., Garcia-Pichel, F., and Muyzer, G. (1997) PCR primers to amplify 16S rRNA genes from cyanobacteria, *Applied and Environmental Microbiology*, **63** (8), 3327–3332.

Schloss, P.D. (2010) The effects of alignment quality, distance calculation method, sequence filtering, and region on the analysis of 16S rRNA gene-based studies, *PLoS Computational Biology*, **6** (7), e1000844.

Schloss, P.D., Gevers, D., and Westcott, S.L. (2011) Reducing the effects of PCR amplification and sequencing artifacts on 16S rRNA-Based Studies, *PLOS ONE*, **6** (12), e27310.

Schloss, P.D., Westcott, S.L., Ryabin, T., *et al.* (2009) Introducing MOTHUR: Open-source, platform-independent, community-supported software for describing and comparing microbial communities, *Applied Environmental Microbiology*, **75** (23), 7537–7541.

Sipos, R., Székely, A., Révész, S., and Márialigeti, K. (2010) Addressing PCR biases in environmental microbiology studies, *Methods in Molecular Biology*, **599**, 37–58.

Tringe, S.G. and Hugenholtz, P. (2008) A renaissance for the pioneering 16S rRNA gene, *Current Opinion in Microbiology*, **11** (5), 442–446.

10.3 Bioinformatic Processing of Amplicon Sequencing Datasets

Maxime Sweetlove[1], Dagmar Obbels[1], Elie Verleyen[1], Igor S. Pessi[2], Annick Wilmotte[2], and Wim Vyverman[1]

[1]*Laboratory of Protistology and Aquatic Ecology, Department of Biology, Ghent University, Ghent, Belgium*

[2]*InBios – Center for Protein Engineering, University of Liège, Liège, Belgium*

10.3.1 Introduction

Amplicon sequencing can be a very powerful approach for detecting toxic cyanobacteria or any other kind of microorganism during monitoring programs. However, owing to the huge size of next-generation sequencing (NGS) datasets (up to several Gb), there is an obvious need for semi-automatic data processing and statistical analysis, as well as visualization of the patterns found. Importantly, raw NGS data contain errors, some of which are easily detected (e.g. too short or low-quality reads), while others remain hidden even after the most stringent quality controls (e.g. chimeras, contaminations, reads with large insertions or deletions, referred to as "indels"). As a consequence, NGS data need to be interpreted with caution, and bioinformatics analysis implementing poor error identification can easily lead to erroneous conclusions. Hence, a crucial step in the analysis of NGS data is the detection and removal of as many erroneous reads as possible. Moreover, bioinformatics involve additional preprocessing steps, including demultiplexing (i.e. grouping reads to samples according to the barcode sequence), deleting non-biological tags together with the adaptors and primer sequences, and removing chimeric sequences. In addition, the bioinformatics pipelines enable the quality-filtered sequences to be clustered into biologically relevant operational taxonomic units (OTUs), which form the basis of the statistical analysis, including the calculation of alpha- and beta-diversity.

10.3.2 Sequencing Platforms

Several platforms are available for amplicon sequencing. Illumina's MiSeq is the most commonly used, with one run quickly yielding up to 50 million high-quality paired-end reads. As recently as 2014, Roche's 454 pyrosequencing platform was widely used, and datasets produced with this technology can still be found. However, because of the high price involved in sequencing and competition with Illumina, support for 454 pyrosequencing has been discontinued. Other platforms such as Ion Torrent from Thermo Fisher Scientific and the SMRT sequencing system of Pacific Biosciences are rarely used for amplicon sequencing studies. The available platforms differ in the amount of data and read length they provide, but also in their speed of analysis and average cost as well as the potential bias related to the PCR reaction (Table 10.1).

Table 10.1 *Overview of sequencing platforms, comparison of the different technologies and platforms currently available for amplicon sequencing*

	Data output	Read length	Speed	Averaged cost per run	PCR
Sanger	96 Mb	700–900 bp	3 h	€250 per 96 well	yes
454 Roche[1]	35–700 Mb	400–700 bp	10–23 h	€6,000	yes
SOLiD[2]	90–120 Gb	2*50 bp (PE)	7–12 days	€4,000	yes
Illumina MiSeq[3]	0.3–15 Gb	2*300 bp (PE)	2–10 days	€2,000–€3,300	yes
Illumina HiSeq[3]	25–500 Gb	2*125 bp (PE)	3–10 days	€3,000–€4,500	yes
Ion Torrent[4]	20 Mb–1 Gb	200 bp	2–4 h	€1,000–€2,000	no
PacBio[5]	1 Gb	3,000 bp	2–3 h	€900	no

[1] roche.com.
[2] appliedbiosystems.com/cms/groups/global_marketing_group/documents/generaldocuments/cms_088662.pdf.
[3] illumina.com/systems/sequencing.ilmn.
[4] appliedbiosystems.com/cms/groups/applied_markets_marketing/documents/generaldocuments/cms_094139.pdf.
[5] files.pacb.com/pdf/PacBio_RS_II_Brochure.pdf.
bp = base pair, Mb = 10^6 base, Gb = 10^9 base, PE = paired end. Each averaged cost per run is based on three quotes from NGS sequencing facilities.

10.3.3 Data Formats

Data output of an Illumina platform is provided in the FASTQ format, in which every sequence read is stored in four lines. The first line starts with an "@" character and contains the header of the read which can include the sequence's geographical X–Y position on the Illumina lane. The sequence string is always in the second line, followed by a third line that starts with a "+", and may or may not contain the header repeated. The last line contains the quality score for each nucleotide, coded in single byte ASCII characters, which represent the estimated probability (p) that the base call on the same position in the sequence string is correct, given by the formula $p = 10^{-Q/10}$, with Q = quality score.

An important point of attention is that starting the header line with an "@" character was a poor choice (Cock *et al.*, 2010), because it is also used in the ASCII coding of the fourth line, rendering it unusable when writing custom scripts to distinguish the start of a sequence record. Also note that Illumina (version 1.8 or higher) uses the Sanger FASTQ notation (ASCII 33, Phred) as a consensus format. Older FASTQ files might differ in which ASCII characters are used as an offset (33 or 64) and the type of Q-score (Phred or Q_{SOLEXA}). For detailed information on older versions of FASTQ see Cock *et al.* (2010).

The standard downstream analysis format in bioinformatics is the FASTA format. Each record in the FASTA file is represented by two lines; the first starts with the ">" character, followed by a header that may contain any character and the second is the sequence string associated with this header. Because of this ">" character at the beginning of each record, custom software can easily recognize the start of any record. Note that the use of some characters, like the colon and semicolon, are better avoided in the header because they can be used by the software to add information, leading to confusion when this information needs to be extracted again.

When dealing with sequences coming from the 454 pyrosequencing platform, raw data are usually distributed in the SFF format, which describes the nucleotides and the associated light intensity peaks of the sequencing chromatogram. A chromatogram is created by flowing mononucleotides of each base over the lanes containing the samples in a repeated and fixed order. If there is an incorporation of one or more bases during such a flow, pyrophosphate releases can be detected as light signals of which the measured intensity is proportional to the number of nucleotides that were incorporated. The vast majority of these homopolymers will have a length of one base, but in theory any length is possible when homopolymers are encountered (Margulies *et al.*, 2005; NCBI, 2014). The actual sequence is constructed by concatenating these homopolymers, while the probability of error can be calculated from the difference between the detected signal and an "ideal signal" that represents an integer number of incorporated nucleotides.

10.3.4 Error Associated with NGS Data

Erroneous reads can be categorized into two main groups, namely (1) errors that arose during the sequencing itself and (2) errors that arose during the DNA storage and subsequent extraction and PCR steps preceding the NGS sequencing.

10.3.4.1 Sequencing Errors

A first aim of a bioinformatics procedure is to provide a quality control in which obvious sequencing errors are removed based on the quality scores. This step will reduce the dataset to sequences with a higher probability of being correct. Homopolymers form a special type of sequencing error and occur frequently on the 454 pyrosequencing platform (Huse *et al.*, 2007; Gilles *et al.*, 2011).

10.3.4.2 PCR Errors and Others

After the first quality control, artifacts that arose during PCR amplification are still present. Importantly, these errors are difficult to identify. Singling out and eliminating this kind of error forms a second aim of any bioinformatic procedure.

A major group of such false sequences comprises chimeras. These artifacts originate from the combination of two phylogenetically distinct parent sequences. Most chimeras are formed during PCR, when an incompletely extended DNA strand will serve as a primer and anneals to a different biologically unrelated strand because they share a conserved region at one 3′ extremity. This sequence can then be completed and copied during the following PCR cycles (Liesack *et al.*, 1991; Kopczynski *et al.*, 1994; Hugenholtz and Huber, 2003). These chimeras can form the basis of so-called spurious or phantom OTUs during classification (Kunin, 2010). Special algorithms are available to detect chimeras, like UCHIME (Edgar *et al.*, 2011), which is based on constructing three-way alignments between potential chimeras and their most probable parent sequences. Chimera Slayer (Haas *et al.*, 2011) requires a chimera-free template reference dataset to which the NGS sequences are compared. If a sequence appears to be a potential chimera and parent sequences can be discovered, a scoring algorithm is used to decide whether the sequences are discarded. Importantly, for *de novo* sequencing of environmental samples, it is recommended to avoid dependence on databases for chimera detection, because novel or rare genera that are not included in the database can wrongly be discarded as being chimeric.

Other errors include indels. During clustering, this type of error can easily make a sequence fall below the OTU similarity threshold, thereby forming the basis of a new cluster. As a rule of thumb, the fewer sequences that are clustered into one OTU, the more cautiously this OTU should be treated. Single nucleotide mis-incorporations (non-natural mutations that occurred during PCR) and misreads can also be present, but they are of lower concern as they are less likely to accumulate to such a degree that they can form the basis of spurious OTUs.

10.3.5   OTU Delineation: Choosing a Similarity Threshold

The choice of a similarity threshold for grouping sequences into OTUs is a very critical but arbitrary decision, and is aimed at reducing sequence variation originating from intra-species population differences and point mutation sequencing errors. Clustering thresholds must be chosen so that OTUs ideally approximate the species level. However, depending on the taxonomic group and its evolutionary rate and the change accumulated in the marker being sequenced, some OTUs may represent anything in the range of strains to genera and families. Most studies use a cutoff level of 97% similarity for the 16S SSU rRNA marker gene (Stackebrandt and Goebel, 1994), although for cyanobacteria, 97.5% similarity has been proposed to be a better cutoff level (Taton *et al.*, 2003; Pessi *et al.*, 2016). The amount of PCR and sequencing errors in the reads and the type of clustering and alignment algorithm can also affect the selection of the most appropriate similarity cutoff level (Huse *et al.*, 2010). Cutoff percentages of more than 5% for species-level identification with the 16S rRNA gene are strongly dissuaded. Whichever cutoff is chosen, results must still be cautiously interpreted as some errors can still form the basis of spurious OTUs.

10.3.6   Conclusions

Illumina's MiSeq is currently the main NGS platform used for amplicon sequencing, although 454 pyrosequencing is still widely used. Raw data output cannot immediately be interpreted because of sequencing and PCR errors. To address this problem, we provide an example of a bioinformatics standard operating procedure (see SOP 10.2) to process Illumina amplicon sequencing data and reduce the errors as much as possible, after which sequences are clustered into OTUs that can be used in downstream alpha- and beta-diversity analyses.

10.4   References

Cock, P.J.A., Fields, C.J., Goto, N., et al. (2010) The Sanger FASTQ file format for sequences with quality scores, and the Solexa/Illumina FASTQ variants, *Nucleic Acids Research*, **38** (6), 1767–1771.

Edgar, R.C., Haas, B.J., Clemente, J.C., et al. (2011) UCHIME improves sensitivity and speed of chimera detection, *Bioinformatics*, **27** (16), 2194–2200, doi: 10.1093/bioinformatics/btr381.

Gilles, A., Meglécz, E., Pech, N., et al. (2011) Accuracy and quality assessment of 454 GS-FLX Titanium pyrosequencing, *BMC Genomics*, **12**, 245, doi: 10.1186/1471-2164-12-245.

Haas, B., Gevers, D., Earl, A., et al. (2011) Chimeric 16S rRNA sequence formation and detection in *Sanger and 454-pyrosequenced PCR amplicons*, **21** (3), 494–504, doi: 10.1101/gr.112730.110.

Hugenholtz, P. and Huber, T. (2003) Chimeric 16S rDNA sequences of diverse origin are accumulating in the public databases, *International Journal of Systematic and Evolutionary Microbiology*, **53**, 289–293.

Huse, S.M., Huber, J.A., Morrison, H.G., et al. (2007) Accuracy and quality of massively parallel DNA pyrosequencing, *Genome Biology*, **8**, R143, doi: 10.1 186/gb2007-8-7-r143.

Huse, S.M., Welch, D.M., Morrison, H.G., and Sogin, M.L. (2010) Ironing out the wrinkles in the rare biosphere through improved OTU clustering, *Environmental Microbiology*, **12**, 1889–1898.

Kopczynski, E.D., Bateson, M.M., and Ward, D.M. (1994) Recognition of chimeric small-subunit ribosomal DNAs composed of genes from uncultivated microorganisms, *Applied Environmental Microbiology*, **60**, 746–748.

Kunin, V. (2010) Wrinkles in the rare biosphere: Pyrosequencing errors can lead to artificial inflation of diversity estimates, Lawrence Berkeley National Laboratory, LBNL Paper LBNL-2793E.

Liesack, W., Weyland, H., and Stackebrandt, E. (1991) Potential risks of gene amplification by PCR as determined by 16S rDNA analysis of a mixed-culture of strict barophilic bacteria, *Microbial Ecology*, **21**, 191–198.

Margulies, M., Egholm, M., Altman, W.E., et al. (2005) Genome sequencing in microfabricated high-density picolitre reactors, *Nature*, **437**, 376–380.

NCBI (2014) SFF file format, http://www.ncbi.nlm.nih.gov/Traces/trace.cgi?cmd=show&f=formats&m=doc&s=format#sff, accessed 21 February 2017.

Pessi, I.S., Maalouf, P.C., Laughinghouse IV,, H.D., et al. (2016) On the use of high-throughput sequencing for the study of cyanobacterial diversity in Antarctic aquatic mats, *Journal of Phycology*, **52**, 356–368.

Stackebrandt, E. and Goebel, B.M. (1994) Taxonomic note: A place for DNA-DNA reassociation and 16sRNA sequence analysis in the present species definition in bacteriology, *International Journal of Systematic Bacteriology*, **44**, 846–849.

Taton, A., Grubisic, S., Brambilla, E., et al. (2003) Cyanobacterial diversity in natural and artificial microbial mats of Lake Fryxell (McMurdo Dry Valleys, Antarctica): A morphological and molecular approach, *Applied and Environmental Microbiology*, **69** (9), 5157–5169, doi: 10.1128/AEM.69.9.5157-5169.

SOP 10.1

Standard Technique to Generating 16S rRNA PCR Amplicons for NGS

Li Deng[1,2*], *Stephan Blank*[1], *Guntram Christiansen*[1,3], *and Rainer Kurmayer*[1]

[1]*Research Institute for Limnology, University of Innsbruck, Mondsee, Austria*
[2]*Institute of Virology, Helmholtz Zentrum Munich, Munich, Germany*
[3]*Miti Biosystems GmbH, Max F. Perutz Laboratories, Vienna, Austria*

SOP 10.1.1   Introduction

Next-generation sequencing (NGS) methods allow for the characterization of microbial communities, including toxic cyanobacteria, at an unprecedented depth with minimal cost and labor. We detail here in this SOP our approach to amplify 16S rRNA amplicons using DNA extracted from environmental samples (see SOP 5.3) to be sequenced by NGS technique. Our method is particularly adapted for relative long amplicon size (~800 bp) in order to achieve more accurate taxonomy assignment. The resulting sequencing data can be analyzed using various bioinformatic pipelines, such as in SOP 10.2.

SOP 10.1.2   Experimental

SOP 10.1.2.1   Materials

- Proof reading polymerase and buffer, dNTPs.
- Fusion primers with different barcodes to identify individual samples. Our following primer set is targeting the V3 to V6 regions within the 16S rRNA, resulting relative

*Corresponding author: li.deng@helmholtz-muenchen.de

long amplicons (~800 bp) to be sequenced: 338F: 5′–<u>CGTATCGCCTCCCTCGCGCCA TCAG</u> <u>ACGAGTGCGT</u> ACTCCTACGGGAGGCAGCAG–3′ and 1046R: 5′–<u>CTATGC GCCTTGCCAGCCCGC</u> <u>TCAG</u> <u>ACGAGTGCGT</u> CGACAGCCATGCANCACCT–3′. The sequences underlined and separated by a space represent the adaptor, key, and barcode sequence, respectively, resulting in ~800 bp of product size (90 bp of adaptor, key and barcodes and 726 bp of 16S rRNA products: *E. coli*, Access. No. U00096).

- Scalpel blade to cut the PCR band of interest for gel purification.
- Gel purification kit (e.g. QIAquick® Gel Extraction Kit).
- PicoGreen dsDNA quantitation kit (for DNA quantification).
- Reaction tubes (0.2 mL, 1.5 mL, 2 mL).
- Sterile Milli-Q® water.

SOP 10.1.2.2   Equipment

- Micropipettes (1 μL, 20 μL, 200 μL, 1000 μL).
- PCR thermal cycler.
- Tabletop centrifuge.
- Agarose gel electrophoresis.
- Heat incubator (for gel purification).
- Fluorescence plate reader or qPCR cycler (for DNA quantification).

SOP 10.1.3   Procedure

1. 50 ng of extracted DNA is used for the amplification of 16S rRNA amplicons in 50 μL containing one unit of a proof-reading polymerase (e.g. Phusion Hot Start High-Fidelity DNA Polymerase, Thermo Fisher Scientific), 10 μL buffer HF (5×), 1 μL of 10 mM of each dNTP, and 2.5 μL of primers (10 pmol μL^{-1}). Perform 3–5 replicates of PCR reactions using aliquots of the extracted DNA as a template.
2. PCR program includes initial denaturation at 95°C for 5 min; 20 cycles of 95°C for 30 sec, 67.8°C for 30 sec and 72°C for 45 sec; final elongation at 72°C for 10 min.
3. For gel purification, PCR amplicons from the same sample are combined and loaded onto a 2% agarose gel using the largest combs available. Add 5 μL of 6× gel loading dye to the reaction and load on the gel. We recommend loading amplicons from different samples in separate gels or separating each sample by an empty well to avoid cross-contamination between samples. Run the gel at 70 V for 50 min, and inspect using gel electrophoresis. To protect sequences from damage by UV light inspection, longwave UV light (e.g. by means of a hand lamp) should be used. PCR products (expected size ~800 bp) are cut out from gel using a clean scalpel and purified from agarose and nucleic acids stain (with e.g. QIAquick® Gel Extraction Kit) following standard procedures.
4. For each sample purified PCR amplicons are combined and quantified using PicoGreen dsDNA quantification assay (e.g. Invitrogen) according to the manufacturer's instruction. Dilute samples to obtain a stock solution at 5×10^9 molecules μL^{-1} (approximately 1 ng μL^{-1}). If some samples show lower amplification, adjust the dilutions to a lower concentration, keeping in mind that the lowest concentration suitable for subsequent procedures is around 1×10^7 molecules μL^{-1} (~20 pM).

5. Pool 10 µL of each diluted amplicon to obtain an equimolar pool of samples. Prepare one separate pool for each of the planned sequencing runs. The pooled PCR products can be kept in the freezer until they are used.

SOP 10.1.4   Notes

We recommend the SOP 5.3 for quantitative DNA isolation from environmental samples where the glass fiber GF/C filters were employed for quantitatively cell filtration.

The primer set listed here is designed specifically for Roche 454. Adaptor, barcodes, and 16S rRNA primer sequencers should be subject to changes according to other sequencing technologies (see Chapter 10).

Amplicons, primer dimers, and PCR artifacts can be also eliminated by size selection using other methods available in the market (e.g. AMPure magnetic beads (Invitrogen)). This method is recommended especially when handling a large number of samples.

Precise DNA quantification is crucial for pooling multiple barcoded amplicons and the downstream sequencing steps, thus we recommend Agilent Bioanalyzer (Agilent Technologies) as an alternative method.

We recommend keeping the number of PCR cycles as low as possible to avoid chimera formation and to decrease amplification biases, keeping in mind that all samples should be amplified using the same number of cycles.

In general, several bands of PCR products are visible on the gel when amplifying 16S rRNA under the specified conditions. Since unspecific PCR products unequal in length reduce the sequence output during NGS, the larger distinct band is cut out and purified. However, cutting out DNA bands is considered unnecessary if only one clear distinct PCR band is observed and DNA can be purified directly.

After gel purification, with 20 cycles of PCR amplification, the amplicons are usually at a concentration of 0.1–10.0 ng in 50 µL of water. This may vary widely depending on the starting DNA concentration, the type of sample, and the purification kit used.

When adapting the described method for other (smaller-sized) prokaryotes, membrane filters with a smaller pore size should be considered.

Please see the following protocols to generate 16S rRNA amplicons for other sequencers: Illumina (Caporaso *et al.*, 2012; Kozich *et al.*, 2013) and Ion Torrent (Sanschagrin and Yergeau, 2014).

SOP 10.1.5   References

Caporaso, J.G., Lauber, C.L., Walters, W.A., *et al.* (2012) Ultra-high-throughput microbial community analysis on the Illumina HiSeq and MiSeq platforms, *The ISME Journal*, **6** (8), 1621–1624.

Kozich, J.J., Westcott, S.L., Baxter, N.T., *et al.* (2013) Development of a dual-index sequencing strategy and curation pipeline for analyzing amplicon sequence data on the MiSeq Illumina Sequencing platform, *Applied and Environmental Microbiology*, **79** (17), 5112–5120.

Sanschagrin, S. and Yergeau, E. (2014) Next-generation sequencing of 16S ribosomal RNA gene amplicons, *Journal of Visualized Experiments*, **90**, e51709.

SOP 10.2

Bioinformatics Analysis for NGS Amplicon Sequencing

Maxime Sweetlove[1]*, Dagmar Obbels*[1]*, Elie Verleyen*[1]*, Igor S. Pessi*[2]*, Annick Wilmotte*[2]*, and Wim Vyverman*[1]

[1]*Laboratory of Protistology and Aquatic Ecology, Department of Biology, Ghent University, Ghent, Belgium*
[2]*InBios – Center for Protein Engineering, University of Liège, Liège, Belgium*

SOP 10.2.1   Introduction

The main aim of bioinformatics analysis of NGS data is to refine and reconstruct a sample's biologic reality. This is done by detecting and removing errors and clustering sequences into operational taxonomic units (OTUs). Below, we describe and explain all the steps needed to extract the raw sequencing reads and cluster them into OTUs, which can be subsequently used in statistical analyses. Assuming paired-end Illumina data in the FASTQ format, we focus on the use of UPARSE (Edgar, 2013) for the major processing steps, as it has been recently shown fit for the study of cyanobacterial diversity using cyanobacteria-specific primers (Pessi *et al.*, 2016). However, when working with single end 454 data, different pre-processing steps will be needed, for instance using the MOTHUR (Schloss *et al.*, 2009) program, which is not discussed here. However, information and tutorials for MOTHUR can be found at www.mothur.org. Alternatively, 454 data can be converted to FASTQ or FASTA data after the first quality-control steps, which can be further processed following the SOP described below. (A standard workflow is shown in Fig. 10.2.)

SOP 10.2.2   Experimental

SOP 10.2.2.1   Equipment

Most of the programs that are designed to handle amplicon sequencing data require quite a lot of computing time and random access memory (RAM). Therefore, a sufficiently powerful computer is needed that can handle these tasks. Although a duo-core system with 8 GB of RAM should suffice, the better the computer, the faster it can process the

*Corresponding author: Maxime.Sweetlove@UGent.be

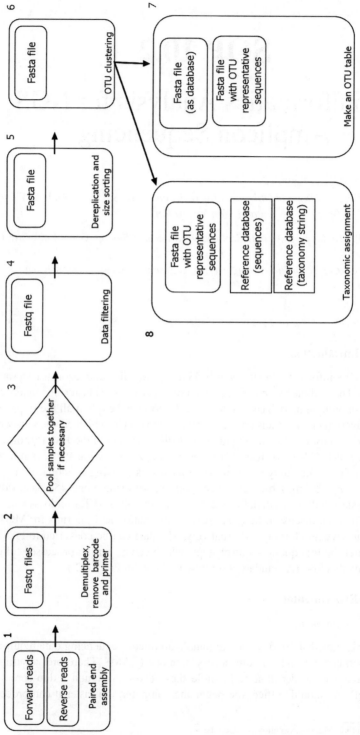

Figure 10.2 *Standard workflow for processing NGS data with USEARCH. Raw 454 data can be converted to a FASTQ format, and join at Step 2. This, however, comes at the cost of quality information loss, and specialized quality filtering algorithms and programs for pyrosequencing data (e.g. MOTHUR) that output FASTA files will yield better results. These, as well as other quality filtered FASTA files, can start at Step 5.*

data. Data storage is generally not an issue; a typical Illumina MiSeq output is about 5–10 GB large. Note that most programs discussed in the SOP below require a UNIX-based operating system such as Linux or MacOS.

To our knowledge, there are no commercial software packages to analyze amplicon sequencing data better than any available freeware. Therefore, all the programs used are open-source software. They have also the advantage of being more transparent, because the source code is available and the algorithms used are detailed in the accompanying scientific literature.

SOP 10.2.2.2　Procedure

UPARSE is a simple pipeline that, because of its greedy clustering algorithm, tends to produce relatively less spurious OTUs than other programs available (Edgar, 2013; Pessi *et al.*, 2016). It can be freely downloaded from http://drive5.com/uparse/, and it is run from the command line. Extensive documentation on parameters and options are given on the website. The example below assumes sequences that were bidirectional reads. Hence, we start with an additional step to merge the paired ends.

SOP 10.2.2.3　Merging Paired-End Data

With paired-end reads, the first step is to assemble forward and reverse reads to make contiguous sequences. There are several programs available that can do this, such as PandaSeq (Masella *et al.*, 2012), MOTHUR (Schloss *et al.*, 2009), USEARCH (Edgar, 2010), and PEAR (Zhang *et al.*, 2013). The latter differs from the others, owing to its simplicity while providing several interesting quality-control parameters, which allow filtering out low-quality overlaps. PEAR also needs to be run from the Linux command line and the most recent version can best be compiled from the source code freely available on github .com (Table 10.2).

```
# *****PEAR*****
# merge paired end reads
> pear -f file-1.fastq -r file-2.fastq -o file.fastq
-v 50 -q 20
```

Table 10.2　*Most important parameters and their meaning*

Parameter	Type	Meaning
-f	obligatory	forward file
-r	obligatory	reverse file
-o	obligatory	name of the output file
-v	optional	number of bases the reads should overlap
-q	optional	quality truncation value
-n	optional	minimal assembly length

SOP 10.2.2.4   Reducing the Sequencing Error in the Data

SOP 10.2.2.4.1   *Removing Non-Biological Tags and Demultiplexing*

Illumina data provided by the sequencing facility are generally demultiplexed into different files representing the barcodes (samples), while primer sequences are still present. The reason for this is derived from the technical properties of Illumina sequencing, in which barcodes are read in a separate index-reading step. However, if barcodes or parts of the adaptor and primers sequences are still present (e.g. if reads were derived from converted 454 data), these can be removed with a Python script provided by UPARSE. It requires a FASTA file that lists the barcodes, while the forward primer should be provided at the command line. This command also adds a prefix to each read label.

```
# demultiplex reads and remove barcodes
python fastq_strip_barcode_relabel2.py file.fastq
PRIMERSEQUENCE barcodes.fasta prefix > new_file.fastq
```

SOP 10.2.2.4.2   *Merging All the Sequences into One File*

Most clustering algorithms have a better overall performance if sequences are merged into one file (i.e. sample pooling). In UPARSE, this also enables a more efficient detection of chimeras, because these are more difficult to detect if the parent sequences have a low abundance. Sample pooling will also save a considerable amount of time, as commands described below need to be typed in and run only once.

This step, however, has not yet been implemented in the UPARSE pipeline. A simple script like the rudimentary python example below will pool all the samples together if a file is provided containing all the sample names. The script should be executed from the directory in which the files are stored.

```
#Python 3.3
#prints all fastq to one file, and changes names

file_files = input("please enter file contain-
ing the names in all sample-files:")
samples = open(file_files, "r")
out = input("what do you want to call your output file?")
output = open(out, "w")

samples_list = []
for sample in samples: #print all sample names to a list
new_sample = sample.strip('\n')
samples_list.append(new_sample)
n = len(samples_list)
seq_num = 1
```

```
print()
print("processing...")

i = j = 0
for sample in sample list: #loops over all the sample
one_sample = open(sample, "r")
sample_name = str(sample)
numlines = 4
reccord = []
for line in one_sample: #loops over all the records in
one sample
reccord.append(line)
if len(reccord) == numlines:
new_line = "@num_" + str(seq_num) + ";barcodela-
bel=" + sample_name + '\n'

output.writelines(new_line) #write sequence header
seq_num += 1
if j%1000 == 0:
print(j)
j += 1
output.writelines(reccord[1]) #write sequence
output.writelines(reccord[2]) #write quality header
output.writelines(reccord[3]) #write quality string
reccord = []
num = str(i+1)
print(num+"th sample complete")
i = i + 1

samples.close()
output.close()
```

Note: the bulky sequence name is replaced by a number, followed by ";barcodelabel=" and the sample name. This is necessary in USEARCH to demultiplex the reads and make an OTU table later in the process. Let us assume that, at the end of this process, we have obtained a file with the name file.assembled.fastq.

SOP 10.2.2.4.3   *Quality Control and Converting to FASTA Format*

Quality filtering of Illumina data is based on the Phred scores provided by the sequencing machine, and is aimed at removing sequence reads with a high amount of erroneous base calls. The maximum expected error (maxee parameter) is calculated as the sum of the probabilities that the base calls are wrong (coded in the Phred scores), and can be seen as an estimated measure of the actual number of errors in the sequence. This value does not have to be an integer number larger than one (Edgar, 2013; see: http://drive5.com/usearch/

manual/expected_errors.html). Before applying the maxee parameter, reads are trimmed to a certain length (300 bp in this example).

```
#Quality filtering
usearch -fastq_filter file.assembled.fastq -fastq_maxee
0.5 -fastq_minlen 300 -fastaout file_filtered.fasta
```

Tip: UPARSE has a command that provides basic statistics for the data. The output from this can be used to determine the optimal values for some parameters, like truncation length (see http://drive5.com/usearch/manual/fastq_choose_filter.html for more information).

SOP 10.2.2.5   OTU Clustering with UPARSE

SOP 10.2.2.5.1   *Dereplication*
During the dereplication step, identical sequences are collapsed into "unique sequences," on which further analysis is performed. This will reduce the computational demand of the data in further steps, and also provides the information on the abundance of each unique sequence which is crucial for the OTU clustering step. To make a clear distinction with truly unique sequences (e.g. found only once in the complete dataset) and to avoid confusion, we refer to the "unique sequences" outputted from the dereplication step as uni-identical (unique-identical) sequences. Please note that the term "unique sequences" refers to a group of identical reads (i.e. the output of the dereplication step). Sequences that appear only once in the complete dataset are called "singletons." The unique sequences are subsequently sorted by the abundance of the effective sequences they represent.

```
#Dereplication
#sizeout: keep track of the sequence abundance data
usearch -derep_fulllength file_filtered.fasta -output
file_derep.fasta -sizeout

#Size sorting
#for singleton removal, add -minsize 2
usearch -sortbysize file_derep.fasta -output file_
sorted.fasta
```

SOP 10.2.2.5.2   *OTU Clustering*
Most OTU-picking algorithms available will overestimate the number of OTUs, which will inflate the estimated diversity, and potentially bias statistical analyses. The reason for this has more to do with shortcomings of the data than with the clustering algorithms themselves. Notably, this problem is considerably reduced by the UPARSE clustering algorithm, which will generally produce fewer OTUs, owing to the implementation of a greedy clustering algorithm combined with immediate chimera-checking. First, the most abundant unique sequence is added to an OTU database as OTU1. In decreasing order of abundance, unique

sequences are compared to the current database. If a sequence is within the threshold set by the user in comparison to the existing OTUs (97.5% in this example), it will be added to this cluster. If not, the algorithm investigates whether the sequence is chimeric. Only if this is not the case, it can become a representative sequence of a new OTU, and is added to the database (Edgar *et al.*, 2013). After clustering, OTU representative sequences are renamed.

```
#OTU clustering and chimera checking
#otu_radius = dissimilarity cut-off percentage
usearch -cluster_otus file_sorted.fasta -otus file_otu.
fasta -otu_radius_pct 2.5 -relabel OTU_
```

SOP 10.2.2.5.3   *Making an OTU Table*
In a final step, an OTU table is constructed by mapping all sequences from the under-replicated dataset to the OTUs that were formed. The result is a table listing the sequence abundance of each OTU in each sample.

```
#making an OTU-table
usearch -usearch_global file_filtered.fasta -db file_otu.
fasta -strand plus -id 0.975 -uc file_map.uc -otutabout
file_table.txt
```

SOP 10.2.2.6   *Taxonomic Assignment*

An important constraint of the UPARSE pipeline is that it stops after OTU clustering and constructing the OTU table. Further analysis can be done in MOTHUR, for example using the classify.seqs command to obtain a taxonomic assignment for representative sequences of the OTUs. This command uses the RDP naive Bayesian classifier algorithm (Wang *et al.*, 2007), and requires a taxonomy database which consists of a fasta template and a corresponding taxonomy file. There are different alternative databases available to annotate a taxonomy to 16S rRNA amplicons, including Greengenes (DeSantis *et al.*, 2006) and RDP (Wang *et al.*, 2007). Here trainset9_032012.pds is used, the most recent version of which can be downloaded from the RDP website.

```
*****MOTHUR*****
#taxonomic classification
#cutoff = bootstrap value for assigning a taxonomy
mothur
classify.seqs(fasta=file_otu.fasta, tem-
plate=trainset9_032012.pds.fasta, taxon-
omy=trainset9_032012.pds.tax, cutoff=80, processors=2)
```

An interesting alternative to MOTHUR for taxonomic assignment is CREST (Lanzén *et al.*, 2012), which can be found at http://apps.cbu.uib.no/crest.

SOP 10.2.3   Practical Tips and Alternatives for Quality Filtering

Any clustering algorithm will tend to overestimate the diversity by producing more OTUs than actually exist (Kunin, 2010). This is due to the sequencing procedure itself, PCR errors (mutations, indels, chimeras,...), and the presence of multiple copies of the investigated gene (which could be slightly different). The inflation of diversity potentially has a large influence on the patterns observed, and the only solution up to now is to critically investigate any suspicious OTU (e.g. singletons) and use mock communities to assess the amount of bias.

Amplicon sequencing can be subject to PCR bias (Acinas, 2005). Hence, relative abundance data should be treated with caution, as it has been shown recently for artificial (mock) cyanobacterial communities (Pessi *et al.*, 2016).

In addition to the maximum expected error, two alternatives for quality filtering are available in UPARSE. The first one involves truncating reads at a certain Phred score, by cutting off any base (and all consecutive ones) as soon as its quality is below a specified threshold. As with the maxee parameter, this is done in combination with a minimal length parameter. After reads are truncated at the first Phred score below the specified threshold, resulting reads shorter than those specified are discarded. The second strategy is based on the averaged Phred score. This approach, however, is strongly dissuaded because very low-quality base calls in a read are masked from the algorithm by averaging out the Phred scores (Edgar, 2014).

Sometimes the final FASTA output (obtained from script in section 10.2.2.6) has line breaks in the sequences. The following Linux command can deal with this.

```
awk '!/^>/ { printf "%s", $0; n = "\n" } /^>/ { print
n $0; n = "" } END { printf "%s", n } ' input.fasta >
output.fasta
```

SOP 10.2.4   References

Acinas, S.G., Sarma-Rupavtarm, R., Klepac-Ceraj, V., and Polz, M.F. (2005) PCR-induced sequence artifacts and bias: Insights from comparison of two 16S rRNA clone libraries constructed from the same sample, *Applied Environmental Microbiology*, **71**, 8966–8969.

DeSantis, T.Z., Hugenholtz, P., Larsen, N., *et al.* (2006) Greengenes: A chimera-checked 16S rRNA gene database and workbench compatible with ARB, *Applied Environmental Microbiology*, **72**, 5069–5072.

Edgar, R.C. (2010) Search and clustering orders of magnitude faster than BLAST, *Bioinformatics*, **26** (19), 2460–246, doi: 10.1093/bioinformatics/btq461.

Edgar, R.C. (2013) UPARSE: Highly accurate OTU sequences from microbial amplicon reads, *Nature Methods*, **10** (10) 996–998, doi: 10.1038/nmeth.2604.

Edgar, R.C. (2014) Average Q Scores, http://drive5.com/usearch/manual/avgq.html, accessed 21 February 2017.

Kunin, V. (2010) Wrinkles in the rare biosphere: Pyrosequencing errors can lead to artificial inflation of diversity estimates, *Lawrence* Berkeley National Laboratory, LBNL Paper LBNL-2793E.

Lanzén, A., Jørgensen, S.L., Huson, D., *et al.*(2012) CREST: Classification resources for environmental sequence tags, *PLOS ONE*, **7**, e49334.

Masella, A.P., Bartram, A.K., Truszkowski, J.M., *et al.* (2012) *PANDAseq: Paired-end assembler for Illumina sequences*, **13** (1), doi: 10.1186/1471-2105-13-31.

Pessi, I.S., Maalouf, P.C., Laughinghouse IV, H.D., *et al.* (2016) On the use of high-throughput sequencing for the study of cyanobacterial diversity in Antarctic aquatic mats, *Journal of Phycology*, **52**, 356–368.

Schloss, P.D., Westcott, S.L., Ryabin, T., *et al.* (2009) Introducing MOTHUR: Open-source, platform-independent, community-supported software for describing and comparing microbial communities, *Applied Environmental Microbiology*, **75** (23), 7537–7541.

Wang, Q., Garrity, G.M., Tiedje, J.M., and Cole, J.R. (2007) Naive Bayesian classifier for rapid assignment of rRNA sequences into the new bacterial taxonomy, *Applied Environmental Microbiology*, **73** (16), 5261–5267.

Zhang, J., Kobert, K., Flouri, T., and Stamatakis, A. (2013) PEAR: A fast and accurate Illumina paired-end read merger, *Bioinformatics*, **30** (5), 614–620, doi: 10.1093/bioinformatics/btt593.

11

Application of Molecular Tools in Monitoring Cyanobacteria and Their Potential Toxin Production

Vitor Ramos[1,2], *Cristiana Moreira*[1], *Joanna Mankiewicz-Boczek*[3,4], *and Vitor Vasconcelos*[1,2]*

[1]*Interdisciplinary Centre of Marine and Environmental Research (CIIMAR/CIMAR), University of Porto, Matosinhos, Portugal*
[2]*Faculty of Sciences, University of Porto, Porto, Portugal*
[3]*European Regional Centre for Ecohydrology of the Polish Academy of Sciences, Łódź, Poland*
[4]*Department of Applied Ecology, Faculty of Biology and Environmental Protection, University of Lodz, Łódź, Poland*

11.1 Introduction

Soon after the advent of PCR-based methods, molecular tools for the study and monitoring of toxigenic cyanobacteria in the environment began to be developed. In fact, these methods have been applied for that purpose for almost two decades, and still are, as can be easily deduced (albeit indirectly) from Fig. 11.1. Before the genetics behind the biosynthesis of the different cyanotoxins started to be unraveled, a common molecular-based approach to the detection of putative toxic cyanobacteria in environmental samples consisted of performing DNA fingerprinting analysis to discriminate between species or strains/genotypes.

*Corresponding author: vmvascon@fc.up.pt

Molecular Tools for the Detection and Quantification of Toxigenic Cyanobacteria, First Edition.
Edited by Rainer Kurmayer, Kaarina Sivonen, Annick Wilmotte and Nico Salmaso.
© 2017 John Wiley & Sons Ltd. Published 2017 by John Wiley & Sons Ltd.

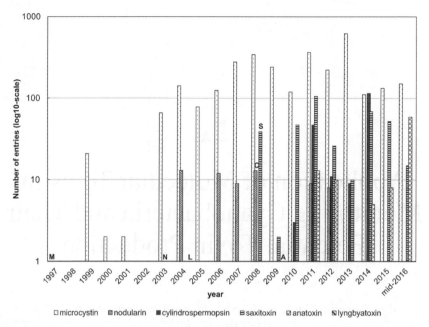

Figure 11.1 *Evolution of the number of new sequences, deposited in the GenBank nucleotide database, for genes involved in the biosynthesis of cyanotoxins. Example of a Boolean search string used: (((microcystin[Title]) OR mcy*[Title]) AND cyanobacter*[Organism]) AND ("1997"[Publication Date] : "1997"[Publication Date]). The first (capitalized) letter of each toxin name indicates the release date of the first annotated gene sequence for this toxin (see text to an accession number and publication reference). (See color plate section for the color representation of this figure.)*

The genetic markers used were not related to (or correlated with) toxin synthesis, and so obtaining a clear and undeniable relationship between those fingerprint profiles and toxicity was difficult if not impossible (e.g. Nishihara *et al.*, 1997; Neilan *et al.*, 2003; see also Chapters 4, 6, and 9). Another approach is to use phylogenetically useful genes, also known as "taxonomic markers," to detect and identify cyanobacterial taxa. These markers, especially the 16S rRNA gene, are still extensively used. Not surprisingly, when cyanotoxin gene clusters are discovered and characterized, and their annotated sequences are made publicly available, the genes involved in the biosynthetic pathway of the different cyanotoxins rapidly become the preferred target region for the development of molecular tools to study and monitor toxigenic cyanobacteria. As can be seen in Fig. 11.1, the microcystin gene cluster was the first to be identified and (partially) sequenced (Dittmann *et al.*, 1997; GenBank sequence accession number, U97078.1), followed by the gene clusters of nodularin (Moffitt and Neilan, 2004; AY210783.2), lyngbyatoxin (Edwards and Gerwick, 2004; AY588942.1), cylindrospermopsin (Mihali *et al.*, 2008; EU140798.1), saxitoxin (Kellmann *et al.*, 2008a; DQ787200.1), and finally the gene cluster for anatoxin biosynthesis (Méjean *et al.*, 2009; FJ477836.1). As a result, the number of available gene sequences of the several cyanotoxins in the GenBank database has increased exponentially since the late 1990s, particularly for microcystin (see Fig. 11.1). This is also in agreement with the higher

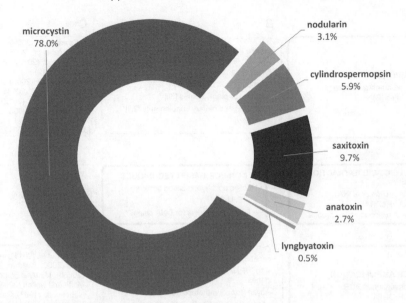

Figure 11.2 *Percentage of publications on toxic cyanobacteria that used molecular methods, until mid-2016; by cyanotoxin. Example of a Boolean search string used: TITLE-ABS-KEY ((cylindrospermopsin* AND (cyr* OR aoa*)).*

number of publications on this cyanotoxin when compared with the others (Fig. 11.2), as retrieved by a literature search on the Scopus database conducted in early September 2016. Remarkably, the biosynthetic gene cluster of the dermatotoxin lyngbyatoxin – known to cause the so-called swimmer's itch – has earned scant attention from molecular ecotoxicologists, so far (Fig. 11.1 and Fig. 11.2).

11.2 Possible Applications

The availability of cyanotoxin gene cluster sequences led to the possibility of applying different molecular methods and approaches for monitoring toxigenic cyanobacteria directly in the environment. Fig. 11.3 summarizes the various applications and different molecular methods being used for the study of naturally occurring (toxic) cyanobacteria. It shows the main stages that need to be taken during molecular-based approaches, pointing out the outcomes that may be achieved at the various stages, while providing some relevant bibliography for each specific issue. A more complete, although not exhaustive, list of publications is presented in Table 11.1.

Despite the wide use of molecular tools for the detection of toxic cyanobacteria by the scientific community (which can be inferred from Table 11.1), very little is known about their use (especially on a regular basis) by other health- or environmental-related entities, namely by regulatory and monitoring agencies or by companies. For instance, Wood *et al.* (2013) stated that, in New Zealand, "despite research and validation demonstrating their potential, the application of these tools by monitoring agencies has been limited." In fact, the attempt to develop molecular tools and to transfer this technology from academia

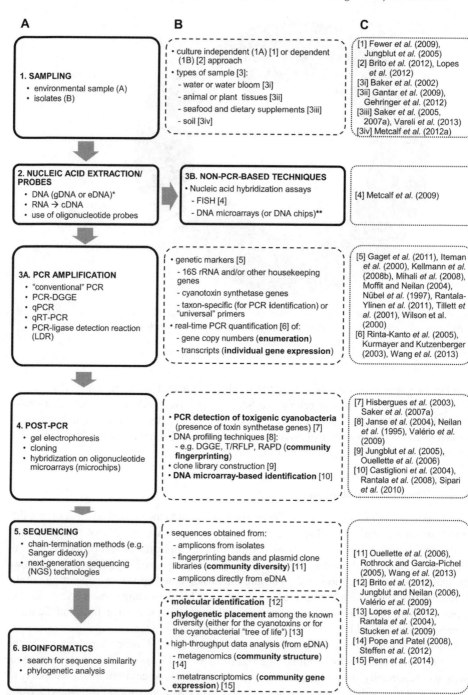

Figure 11.3 *Overview of the main applications of molecular-based methods for cyanobacterial (and/or their toxins) detection and identification. (A) General flow diagram showing main stages. (B) Methodological notes or applications description; in bold are possible outcomes or achievements, at this particular stage. (C) Examples of relevant literature for topics identified in (B), within square brackets. * gDNA and eDNA regard to isolate genomic DNA and environmental DNA, respectively; ** in the scarce studies on cyanobacteria, hybridization is preceded by PCR (see stage 4). Source: Reproduced from Moreira* et al. *(2014) with permission from Springer.*

Table 11.1 List of studies showing the main applications (and some methodological features) in which the detection/identification of tox(igen)ic cyanobacteria have been pursued, by molecular-based method used

Method[1]	Application	Type of sample	Targeted molecular marker(s)[2]	Cyanotoxin biosynthesis genes[3]	Complementary evaluation/ validation[4]	References
Conventional PCR (Chapter 6)	aquatic ecosystems	isolate(s)	16S rRNA gene	n.d.	ELISA (+, for MC), LC-MS/MS (+, for MC)	Carmichael and Li (2006)
	aquatic ecosystems	water	mcyA; mcyB; mcyC; mcyD; mcyE; mcyG; ndaF; cyrB; cyrC; sxtI	MC (+), NOD (+), CYL (+), SXT (+)	ELISA (−), MALDI-TOF/MS (−)	Lopes et al. (2012)
	aquatic ecosystems	isolate(s)	aoaA; aoaB; aoaC	CYL (−)	ELISA (−), HPLC (−)	Fathalli et al. (2011)
	aquatic ecosystems	water	mcyA; mcyB; mcyC; mcyD; mcyE	MC (+)	ELISA (+)	Te and Gin (2011)
	aquatic ecosystems	isolate(s)	mcyB	MC (+)	MALDI-TOF (+)	Kurmayer et al. (2002)
	aquatic ecosystems	isolate(s)	sxtA; sxtG; sxtH; sxtI; sxtX	SXT (+)	ELISA (+) and reported in previous studies	Casero et al. (2014)
	aquatic ecosystems	water	mcyA; mcyB; mcyD; mcyE	MC (+)	HPLC (+), PPIA (+)	Mankiewicz-Boczek et al. (2006)
	aquatic ecosystems	water	cyrJ	CYL (+)	HPLC (+)	Mankiewicz-Boczek et al. (2012a)
	aquatic ecosystems	isolate(s) and water	cyrI, pks	CYL (+)	ELISA (+), HPLC (+), HPLC-MS/MS (+)	Kokociński et al. (2013)

(Continued)

Table 11.1 (Continued)

Method[1]	Application	Type of sample	Targeted molecular marker(s)[2]	Cyanotoxin biosynthesis genes[3]	Complementary evaluation/validation[4]	References
	aquatic ecosystems	isolate(s)	aoaB; aoaC	CYL (−)	HPLC (−)	Valério et al. (2005)
	aquatic ecosystems	isolate(s) and water	mcyA; mcyB; mcyE; aoaA; aoaB; aoaC	MC (+)	ELISA (+)	Martins et al. (2010)
	aquatic ecosystems	water	mcyA	MC (+)	ELISA (+)	Saker et al. (2007 b)
	aquatic ecosystems	water	mcyA; mcyB	MC (+)	PPIA (+), LC-MS (+)	Allender et al. (2009)
	aquatic ecosystems	water (from food fish farming)	mcyB; 16S rRNA gene	MC (+)	HPLC (+)	Nonneman and Zimba (2002)
	aquatic ecosystems	bloom	mcyE; ndaF; sxtA; cyrA; 16S rRNA gene	MC and/or NOD (+), SXT (+), CYL (+)	ELISA (−;+;+)	Al-Tebrineh et al. (2012a)
	aquatic ecosystems	bloom	mcyA; mcyC	MC (+)	Mouse bioassay (+), PPIA (+)	Baker et al. (2001)
	aquatic ecosystems	bloom	mcyA; 16S rRNA gene	MC (+), SXT (+)	HPLC (+;+)	Baker et al. (2002)
	aquatic ecosystems	water	cyrC; mcyA; mcyC; sxtI	CYL (+), MC (−), SXT (−)	HPLC-MS/MS (+)	McGregor et al. (2011)
	aquatic ecosystems	water	mcyE	MC (+)	HPLC (+)	Belykh et al. (2011)
	aquatic ecosystems	bloom	mcyB	MC (+)	ELISA (+)	Bittencourt-Oliveira (2003)

Context	Sample	Target genes	Toxins	Methods	Reference
aquatic ecosystems	isolate(s)	cyrA; cyrB; cyrC; cyrI; sxtA; sxtB; sxtI	CYL (–), SXT (+)	LC-MS (–;+)	Hoff-Risseti et al. (2013)
aquatic ecosystems	mat	mcyE	MC (+)	LC-MS (+)	Wood et al. (2008)
aquatic ecosystems	isolate(s)	mcyA; mcyB; mcyC; ndaF; CYL biosynthesis-related genes; 16S rRNA gene	MC (+), CYL (–)	Artemia salina bioassay (+), MALDI-TOF/MS (–)	Frazão et al. (2010)
aquatic ecosystems	isolate(s)	16S rRNA gene	n.d.	PPIA (+), ELISA (+)	Gantar et al. (2009)
aquatic ecosystems	mat (coral tissue)	mcyA, mcyE; ndaF; cyrI; sxtI	MC (+), NOD (–), CYL (–), SXT (–)	LC-MS (–;–;–;–)	Glas et al. (2010)
aquatic ecosystems	herbarium specimens	mcyD	MC (+)	HPLC (+), ELISA (+), MALDI-TOF (+)	Metcalf et al. (2012b)
food (additive) surveillance	dietary supplements	mcyE; ndaF; cpcBA-IGS	MC and/or NOD (+)	PPIA, ELISA, and LC-MS (+, for MC); LC-MS (–, for CYL, ANA, and SXT/PSPs)	Heussner et al. (2012)
food (additive) surveillance	dietary supplements	mcyA; mcyE; ndaF; cpcBA-IGS	MC (+), MC and/or NOD (+)	ELISA (+, for MC)	Saker et al. (2005)
food web bioaccumulation	bloom and grazers (gut content)	mcyA; mcyB	MC (+)	PPIA (+), ELISA (+)	Oberholster et al. (2006)

(Continued)

Table 11.1 (Continued)

Method[1]	Application	Type of sample	Targeted molecular marker(s)[2]	Cyanotoxin biosynthesis genes[3]	Complementary evaluation/validation[4]	References
	terrestrial ecosystems	cryptogamic (soil) crust	mcyD; aoaB; aoaC	MC (+), CYL (−)	HPLC (+;−), ELISA (+, for MC)	Metcalf et al. (2012a)
	terrestrial ecosystems	cyanobiont isolate(s) and in planta	mcyE; ndaA; ndaB; ndaF; 16S rRNA gene	NOD (+), MC, and/or NOD (+)	HPLC HESI-MS/MS (+, for NOD) (−, for MC)	Gehringer et al. (2012)
	terrestrial ecosystems	cyanobiont isolate(s) and lichen tissue	mcyE	MC (+)	LC-MS (+, for MC and NOD)	Kaasalainen et al. (2012)
	terrestrial ecosystems	cyanobiont isolate(s)	mcyA; mcyD; mcyE; rpoC1; 16S rRNA gene	MC (+)	PPIA (+), ELISA (+), HPLC-MS/MS (+)	Oksanen et al. (2004)
Multiplex-PCR (Chapter 6)	aquatic ecosystems	water	mcyA; mcyB; mcyC; mcyD; mcyE; mcyG; mcyI; cyrC	MC (+), CYL (+)	HPLC-MS/MS (+;+)	Zhang et al. (2014)
	aquatic ecosystems	bloom	mcyA; mcyB; mcyC; mcyD; mcyE; mcyG	MC (+)	MALDI-TOF (+), HPLC (+)	Ouahid and Fernández del Campo (2009)
	aquatic ecosystems	water	mcyB; mcyD	MC (+)	n.d.	Ouellette et al. (2006)
	food (additive) surveillance	dietary supplements	mcyA; cpcBA-IGS	MC (+)	MALDI-TOF/MS (+)	Saker et al. (2007 a)

Method	Sample	Gene target	Toxin	Analytical method	Reference
Multiplexed-tandem-PCR (Chapter 6)	aquatic ecosystems	bloom	mcyE; ndaF; cyrC; sxtI	MC (+), NOD (+), CYL (+), SXT (+)	ELISA (+;+;+;+)	Baker et al. (2013)
Quantitative PCR (Chapter 7)	aquatic ecosystems	water	mcyD; cpcBA-IGS	MC (+)	n.d.	Li et al. (2012)
	aquatic ecosystems	water	mcyA	MC (+)	HPLC (+), ELISA (+)	Yoshida et al. (2005)
	aquatic ecosystems	water	mcyA	MC (+)	n.d.	Yoshida et al. (2007)
	aquatic ecosystems	water	mcyA	MC (+)	ELISA (+)	Ha et al. (2011)
	aquatic ecosystems	water	mcyA, 16S rRNA gene	MC (+)	HPLC (+), PPIA (+)	Gągała et al. (2014)
	aquatic ecosystem	water	cyrA	CYL (+)	ELISA (+), HPLC (+), HPLC-MS/MS (+)	Kokociński et al. (2013)
	aquatic ecosystems	water	mcyA; mcyB; mcyC; mcyD; mcyE; mcyG; mcyI; cyrC	MC (+), CYL (+)	HPLC-MS/MS (+;+)	Zhang et al. (2014)
	aquatic ecosystems	water	mcyE	MC (+)	ELISA (+)	Te and Gin (2011)
	aquatic ecosystems	water	mcyJ	MC (+)	HPLC (+)	Joung et al. (2011)
	aquatic ecosystems	bloom	mcyE; ndaF	MC (+), NOD (+)	HPLC (+;+)	Al-Tebrineh et al. (2011)

(Continued)

Table 11.1 (Continued)

Method[1]	Application	Type of sample	Targeted molecular marker(s)[2]	Cyanotoxin biosynthesis genes[3]	Complementary evaluation/validation[4]	References
	aquatic ecosystems	water	mcyE	MC (+)	n.d.	Vaitomaa et al. (2003)
	aquatic ecosystems	sediment	mcyB	MC (+)	PPIA (+)	Misson et al. (2012)
	aquatic ecosystems	water	mcyA; cpc operons	MC (+)	PPIA (+)	Briand et al. (2008)
	aquatic ecosystems	water	mcyB; cpc operons	MC (+)	n.d.	Kurmayer and Kutzenberger (2003)
	aquatic ecosystems	water	mcyA	MC (+)	HPLC (+)	Ostermaier and Kurmayer (2010)
	aquatic ecosystems	water	aoaA; aoaB; aoaC	CYL (+)	HPLC LC-MS (+)	Rasmussen et al. (2008)
	food (additive) surveillance	dietary supplements	mcyB; cpcBA-IGS; 16S rRNA gene	MC (+)	ELISA (+), LC-MS/MS (+)	Vichi et al. (2012)
	food web bioaccumulation	gut content of grazers and edible fish	16S rRNA gene	n.d.	ELISA (+, for MC), HPLC/PDA (+, for MC)	Sotton et al. (2014)
	food web bioaccumulation	bloom	mcyA; mcyB	MC (+)	PPIA (+), ELISA (+)	Oberholster et al. (2006)
Multiplex real-time PCR (Chapter 7)	aquatic ecosystems	isolate(s)	mcyE; ndaF; sxtA; cyrA	MC and/or NOD (+), SXT (+), CYL (+)	n.d.	Al-Tebrineh et al. (2012b)
	aquatic ecosystems	bloom	mcyB; cpc operons	MC (+)	n.d.	Briand et al. (2009)
	aquatic ecosystems	bloom	mcyE; ndaF; sxtA; cyrA	MC and/or NOD (+), SXT (+), CYL (+)	ELISA (-;+;+)	Al-Tebrineh et al. (2012a)

Microarrays (Chapter 8)	aquatic ecosystems	isolate(s) and water	*mcyE*; 16S rRNA gene	MC (+)	HPLC-MS (+), LC-MS (+)	Sipari et al. (2010)
	aquatic ecosystems	isolate(s)	HIP1; putative Na+-dependent transporter gene	n.d.	toxigenicity of strains (SXT-producer) reported in previous studies	Pomati et al. (2004)
	aquatic ecosystems	isolate(s) and water	*mcyE*; *ndaF*	MC and/or NOD (+)	n.d.	Rantala et al. (2008)
DGGE/clone libraries/DNA typing methods (Chapter 9)	aquatic ecosystems	isolate(s)	*cpc*BA-IGS; rRNA ITS	n.d.	toxigenicity of strains reported in previous studies	Neilan (2002)
	aquatic ecosystems	water	*mcyE*; *ndaF*; 16S rRNA gene	MC and/or NOD (+)	ELISA (-;-)	Amer et al. (2009)
	aquatic ecosystems	isolate(s)	RAPD markers	n.d.	toxigenicity of strains (MC-producer) reported in previous studies	Nishihara et al. (1997)
	aquatic ecosystems	sediment	*mcyB*; 16S rRNA gene; rRNA ITS; *rpoC1*	MC (–)?	n.d.	Innok et al. (2005)

(Continued)

Table 11.1 (Continued)

Method[1]	Application	Type of sample	Targeted molecular marker(s)[2]	Cyanotoxin biosynthesis genes[3]	Complementary evaluation/validation[4]	References
	aquatic ecosystems	isolate(s)	STRR sequences; 16S rRNA gene	n.d.	HPLC-MS (+, for CYL)	Chonudomkul et al. (2004)
	aquatic ecosystems	sediment	mcyB; rRNA ITS	MC (+)	PPIA (+)	Misson et al. (2012)
	aquatic ecosystems	isolate(s) and water	mcyA	MC (+)	ELISA (+), MALDI-TOF (+)	Hisbergues et al. (2003)
	aquatic ecosystems	water	rRNA ITS	n.d.	HPLC (+, for MC)	Kardinaal et al. (2007)
	aquatic ecosystems	isolate(s) and water	mcyB; mcyD; 16S rRNA gene	MC (+)	toxigenicity of strains (MC-producer reported in previous studies	Ouellette et al. (2006)
	aquatic ecosystems	isolate(s)	uncharacterized PS and PKS genes, 16S rRNA gene	PS/PKS (+)	HPLC (+, for CYL), (–, for SXT (i.e. PSP))	Valério et al. (2005)
	aquatic ecosystems	mat	mcyE; ndaF; rRNA ITS; 16S rRNA gene	MC and/or NOD (+)	LC-MS (+, NOD), LC-MS (–, for MC and ANA)	Wood et al. (2012a)
	aquatic ecosystems	sediment	rRNA ITS	n.d.	LC-MS (+, for ANA)	Wood et al. (2009)
	aquatic ecosystems	isolate(s)	mcyE; ndaF; STRR sequences; 16S rRNA gene	MC and/or NOD (+)	PPIA (+); ELISA (+)	Lyra et al. (2005)

aquatic ecosystems	isolate(s)	HIP1; 16S rRNA gene	n.d.	Mouse bioassay (+), HPLC (+, for CYL)	Neilan et al. (2003)
aquatic ecosystems	naturally occurring colonies and water	rRNA ITS	n.d.	MALDI-TOF/MS (+, for MC)	Janse et al. (2004)
aquatic ecosystems	mat	mcyE; rRNA ITS; 16S rRNA gene	MC (+)	PPIA (+), ELISA (+), LC-MS (+)	Wood et al. (2008)
aquatic ecosystems	isolate(s) and mat (coral tissue)	mcyA; 16S rRNA gene	MC (+)	PPIA (+), ELISA (+), HPLC-MS (+), HPLC-MS (–, for ANA and SXT)	Richardson et al. (2007)
aquatic ecosystems	mat	mcyE; ndaF; ANA biosynthesis-related gene; cyrJ; sxtG; rRNA ITS; 16S rRNA gene	MC and/or NOD (–), ANA-related (–), CYL (–), SXT (+)	n.d.	Villeneuve et al. (2012)
food web bioaccumulation	seawater and edible mussels	mcyA; mcyE; uncharacterized PKS genes; rRNA ITS	MC (+) and PKS (+)	ELISA (+, for MC), HPLC-MS: (+, for MC)	Vareli et al. (2012)

(Continued)

Table 11.1 (Continued)

Method[1]	Application	Type of sample	Targeted molecular marker(s)[2]	Cyanotoxin biosynthesis genes[3]	Complementary evaluation/ validation[4]	References
NGS and/or metagenomics techniques (Chapter 10)	aquatic ecosystems	bloom	mcyA; 16S rRNA gene	MC (+)	PPIA (+), LC-MS (+), LC-MS (−, for CYL, SXT, and BMAA)	Steffen et al. (2012)
	aquatic ecosystems	bloom	Microcystis metagenome	n.d.	toxigenicity of strains (MC-producer) reported in previous studies	Wilhelm et al. (2011)
	aquatic ecosystems	bloom	16S rRNA gene; BAC library	n.d.	n.d.	Pope and Patel (2008)
FISH	aquatic ecosystems	isolate(s) and water	mcyA	MC (+)	ELISA (+), HPLC (+)	Gan et al. (2010)
	aquatic ecosystems	isolate(s) and water	mcyA	MC (+)	ELISA (+)	Metcalf et al. (2009)

[1] For more information on the method, see the respective Chapter; FISH – fluorescence in situ hybridization.

[2] Biosynthetic gene cluster (cyanotoxin): mcy (microcystin), nda (nodularin), cyr ≡ aoa (cylindrospermopsin); gene name as it is in the original study), sxt (saxitoxin).
Other markers: rRNA ITS – 16S-23S rRNA internal transcribed spacer region; cpc operon – phycocyanin operon; cpcBA-IGS – phycocyanin intergenic spacer and flanking regions; rpoC1 – RNA polymerase subunit beta; PS – nonribosomal peptide synthetase; PKS – polyketide synthase.
Other abbreviations: HIP1 – highly iterated palindrome; RAPD – Random Amplified Polymorphic DNA; STRR – short tandemly repeated repetitive; BAC – bacterial artificial chromosome; CYL – cylindrospermopsin; ANA – anatoxin-a; BMAA – β-methyl-amino-L-alanine.

[3] Molecular-based assessment of the presence (+) or absence (−) of biosynthesis genes for the production of the following toxins: MC – microcystin; NOD – nodularin; CYL – cylindrospermopsin; ANA – anatoxin-a; SXT – saxitoxin. n.d. – not determined.

[4] Toxicity/Toxin detection (+) or no detection (−) by additional complementary methods. Biological methods (non-specific): Mouse and brine shrimp (Artemia salina) bioassays; Biochemical methods: ELISA – enzyme-linked immunosorbent assay; PPIA – protein phosphatase inhibition assays; Analytical methods: LC-MS – liquid chromatography–mass spectrometry; HPLC – high-performance liquid chromatography; MALDI-TOF – matrix-assisted laser desorption/ionization time-of-flight (several variants of these techniques exist). n.d. – not determined.

Note: a full form of this table is made available in the Appendix, as Table S11.1.

to society has been a recurrent challenge during the last decade. Several transnational, regional, or national projects, with different types of stakeholders, were or still are funded in Europe (e.g. CYANOCOST; B-BLOOMS2), North America (e.g. MERHAB-LGL; Water Research Foundation, 2007a) and Oceania (e.g. Water Research Foundation, 2007b; Wood *et al.*, 2013), with the purpose of, or having as a specific goal, developing and establishing cost-effective molecular-based tools, with quick turnaround times able to assist monitoring decision processes. However, constraints to their adoption and implementation by potential stakeholders include (lack of) legislative requirements, perceived costs, and unwilling-ness to change procedures and practices (Water Research Foundation, 2007b; Wood *et al.*, 2013).

11.3 Checklist of Publications, Applications and Lessons from Practice

11.3.1 Molecular-Based Studies on (Toxic) Cyanobacteria: Overview of Methods Being Used, and Generic Findings and Concerns

Table 11.1 has an extensive yet not full list of different molecular-based studies dealing with (possible) toxic cyanobacteria. The effective or potential toxicity of cyanobacteria deter-mined in these studies was achieved either by molecular methods or other techniques (such as ELISA, HPLC, LC-MS), or both. Owing to being larger in number, studies based on conventional or quantitative PCR techniques included in Table 11.1 are intentionally under-represented. Nonetheless, the selected examples intend to embrace as much as possible the different aspects of the overall studies/applications.

Thus, although incomplete, this table in conjunction with its full form (Table S11.1, which has additional information on the sampling location, type of habitat, and cyanobac-terial identification) allows some general conclusions to be made:

- The large majority of the studies are being applied to aquatic environments, particularly surface freshwater (including planktonic bloom) (Table S11.1).
- Other applications include food/additive surveillance (which will receive particular attention further ahead), as well as monitoring of terrestrial environments and of bioaccumulation through food web.
- In addition, to water or bloom samples, other types of samples include tissues from dif-ferent organisms (lichens, plants, or animals), sediment or soil, as well as cyanobacterial isolates (either live cultures or dried material; for the latter, see Metcalf *et al.* (2012a, 2012b).
- DNA samples to be used in PCR-based techniques are obtained either by direct DNA isolation from environmental samples (eDNA) (culture-independent techniques; see Fig. 11.3) or through genomic DNA extracted from strains previously isolated (culture-dependent techniques; see Fig. 11.3).
- Conventional PCR is the most employed technique, while qPCR is becoming more prevalent in recent years. With the exception of the methods usually used to profile the diversity of a community (especially DGGE), other molecular techniques (e.g. DNA chips, NGS, FISH) are still scarcely used.
- The most preferred molecular markers are genes from the cyanotoxins' biosynthetic pathways. Nevertheless, indirect markers for identifying toxigenic cyanobacteria, such as the 16S rRNA gene, are still being used.

- Most of the researches are designed to look for a specific cyanotoxin. In this particular, microcystins seem to be the preferred target, which is in accordance with Fig. 11.1 and Fig. 11.2.
- Related with it, the most studied taxon is *Microcystis* (Table S11.1).
- Molecular-based studies are being preponderantly performed by (and in) economically more developed countries (Table S11.1).

11.3.2   The Need for Complementary Approaches

Owing to their high sensitivity, fairly rapid procedure time, and obtainment of reliable results molecular-based methods are widely and routinely used in the analysis of natural samples from potentially toxic environments (Fig. 11.3 and Table 11.1), either through simple (i.e. conventional) PCR amplification, qPCR quantification, or by using oligonucleotide-arrays. In fact, they allow a fast screening for the presence of toxin-related genes or other markers for toxic cyanobacteria, having the advantage of giving a prompt alarm if there is toxicity potential. But since they generate qualitative or indirect results, complementary approaches for assessing the presence of cyanotoxins in the samples are further required to corroborate or fully validate the molecular data. These include screening biochemical assays developed for detecting (ELISA) or toxicity determination of specific cyanotoxins (PPIA), and stricter analysis obtained by chemical (analytical) techniques, such as HPLC, LC-MS, etc. (Moreira *et al.*, 2014). Notice that, in comparison to the molecular approach, these chemical methods are relatively expensive, and may require more specialized technicians and equipment. Preparation of the samples for chemical analyses is also more time consuming.

11.3.3   Interpreting Results

Possible false negative results may occur owing to improper sample collection/nucleic acid extraction or from poor handling of nucleic acid samples before testing (see Chapter 2 and 5). They may also originate from following procedural steps, for instance owing to a wrong decision on which molecular tools to use (e.g. by choosing a PCR primer set or a probe not suitably designed for the genetic diversity present in the sample) (see Chapters 6–8). False positive PCR results for the detection of cyanotoxin genes (indeed, as happens with other molecular markers) are likely to occur (Nonneman and Zimba, 2002; Villeneuve *et al.*, 2012). Thus, sequencing confirmation is highly recommended, or mandatory if a PCR-based technique is the only method being used. Even so, one should remember that the presence of these genes by itself does not confirm the existence of toxin production. Consequently, contrasting results are expected between molecular-derived data and those from complementary methods. One of these common conflicts is the evidence of a toxigenic potential inferred by molecular analysis, as deduced by the presence of known cyanotoxin genes (often confirmed by sequencing), which is not corroborated by data from other methods. Assuming that no procedural errors were made, this may be due to the fact that (a) the gene cluster is present but inactive (Christiansen *et al.*, 2008) or (b) the toxin value present in the sample is below the detection level of the biochemical or analytical methods applied (Dietrich and Hoeger, 2005). As an illustration of the latter, Saker *et al.* (2007a) were only able to detect microcystins in food supplements in a second round of MALDI-TOF analyses, after concentrating the samples.

Yet, looking at Table 11.1, it can be observed that the presence/absence of genes involved in the biosynthesis of cyanotoxins seems to broadly correlate positively with the results from biochemical and analytical methods. This general observation may derive from a positive-outcome bias (a type of publication bias) since congruent findings between molecular and biochemical or analytical data are more prone to be published than apparently incongruent results between them. Still, there are exceptions (see Table 11.1) to the aforementioned harmony between results from different types of data (Amer *et al.*, 2009; Al-Tebrineh *et al.*, 2012a), namely in those obtained from coastal environments (Frazão *et al.*, 2010; Glas *et al.*, 2010; Lopes *et al.*, 2012; Vareli *et al.*, 2012; see also Vareli *et al.*, 2013 for a review on the presence of microcystin in marine ecosystems).

In brief, it is important to be aware of the possibility of false (positive or negative) results, to know how to interpret the data, and to recognize which are the limitations of the method(s).

11.3.4 Choice of Molecular Tools for Toxigenicity Assessment

Several group- and species-specific PCR primer sets have already been developed, using taxonomic markers such as the 16S rRNA gene, *gyr*A or *rpo*B, with the goal of distinguishing toxic cyanobacteria (see Chapters 4 and 6). But actually, species identification is not conclusive in ascertaining whether a cyanobacterium is a potential cyanotoxin producer (this is true whether the identification has been achieved either by molecular- or by morphological-based methods). At most, it only indirectly indicates such a capability, and only if it undoubtedly fit in a known phylo- or morphotype (i.e. a taxon defined by a gene marker or by morphological features, respectively) for which toxigenicity has been already proven. Though, even within these taxa, there are some genotypes (i.e. strains) unable to produce toxins. For example, only 10 to 50% of *Microcystis* cells in natural systems have the genetic capability to produce microcystin (Joung *et al.*, 2011; Rinta-Kanto *et al.*, 2009). DNA profiling techniques using short tandem repeat (STR) analyses also fail to distinguish between toxic and nontoxic strains (Chonudomkul *et al.*, 2004; Lyra *et al.*, 2005).

Thus, the main tools available to assess toxigenicity by molecular methods are primers or probes designed after the sequencing and characterization of the several cyanotoxins gene clusters (for primers and probes development see also Chapters 6, 7, and 10). However, given the foregoing (see previous subsection), the choice of which is the best genetic marker (i.e. gene region within the cluster) that corroborates the existence of toxicity is an issue that is still being debated. For instance, in microcystins, some authors claim that the best candidate marker that confirms the presence of this cyanotoxin is the *mcy*E (Mankiewicz-Boczek *et al.*, 2006; Rantala *et al.*, 2006) while others refer to the *mcy*G (Tanabe *et al.*, 2007). Both markers showed to have had the best correlation with the chemical data, as their detection by PCR always returns a detection of microcystins by analytical methods. In the same way, if it is not detected by PCR, neither it is by analytical techniques. For cylindrospermopsin the *cyr*J appears to be the best candidate marker (Mankiewicz-Boczek *et al.*, 2012a, Hoff-Risseti *et al.*, 2013). For saxitoxin, *sxt*I seems to be useful to discriminate between SXT producers and non-producers species in the genus *Aphanizomenon* (Casero *et al.*,

2014). However, this does not apply to other SXT-producing genera (Casero *et al.*, 2014). Regarding anatoxin-a (including its analog homoanatoxin-a), *ana*C shown to not be able to discriminate between all ANA-producing and non-producing strains of *Anabaena* spp. evaluated by Rantala-Ylinen *et al.* (2011), giving also a false positive for *Aphanizomenon* sp. On the other hand, it was successful to distinguish producers from non-producers among 11 tested strains of *Oscillatoria* spp. (Rantala-Ylinen *et al.*, 2011). PCR results for *ana*F, another gene from the *ana* gene cluster, were generally congruent with the detection of ANA production in *Phormidium* spp. (Wood *et al.*, 2012b) or *Tychonema* spp. (Shams *et al.*, 2015; Salmaso *et al.*, 2016). The only exception was obtained for *T. bornetii* NIVA-CYA 60, which has tested positive by PCR, but negative for ANA production by the receptor-binding assay (Shams *et al.*, 2015).

11.3.5   Common and Possible Applications of Molecular Tools

11.3.5.1   Examples from Food (Additive) Surveillance

The consumption of cyanobacteria, either as direct food or as an additive/supplement, represents a not-so-well-known risk for human or animal health despite several evidences already reported (Saker *et al.*, 2005; Heussner *et al.*, 2012; Vichi *et al.*, 2012), including some obtained by means of molecular techniques (Table 11.1). A special focus will be given to this issue, exploring how molecular techniques can be helpful to the control of food safety and quality. For that reason, a short summary on this topic will be given below. Moreover, three interlinked SOPs were conceived for "blue–green algae" food supplements (BGAS), where one can also find additional information and advices. The first (see SOP 2.4) regards to sampling and processing of food supplement samples, the second (see SOP 5.6) concerns to methodological steps related with DNA extraction from those samples, while the third (see SOP 6.9) comprises a multiplex-PCR protocol previously developed by Saker *et al.* (2007a) for the detection of microcystin. All SOPs give methodological guidance, alternatives, and suggestions.

11.3.5.1.1   Cyanobacteria as Food and Related Concerns

Common edible "microalgae" include cyanobacteria as *Arthrospira* spp. (which may also be known as *Spirulina* spp.), *Nostoc* spp., and *Aphanizomenon flos-aquae* or the green algae *Chlorella* spp. Usually, they are commercially distributed worldwide as (blue–green) algae dietary supplements, but may also be used as a whole food (some *Nostoc* species; see Gao 1998; But *et al.*, 2002; Johnson *et al.*, 2008). The food supplements can be found in the marketplace in the form of tablets, capsules, flakes, or in powder form, and are commercialized as being one hundred percent pure, as a varying mixture of the aforementioned organisms or as a mixture with other foodstuffs (Heussner *et al.*, 2012). Besides the (nutraceutical) food sector, edible microalgae are also used as feed supplement in the aquaculture, fish-keeping, and poultry industries. A typical commercial system for algal biomass production involves two main stages (for a review see Takenaka and Yamaguchi, 2014). The inoculums production stage (starter cultures) usually occurs in closed systems (e.g. small-scale photobioreactors). The second stage is the biomass production, which is usually conducted in large open ponds, normally in a batch process. It is in this stage that contaminations with other organisms (Görs *et al.*, 2010) are more prone to occur. Some of them can be toxic cyanobacteria, as detected in some BGAS supplements (Saker *et al.*, 2005; Vichi *et al.*, 2012). Moreover, the source of the edible microalgae (i.e. the origin of the strain used as a starter culture) is very likely to be different between different manufacturers/producers (Habib *et al.*, 2008). Therefore, even the species identification may often be disputable (Görs *et al.*, 2010), and consequently it should be inspected. All these subjects are likely to be monitored by means of molecular tools (see SOP 5.6 and SOP 6.9).

Another subject that may represent a concern is the possible production of secondary metabolites other than cyanotoxins. This applies not only to potential contaminants but also to BGAS strains itself. This finding was exposed by Shih *et al.* (2013), who showed that some strains assigned to *Arthrospira* and *Spirulina* spp. possess gene clusters for the synthesis of bacteriocins, cyanobactins, and terpenes. Remarkably, two of those strains are known for being used as BGAS (*Arthrospira* sp. PCC 8005 and *Arthrospira platensis* NIES-39; Habib *et al.*, 2008 and Janssen *et al.*, 2010, respectively). Their effects on humans can be either beneficial or adverse, depending furthermore on the substance itself, the dosage, its bioavailability, the frequency, and the duration of exposure, etc. Until these questions are investigated and satisfactorily clarified, the intake of such secondary metabolites must also raise concern for health, owing to the chronic exposure that may incur frequent consumers (see also Gantar and Svirčev, 2008). A molecular-based monitoring may also be applied in this case.

11.3.5.2   Examples from Aquatic Ecosystems: The Particular Cases of Poland and Portugal

Unsurprisingly, molecular methods are above all applied to aquatic environments, especially in monitoring freshwater ecosystems, as it can be perceived by the number of related studies published (Table 11.1 and Table S11.1, Appendix). Since it is not feasible to cover them all in detail, below are some possible outcomes to be obtained through the application of molecular-based techniques, taking as examples studies performed in Polish and Portuguese freshwater aquatic systems.

In Poland, by routinely obtaining PCR products from water samples, it was possible to determine the occurrence of toxic genotypes responsible for synthesis of microcystins (target genes: *mcy*A, *mcy*B, *mcy*D, or *mcy*E) or cylindrospermopsin (target genes: *cyr*J or *pks*) in different water bodies (Mankiewicz-Boczek *et al.*, 2006, 2011a, 2012a; Kokociński *et al.*, 2013). Molecular analysis indicated that *mcy*E gene – which takes part in the process of addition of the ADDA moiety, responsible for toxic properties to the microcystin molecule – was a very sensitive molecular marker for the determination of potential hepatotoxicity of cyanobacteria in different environmental samples, even if the cyanobacterial biomass in water was below 0.1 mg L^{-1} (Mankiewicz-Boczek *et al.*, 2006, 2011b). Also through the utilization of PCR, and, as stated before, *cyr*J – the sulfotransferase gene required for tailoring reaction to complete the biosynthesis of cylindrospermopsin – was shown to be an appropriate genetic marker for the detection of cylindrospermopsin producers in Polish lakes (Mankiewicz-Boczek *et al.*, 2012a; Kokociński *et al.*, 2013). Quantitative real-time PCR (qPCR) techniques are becoming widely used in the monitoring of toxic cyanobacterial blooms. They combine the qualitative confirmation of cyanobacterial presence along with their quantitative determination, allowing also a comparison with other measurable environmental variables. This, in turn, promotes better understanding of the mechanisms determining the dynamics of blooms, including the detection of toxic genotypes and the evaluation of their potential toxicity to other organisms, and the recognition of key environmental factors for bloom emergence and duration. Such knowledge is important to better understand the interaction between cyanobacteria and environmental factors, and to plan further remediation strategies for a contaminated water body (Mankiewicz-Boczek, 2012b). The application of qPCR (based on 16S rRNA and *mcy*A genes copies, respectively for total *Microcystis* population and its toxic genotypes) allowed to observe that, owing to poor hydrological conditions in a Polish aquatic system (manifested by short water retention time), the total *Microcystis* gene copy number decreased, but the proportion of toxic *Microcystis* genotypes increased in comparison to values previously observed for the same location (Gągała *et al.*, 2014). This led to maintaining a similar level of microcystins concentration as that observed earlier, when the average *Microcystis* quantity was 10 times higher (Gągała *et al.*, 2014).

In Portugal, studies on the use of culture-dependent or -independent molecular methodologies in the screening of cyanotoxins has begun little more than a decade ago. These studies used primers previously published in the literature, based on toxicity markers from the gene clusters of microcystins (*mcy*), cylindospermopsin (*cyr*), saxitoxin (*sxt*), and anatoxin (*ana*) (for a revision, see Moreira *et al.* (2014)). Microcystins are the most widely screened and the most prevalent cyanotoxins in almost every freshwater system studied in Portugal. In two of the works published so far by Saker *et al.* (2007b) and Valério *et al.* (2010), single conventional PCR and multiplex PCR techniques were routinely implemented in field samples. The molecular markers targeted were *mcy*A, in the two studies, and additionally *mcy*B in Valério *et al.* (2010). Both studies came to demonstrate a good correlation between the presence or absence of genes and the detection of the toxin by analytical methods (e.g. ELISA and HPLC), suggesting that these markers can be used in the early warning of cyanotoxins with a good certainty. Recently, through single PCR reactions, Moreira *et al.* (2011a) establish the potential presence of the cyanotoxins microcystins (*mcy*ABE), cylindrospermopsin (*cyr*BC) and saxitoxins (*sxt*1) in the lakes and hot springs of the Azores islands. This study revealed that microcystin genes are the most commonly found in these aquatic systems, and that cylindrospermopsin genes are also present. Similarly, qPCR studies have also been applied in Portuguese freshwater samples, namely for the cyanotoxins microcystins and cylindrospermopsin. Martins *et al.* (2011) used specific marker genes for cyanobacteria, for *Microcystis* spp. and for MC-producing *Microcystis* (*mcy*AB). During this qPCR study, a negative significant correlation between the toxic (with *mcy* genes) and the nontoxic (without *mcy* genes) genotypes ratio and the overall *Microcystis* density was observed. The findings from this study support that qPCR allows evaluating the dynamics of toxic cyanobacterial genotypes in a given freshwater system as well as to establish its seasonal variations in order to be used as a tool in a risk management decision. In other work using qPCR, Moreira *et al.* (2011b) assessed the presence of *Cylindrospermopsis raciborskii* and the *cyr*C gene for cylindrospermopsin in a pond located in the center of Portugal and revealed, for the first time, evidences of the presence of this cyanotoxin in continental waters of this country. Recently, the same authors reported the presence of the toxin itself, by analytical methodologies, in the same previously studied aquatic system (Moreira *et al.* unpublished data).

(See Table 11.1 to get to know the different molecular techniques being applied in studies on aquatic environments or Fig. 11.3 for a complete picture on the main outcomes that may arise from molecular-based studies.)

11.3.5.3 Other Examples

Surveys concerning terrestrial environments and food web bioaccumulation are other important applications of molecular methods for monitoring/detecting the presence of potential toxic cyanobacteria (Table 11.1). For instance, the usage of conventional PCR has exposed the hepatotoxin production capability of *Nostoc* cyanobionts (Table S11.1, Appendix) living in symbioses with lichens (Oksanen *et al.*, 2004; Kaasalainen *et al.*, 2012) and plants (Gehringer *et al.*, 2012). Using a similar methodological approach, Metcalf *et al.* (2012a) revealed the risk that soil dust may represent to human health after discovering that cyanobacteria inhabiting soil crusts commonly found in arid environments are toxic. Conventional PCR (Oberholster *et al.*, 2006), qPCR (Oberholster *et al.*, 2006; Sotton *et al.*, 2014), and PCR-DGGE (Vareli *et al.*, 2012) methodologies have also allowed it to be determined that toxic cyanobacteria may accumulate, sometimes extensively, in different organisms along the food chain, including in edible fishes and mussels (see Sukenik *et al.*, 2015 for a review on bioaccumulation and effects of cyanotoxins on trophic webs).

11.3.5.4 Non-Toxigenicity-Related Examples

Table 11.2 lists some studies that did not have as their main purpose the investigation of toxic cyanobacteria, although it may have been related to them. Hence, the toxigenicity was not determined, neither by molecular techniques nor by other methods. Instead, these are studies that show other/new possibilities of applications of molecular techniques for the study of cyanobacteria, several of them also related to public health and environmental issues. For instance, Loza *et al.* (2013) studied cyanobacterial communities by TGGE (a very similar technique to DGGE, but in this case a temperature gradient rather than a chemical gradient is used to denature the DNA) in order to develop a monitoring tool for water-quality assessment. Likewise, Gomila *et al.* (2006) repeatedly detected the presence of cyanobacteria in hemodialysis water while screening clone libraries obtained from different points of a water treatment system. Other applications include, for example, attempts to DNA barcode potential toxigenic cyanobacteria (Kurobe *et al.*, 2013) and development of a microarray for detecting genomic diversity associated with secondary metabolites in cyanobacteria (Pomati and Neilan, 2004). Also worth noting is the emergence of microbial community studies employing newer molecular methods (e.g. NGS technologies; see Chapter 10).

11.4 General Conclusions

Since the late 1990s, the number of available gene sequences of the several cyanotoxins has increased exponentially in the GenBank database, particularly for the microcystins, which have always prevailed. This finding is supported by a great many publications on this cyanotoxin. The characterization of the biosynthetic gene cluster of this and the other

Table 11.2 *Examples of other applications/studies where molecular methods were applied for the detection/identification of cyanobacteria (including potentially toxic or harmful taxa) and of the microbial communities related to them*

Application/type of study	Method[1] (Chapter number)	Type of sample	Targeted molecular marker(s)	Studied organism(s)	References
ecosystem health and conservation (coral reefs monitoring)	DGGE/clone libraries/DNA typing methods (9)	coral tissue	16S rRNA gene	*Lyngbya*-like species and unidentified cyanobacterium	Sekar *et al.* (2006)
food web bioaccumulation	Conventional PCR (6)	water and grazers	16S rRNA gene	*Nodularia spumigena*	Gorokhova (2009)
genetics of secondary metabolites	Microarrays (8)	isolates	several, genes encoding unknown/putative proteins	*Anabaena circinalis*	Pomati and Neilan 2004
microbial community structure	DGGE/clone libraries/DNA typing methods (9)	tree leaves	16S rRNA gene	several epiphytic cyanobacteria	Rigonato *et al.* (2012)
microbial community structure	NGS and/or metagenomics techniques (10)	water	16S rRNA gene	Cyanobacteria (at the phylum level)	Ghai *et al.* (2011)
microbial community structure	NGS and/or metagenomics techniques (10)	water	16S rRNA gene	*Cyanobium, Merismopedia, Synechococcus, Pseudanabaena*	Ghai *et al.* (2012)
microbial community structure	NGS and/or metagenomics techniques (10)	bloom	16S rRNA gene	*Microcystis* sp.	Chen *et al.* (2011)

Application	Method	Sample	Target gene	Organisms	Reference
microbial community structure	NGS and/or metagenomics techniques (10)	bloom	several, including putative genes of MC-LR cleavage pathway (*mlr*) and xenobiotic metabolisms (GST genes)[2]	transforming and detoxifying microcystins microbiota	Mou et al. (2013)
microbial community structure	NGS and/or metagenomics techniques (10)	bloom	23S rRNA gene	Cyanobacteria (at the order level) and eukaryotic plastids (at the phylum level)	Steven et al. (2012)
taxonomy	Conventional PCR (6)	herbarium specimens	16S rRNA gene	several cyanobacteria	Palinska et al. (2006)
taxonomy (DNA barcoding of toxigenic cyanobacteria)	DGGE/clone libraries/DNA typing methods (9)	water	16S rRNA gene	several potential toxin-producing cyanobacteria	Kurobe et al. (2013)
water quality (cyanobacterial ecotypes as bioindicators)	DGGE/clone libraries/DNA typing methods (9)	epilithic biofilms	16S rRNA gene	several epilithic cyanobacteria	Loza et al. (2013)
water quality (for hemodialysis)	DGGE/clone libraries/DNA typing methods (9)	hemodialysis water	16S rRNA gene	several unidentified cyanobacteria	Gomila et al. (2006)

[1] For more information on the method, see the respective chapter.
[2] MC-LR - microcystin-LR; GST - glutathione S-transferase.
The full version of this table is available in the Appendix as Table S11.2.

cyanotoxins gave rise to the possibility of new applications and methodological improvements. In fact, since the detection of genes involved in cyanotoxins synthesis allows the early recognition of cyanobacterial threats, soon different molecular methods were adopted and developed to become a valuable tool for monitoring toxigenic cyanobacteria in the environment. Nevertheless, genetic information has to be seen as a primary assessment of toxicity, and so the presence of toxins in a given sample should be validated analytically.

Owing to its specificity, from all the applications discussed above food supplement surveillance has deserved special attention. Also, the lack of regulatory standards for the commercialization of BGAS (Gilroy *et al.*, 2000; Habib *et al.*, 2008) gives further importance to the monitoring of these products. As a result, three interlinked SOPs regarding the monitoring of BGAS were developed and made available (see SOP 2.4, SOP 5.6, and SOP 6.9).

11.5 Acknowledgments

This work was partially supported by COST Action CYANOCOST, the Structured Program of RandDandI INNOVMAR – Innovation and Sustainability in the Management and Exploitation of Marine Resources (reference NORTE-01-0145-FEDER-000035, Research Line NOVELMAR), funded by the Northern Regional Operational Program (NORTE2020) through the European Regional Development Fund (ERDF) and project FCT UID/Multi/04423/2013. VR would also like to acknowledge financial support from Fundação para a Ciência e a Tecnologia (FCT) fellowship SFRH/BD/80153/2011.

11.6 References

Allender, C.J., LeCleir, G.R., Rinta-Kanto, J.M., *et al.* (2009) Identifying the source of unknown microcystin genes and predicting microcystin variants by comparing genes within uncultured cyanobacterial cells, *Applied and Environmental Microbiology*, **75** (11), 3598–3604.

Al-Tebrineh, J., Gehringer, M.M., Akcaalan, R., and Neilan, B.A. (2011) A new quantitative PCR assay for the detection of hepatotoxigenic cyanobacteria, *Toxicon*, **57** (4), 546–554.

Al-Tebrineh, J., Merrick, C., Ryan, D., et al. (2012a) Community composition, toxigenicity, and environmental conditions during a cyanobacterial bloom occurring along 1,100 kilometers of the Murray River, *Applied and Environmental Microbiology*, **78** (1), 263–272.

Al-Tebrineh, J., Pearson, L.A., Yasar, S.A., and Neilan, B.A. (2012b) A multiplex qPCR targeting hepato- and neurotoxigenic cyanobacteria of global significance, *Harmful Algae*, **15**, 19–25.

Amer, R., Díez, B., and El-Shehawy, R. (2009) Diversity of hepatotoxic cyanobacteria in the Nile Delta, Egypt, *Journal of Environmental Monitoring*, **11** (1), 126–133.

Baker, J.A., Entsch, B., Neilan, B.A., and McKay, D.B. (2002) Monitoring changing toxigenicity of a cyanobacterial bloom by molecular methods, *Applied and Environmental Microbiology*, **68** (12), 6070–6076.

Baker, J.A., Neilan, B.A., Entsch, B., and McKay, D.B. (2001) Identification of cyanobacteria and their toxigenicity in environmental samples by rapid molecular analysis. *Environmental Toxicology*, **16** (6), 472–482.

Baker, L., Sendall, B.C., Gasser, R.B., et al. (2013) Rapid, multiplex-tandem PCR assay for automated detection and differentiation of toxigenic cyanobacterial blooms, *Molecular and Cellular Probes*, **27**, 208–214.

Belykh, O.I., Sorokovikova, E.G., Fedorova, G.A., et al. (2011) Presence and genetic diversity of microcystin-producing cyanobacteria (*Anabaena* and *Microcystis*) in Lake Kotokel (Russia, Lake Baikal Region), *Hydrobiologia*, **671** (1), 241–252.

Bittencourt-Oliveira, M.C. (2003) Detection of potential microcystin-producing cyanobacteria in Brazilian reservoirs with a *mcy*B molecular marker, *Harmful Algae*, **2** (1), 51–60.

Briand, E., Escoffier, N., Straub, C., et al. (2009) Spatiotemporal changes in the genetic diversity of a bloom-forming *Microcystis aeruginosa* (Cyanobacteria) population, *The ISME Journal*, **3**, 419–429.

Briand, E., Gugger, M., François, J.C., et al. (2008) Temporal variations in the dynamics of potentially microcystin-producing strains in a bloom-forming *Planktothrix agardhii* (cyanobacterium) population, *Applied and Environmental Microbiology*, **74** (12), 3839–3848.

Brito, Â., Ramos, V., Seabra, R., et al. (2012) Culture-dependent characterization of cyanobacterial diversity in the intertidal zones of the Portuguese coast: A polyphasic study, *Systematic and Applied Microbiology*, **35** (2), 110–119.

But, P.P.H., Cheng, L., Chan, P.K., et al. (2002) *Nostoc flagelliforme* and faked items retailed in Hong Kong, *Journal of Applied Phycology*, **14** (2), 143–145.

Carmichael, W.W. and Li, R. (2006) Cyanobacteria toxins in the Salton Sea, *Saline Systems*, **2**, 5.

Casero, M.C., Ballot, A., Agha, R., et al. (2014) Characterization of saxitoxin production and release and phylogeny of *sxt* genes in paralytic shellfish poisoning toxin-producing *Aphanizomenon gracile*, *Harmful Algae*, **37**, 28–37.

Castiglioni, B., Rizzi, E., Frosini, A., et al. (2004) Development of a universal microarray based on the ligation detection reaction and 16S rRNA gene polymorphism to target diversity of cyanobacteria, *Applied and Environmental Microbiology*, **70** (12), 7161–7172.

Chen, C., Zhang, Z., Ding, A., et al. (2011) Bar-coded pyrosequencing reveals the bacterial community during Microcystis water bloom in Guanting Reservoir, Beijing, *Procedia Engineering*, **18**, 341–346.

Chonudomkul, D., Yongmanitchai, W., Theeragool, G., et al. (2004) Morphology, genetic diversity, temperature tolerance and toxicity of *Cylindrospermopsis raciborskii* (Nostocales, Cyanobacteria) strains from Thailand and Japan, *FEMS Microbiology Ecology*, **48** (3), 345–355.

Christiansen, G., Molitor, C., Philmus, B., and Kurmayer, R. (2008) Nontoxic strains of cyanobacteria are the result of major gene deletion events induced by a transposable element, *Molecular Biology and Evolution*, **25** (8), 1695–1704.

Dietrich, D. and Hoeger, S. (2005) Guidance values for microcystins in water and cyanobacterial supplement products (blue-green algal supplements): A reasonable or misguided approach?, *Toxicology and Applied Pharmacology*, **203** (3), 273–289.

Dittmann, E., Neilan, B.A., Erhard, M., et al. (1997) Insertional mutagenesis of a peptide synthetase gene that is responsible for hepatotoxin production in the cyanobacterium *Microcystis aeruginosa* PCC 7806, *Molecular Microbiology*, **26**, 779–787.

Edwards, D.J. and Gerwick, W.H. (2004) Lyngbyatoxin biosynthesis: Sequence of biosynthetic gene cluster and identification of a novel aromatic prenyltransferase, *Journal of the American Chemical Society*, **126** (37), 11432–11433.

Fathalli, A., Jenhani, A.B., Moreira, C., et al. (2011) Genetic variability of the invasive cyanobacteria *Cylindrospermopsis raciborskii* from Bir M'cherga reservoir (Tunisia), *Archives of Microbiology*, **193**, 595–604.

Fewer, D.P., Köykkä, M., Halinen, K., et al. (2009) Culture-independent evidence for the persistent presence and genetic diversity of microcystin-producing *Anabaena* (Cyanobacteria) in the Gulf of Finland, *Environmental Microbiology*, **11** (4), 855–866.

Frazão, B., Martins, R., and Vasconcelos, V. (2010) Are known cyanotoxins involved in the toxicity of picoplanktonic and filamentous North Atlantic marine cyanobacteria?, *Marine Drugs*, **8** (6), 1908–1919.

Gągała, I., Izydroczyk, K., Jurczak, T., et al. (2014) Role of environmental factors and toxic genotypes in the regulation of microcystins-producing cyanobacterial blooms, *Microbial Ecology*, **67** (2), 465–479.

Gaget, V., Gribaldo, S., and de Marsac, N.T. (2011) An *rpo*B signature sequence provides unique resolution for the molecular typing of cyanobacteria, *International Journal of Systematic and Evolutionary Microbiology*, **61** (1), 170–183.

Gan, N.Q., Huang, Q., Zheng, L.L., and Song, L.R. (2010) Quantitative assessment of toxic and nontoxic *Microcystis* colonies in natural environments using fluorescence *in situ* hybridization and flow cytometry, *Science China Life Sciences*, **53** (8), 973–980.

Gantar, M. and Svirčev, Z. (2008) Microalgae and cyanobacteria: Food for thought, *Journal of Phycology*, **44**, 260–268.

Gantar, M., Sekar, R., and Richardson, L.L. (2009) Cyanotoxins from black band disease of corals and from other coral reef environments, *Microbial Ecology*, **58** (4), 856–864.

Gao, K. (1998) Chinese studies on the edible blue-green alga, *Nostoc flagelliforme*: A review, *Journal of Applied Phycology*, **10** (1), 37–49.

Gehringer, M.M., Adler, L., Roberts, A.A., et al. (2012) Nodularin: A cyanobacterial toxin, is synthesized *in planta* by symbiotic *Nostoc* sp., *The ISME Journal*, **6** (10), 1834–1847.

Ghai, R., Hernandez, C.M., Picazo, A., et al. (2012) Metagenomes of Mediterranean coastal lagoons, *Scientific Reports*, **2**, 490.

Ghai, R., Rodríguez-Valera, F., McMahon, K.D., et al. (2011) Metagenomics of the water column in the pristine upper course of the Amazon River, *PLOS ONE*, **6** (8), e23785.

Gilroy, D.J., Kauffman, K.W., Hall, R.A., et al. (2000) Assessing potential health risks from microcystin toxins in blue-green algae dietary supplements, *Environmental Health Perspectives*, **108** (5), 435–439.

Glas, M.S., Motti, C.A., Negri, A.P., et al. (2010) Cyanotoxins are not implicated in the etiology of coral black band disease outbreaks on Pelorus Island, Great Barrier Reef, *FEMS Microbiol Ecology*, **73** (1), 43–54.

Gomila, M., Gascó, J., Gil, J., et al. (2006) A molecular microbial ecology approach to studying hemodialysis water and fluid, *Kidney International*, **70** (9), 1567–1576.

Gorokhova, E. (2009) Toxic cyanobacteria *Nodularia spumigena* in the diet of Baltic mysids: Evidence from molecular diet analysis, *Harmful Algae*, **8** (2), 264–272.

Görs, M., Schumann, R., Hepperle, D., and Karsten, U. (2010) Quality analysis of commercial *Chlorella* products used as dietary supplement in human nutrition, *Journal of Applied Phycology*, **22** (3), 265–276.

Ha, J.H., Hidaka, T., and Tsuno, H. (2011) Analysis of factors affecting the ratio of microcystin to chlorophyll-a in cyanobacterial blooms using real-time polymerase chain reaction, *Environmental Toxicology*, **26** (1), 21–28.

Habib, M.A.B., Pariv, M., Huntington, T.C., and Hasan, M.R. (2008) *A review on culture, production and use of Spirulina as food for humans and feeds for domestic animals and fish. FAO Fisheries and Aquaculture Circular no. 1034*, Food and Agriculture Organization (FAO) of the United Nations, Rome, Italy.

Heussner, A.H., Mazija, L., Fastner, J., and Dietrich, D.R. (2012) Toxin content and cytotoxicity of algal dietary supplements, *Toxicology and Applied Pharmacology*, **265** (2), 263–271.

Hisbergues, M., Christiansen, G., Rouhiainen, L., et al. (2003) PCR-based identification of microcystin-producing genotypes of different cyanobacterial genera, *Archives of Microbiology*, **180** (6), 402–410.

Hoff-Risseti, C., Dörr, F.A., Schaker, P.D.C., et al. (2013) Cylindrospermopsin and saxitoxin synthetase genes in *Cylindrospermopsis raciborskii* strains from Brazilian freshwater, *PLOS ONE*, **8** (8), e74238.

Innok, S., Matsumura, M., Boonkerd, N., and Teaumroong, N. (2005) Detection of *Microcystis* in lake sediment using molecular genetic techniques, *World Journal of Microbiology and Biotechnology*, **21** (8–9), 1559–1568.

Iteman, I., Rippka, R., de Marsac, N.T., and Herdman, M. (2000) Comparison of conserved structural and regulatory domains within divergent 16S rRNA–23S rRNA spacer sequences of cyanobacteria, *Microbiology*, **146** (6), 1275–1286.

Janse, I., Kardinaal, W.E.A., Meima, M., et al. (2004) Toxic and nontoxic *Microcystis* colonies in natural populations can be differentiated on the basis of rRNA gene internal transcribed spacer diversity, *Applied and Environmental Microbiology*, **70** (7), 3979–3987.

Janssen, P.J., Morin, N., Mergeay, M., et al. (2010) Genome sequence of the edible cyanobacterium *Arthrospira* sp. *PCC 8005, Journal of Bacteriology*, **192** (9), 2465–2466.

Johnson, H.E., King, S.R., Banack, S.A., et al. (2008) Cyanobacteria (*Nostoc commune*) used as a dietary item in the Peruvian highlands produce the neurotoxic amino acid BMAA, *Journal of Ethnopharmacology*, **118** (1), 159–165.

Joung, S.H., Oh, H.M., Ko, S.R., and Ahn, C.Y. (2011) Correlations between environmental factors and toxic and non-toxic *Microcystis* dynamics during bloom in Daechung Reservoir, Korea, *Harmful Algae*, **10**, 188–193.

Jungblut, A.D. and Neilan, B.A. (2006) Molecular identification and evolution of the cyclic peptide hepatotoxins, microcystin and nodularin, synthetase genes in three orders of cyanobacteria, *Archives of Microbiology*, **185** (2), 107–114.

Jungblut, A.D., Hawes, I., Mountfort, D., et al. (2005) Diversity within cyanobacterial mat communities in variable salinity meltwater ponds of McMurdo Ice Shelf, Antarctica, *Environmental Microbiology*, **7** (4), 519–529.

Kaasalainen, U., Fewer, D.P., Jokela, J., et al. (2012) Cyanobacteria produce a high variety of hepatotoxic peptides in lichen symbiosis, *Proceedings of the National Academy of Sciences of the USA*, **109** (15), 5886–5891.

Kardinaal, W.E.A., Janse, I., Kamst-van Agterveld, M. et al. (2007) *Microcystis* genotype succession in relation to microcystin concentrations in freshwater lakes, *Aquatic Microbial Ecology*, **48** (1), 1–12.

Kellmann, R., Michali, T.K., and Neilan, B.A. (2008b) Identification of a saxitoxin biosynthesis gene with a history of frequent horizontal gene transfers, *Journal of Molecular Evolution*, **67** (5), 526–538.

Kellmann, R., Mihali, T.K., Jeon, Y.J., et al. (2008a) Biosynthetic intermediate analysis and functional homology reveal a saxitoxin gene cluster in cyanobacteria, *Applied and Environmental Microbiology*, **74** (13), 4044–4053.

Kokociński, M., Mankiewicz-Boczek, J., Jurczak, T., et al. (2013) *Aphanizomenon gracile* (Nostocales): A cylindrospermopsin-producing cyanobacterium in Polish lakes, *Environmental Science and Pollution Research International*, **20**, 5243–5264.

Kurmayer, R. and Kutzenberger, T. (2003) Application of real-time PCR for quantification of microcystin genotypes in a population of the toxic cyanobacterium *Microcystis* sp., *Applied and Environmental Microbiology*, **69** (11), 6723–6730.

Kurmayer, R., Dittmann, E., Fastner, J., and Chorus, I. (2002) Diversity of microcystin genes within a population of the toxic cyanobacterium *Microcystis* spp. in Lake Wannsee (Berlin, Germany), *Microbial Ecology*, **43** (1), 107–118.

Kurobe, T., Baxa, D.V., Mioni, C.E., et al. (2013) Identification of harmful cyanobacteria in the Sacramento-San Joaquin Delta and Clear Lake, California by DNA barcoding, *SpringerPlus*, **2** (1), 1–12.

Li, D., Kong, F., Shi, X., et al. (2012) Quantification of microcystin-producing and non-microcystin producing Microcystis populations during the 2009 and 2010 blooms in Lake Taihu using quantitative real-time PCR, *Journal of Environmental Sciences*, **24** (2), 284–290.

Lopes, V.R., Ramos, V., Martins, A., et al. (2012) Phylogenetic, chemical and morphological diversity of cyanobacteria from Portuguese temperate estuaries, *Marine Environmental Research*, **73**, 7–16.

Loza, V., Perona, E., and Mateo, P. (2013) Molecular fingerprinting of cyanobacteria from river biofilms as a water quality monitoring tool, *Applied and Environmental Microbiology*, **79** (5), 1459–1472.

Lyra, C., Laamanen, M., Lehtimäki, J.M., et al. (2005) Benthic cyanobacteria of the genus *Nodularia* are non-toxic, without gas vacuoles, able to glide and genetically more diverse than planktonic *Nodularia*, *International Journal of Systematic and Evolutionary Microbiology*, **55** (2), 555–568.

Mankiewicz-Boczek, J. (2012b) Application of molecular tools in Ecohydrology, *Ecohydrology & Hydrobiology*, **12** (2), 165–170.

Mankiewicz-Boczek, J., Gągała, I., Kokociński, M., et al. (2011a) Perennial toxigenic *Planktothrix agardhii* bloom in selected lakes of Western Poland, *Environmental Toxicology*, **26**, 10–20.

Mankiewicz-Boczek, J., Izydorczyk, K., Romanowska-Duda, Z., et al. (2006) Detection and monitoring toxigenicity of cyanobacteria by application of molecular methods, *Environmental Toxicology*, **21**, 380–387.

Mankiewicz-Boczek, J., Kokociński, M., Gągała, I., et al. (2012a) Preliminary molecular identification of cylindrospermopsin- producing Cyanobacteria in two Polish lakes (Central Europe), *FEMS Microbiology Letters*, **326** (2), 173–179.

Mankiewicz-Boczek, J., Palus, J., Gągała, I., et al. (2011b) Effects of microcystins-containing cyanobacteria from a temperate ecosystem on human lymphocytes culture and their potential for adverse human health effects, *Harmful Algae*, **10**, 356–365.

Martins, A., Moreira, C., Vale, M., et al. (2011) Seasonal dynamics of *Microcystis* spp. and their toxigenicity as assessed by qpcr in a temperate reservoir, *Marine Drugs*, **9** (10), 1715–1730.

Martins, J., Peixe, L., and Vasconcelos, V. (2010) Cyanobacteria and bacteria co-occurrence in a wastewater treatment plant: Absence of allelopathic effects, *Water Science and Technology*, **62** (8), 1954–1962.

McGregor, G.B., Sendall, B.C., Hunt, L.T., and Eaglesham, G.K. (2011) Report of the cyanotoxins cylindrospermopsin and deoxy-cylindrospermopsin from *Raphidiopsis mediterranea* Skuja (Cyanobacteria/Nostocales), *Harmful Algae*, **10**, 402–410.

Méjean, A., Mann, S., Maldiney, T., et al. (2009) Evidence that biosynthesis of the neurotoxic alkaloids anatoxin-a and homoanatoxin-a in the cyanobacterium *Oscillatoria* PCC 6506 occurs on a modular polyketide synthase initiated by L-proline, *Journal of the American Chemical Society*, **131** (22), 7512–7513.

Metcalf, J.S., Beattie, K.A., Purdie, E.L., et al. (2012b) Analysis of microcystins and microcystin genes in 60–170-year-old dried herbarium specimens of cyanobacteria, *Harmful Algae*, **15**, 47–52.

Metcalf, J.S., Reilly, M., Young, F.M., and Codd, G.A. (2009) Localization of microcystin synthetase genes in colonies of the cyanobacterium *Microcystis* using fluorescence *in situ* hybridization, *Journal of Phycology*, **45** (6), 1400–1404.

Metcalf, J.S., Richer, R., Cox, P.A., and Codd, G.A. (2012a) Cyanotoxins in desert environments may present a risk to human health, *Science of the Total Environment*, **421–422**, 118–123.

Mihali, T.K., Kellmann, R., Muenchhoff, J., et al. (2008) Characterization of the gene cluster responsible for cylindrospermopsin biosynthesis, *Applied and Environmental Microbiology*, **74** (3), 716–722.

Misson, B., Donnadieu-Bernard, F., Godon, J.J., et al. (2012) Short- and long-term dynamics of the toxic potential and genotypic structure in benthic populations of *Microcystis*, *Water Research*, **46** (5), 1438–1446.

Moffitt, M.C. and Neilan, B.A. (2004) Characterization of the nodularin synthetase gene cluster and proposed theory of the evolution of cyanobacterial hepatotoxins, *Applied and Environmental Microbiology*, **70** (11), 6353–6362.

Moreira, C., Martins, A., Azevedo, J., et al. (2011b) Application of real-time PCR in the assessment of the toxic cyanobacterium *Cylindrospermopsis raciborskii* abundance and toxicological potential, *Applied Microbiology Biotechnology*, **92** (1), 189–197.

Moreira, C., Martins, A., Moreira, C., and Vasconcelos, V. (2011a) Toxigenic cyanobacteria in volcanic lakes and hot springs of a North Atlantic Island (S. Miguel, Azores, Portugal), *Fresenius Environmental Bulletin*, **20** (2a), 420–426.

Moreira, C., Ramos, V., Azevedo, J., and Vasconcelos, V.M. (2014) Methods to detect cyanobacteria and their toxins in the environment, *Applied Microbiology and Biotechnology*, **98**, 8073–8082.

Mou, X., Lu, X., Jacob, J., et al. (2013) Metagenomic identification of bacterioplankton taxa and pathways involved in microcystin degradation in Lake Erie, *PLOS ONE*, **8** (4), e61890.

Neilan, B.A. (2002) The molecular evolution and DNA profiling of toxic cyanobacteria, *Current Issues in Molecular Biology*, **4** (1), 1–11.

Neilan, B.A., Jacobs, D., and Goodman, A.E. (1995) Genetic diversity and phylogeny of toxic cyanobacteria determined by DNA polymorphisms within the phycocyanin locus, *Applied and Environmental Microbiology*, **61** (11), 3875–3883.

Neilan, B.A., Saker, M.L., Fastner, J., et al. (2003) Phylogeography of the invasive cyanobacterium *Cylindrospermopsis raciborskii*, *Molecular Ecology*, **12** (1), 133–140.

Nishihara, H., Miwa, H., Watanabe, M., et al. (1997) Random amplified polymorphic DNA (RAPD) analyses for discriminating genotypes of *Microcystis* cyanobacteria, *Bioscience, Biotechnology and Biochemistry*, **61** (7), 1067–1072.

Nonneman, D. and Zimba, P.V. (2002) A PCR-based test to assess the potential for microcystin occurrence in channel catfish production ponds, *Journal of Phycology*, **38** (1), 230–233.

Nübel, U., Garcia-Pichel, F., and Muyzer, G. (1997) PCR primers to amplify 16S rRNA genes from cyanobacteria, *Applied and Environmental Microbiology*, **63** (8), 3327–3332.

Oberholster, P.J., Botha, A.M., and Cloete, T.E. (2006) Use of molecular markers as indicators for winter zooplankton grazing on toxic benthic cyanobacteria colonies in an urban Colorado lake, *Harmful Algae*, **5** (6), 705–716.

Oksanen, I., Jokela, J., Fewer, D.P. et al. (2004) Discovery of rare and highly toxic microcystins from lichen-associated cyanobacterium *Nostoc* sp. strain IO-102-I, *Applied and Environmental Microbiology*, **70** (10), 5756–5763.

Ostermaier, V. and Kurmayer, R. (2010) Application of real-time PCR to estimate toxin production by the cyanobacterium *Planktothrix sp.*, *Applied and Environmental Microbiology*, **76** (11), 3495–3502.

Ouahid, Y. and Fernández del Campo, F. (2009) Typing of toxinogenic *Microcystis* from environmental samples by multiplex PCR, *Applied Microbiology and Biotechnology*, **85**, 405–412.

Ouellette, A.J., Handy, S.M., and Wilhelm, S.W. (2006) Toxic *Microcystis* is widespread in Lake Erie: PCR detection of toxin genes and molecular characterization of associated cyanobacterial communities, *Microbial Ecology*, **51** (2), 154–165.

Palinska, K.A. (2006) Phylogenetic evaluation of cyanobacteria preserved as historic herbarium exsiccata, *International Journal of Systematic and Evolutionary Microbiology*, **56** (10), 2253–2263.

Penn, K., Wang, J., Fernando, S.C., and Thompson, J.R. (2014) Secondary metabolite gene expression and interplay of bacterial functions in a tropical freshwater cyanobacterial bloom, *The ISME Journal*, **8** (9), 1866–1878.

Pomati, F. and Neilan, B.A. (2004) PCR-based positive hybridization to detect genomic diversity associated with bacterial secondary metabolism, *Nucleic Acids Research*, **32** (1), e7.

Pomati, F., Burns, B.P., and Neilan, B.A. (2004) Identification of an Na^+-dependent transporter associated with saxitoxin-producing strains of the cyanobacterium *Anabaena circinalis*, *Applied and Environmental Microbiology*, **70** (8), 4711–4719.

Pope, P.B. and Patel, B.K.C. (2008) Metagenomic analysis of a freshwater toxic cyanobacteria bloom, *FEMS Microbiology Ecology*, **64** (1), 9–27.

Rantala, A., Fewer, D.P., Hisbergues, M., et al. (2004) Phylogenetic evidence for the early evolution of microcystin synthesis, *Proceedings of the National Academy of Sciences of the United States of America*, **101** (2), 568–573.

Rantala, A., Rajaniemi-Wacklin, P., Lyra, C., et al. (2006) Detection of microcystin-producing cyanobacteria in Finnish lakes with genus-specific microcystin synthetase gene E (*mycE*) PCR and associations with environmental factor, *Applied and Environmental Microbiology*, **72** (9), 6101–6110.

Rantala, A., Rizzi, E., Castiglioni, B., et al. (2008) Identification of hepatotoxin-producing cyanobacteria by DNA-chip, *Environmental Microbiology*, **10** (3), 653–664.

Rantala-Ylinen, A., Känä, S., Wang, H., et al. (2011) Anatoxin-a synthetase gene cluster of the cyanobacterium *Anabaena* sp. strain 37 and molecular methods to detect potential producers, *Applied and Environmental Microbiology*, **77** (20), 7271–7278.

Rasmussen, J.P., Giglio, S., Monis, P.T., et al. (2008) Development and field testing of a real-time PCR assay for cylindrospermopsin-producing cyanobacteria, *Journal of Applied Microbiology*, **104** (5), 1503–1515.

Richardson, L.L., Sekar, R., Myers, J., et al. (2007) The presence of the cyanobacterial toxin microcystin in black band disease of corals, *FEMS Microbiology Letters*, **272**, 182–187.

Rigonato, J., Alvarenga, D.O., Andreote, F.D. et al. (2012) Cyanobacterial diversity in the phyllosphere of a mangrove forest, *FEMS Microbiology Ecology*, **80** (2), 312–322.

Rinta-Kanto, J.M., Konopko, E.A., DeBruyn, J.M., et al. (2009) Lake Erie *Microcystis*: Relationship between microcystin production, dynamics of genotypes and environmental parameters in a large lake, *Harmful Algae*, **8** (5), 665–673.

Rinta-Kanto, J.M., Ouellette, A.J.A., Boyer, G.L., et al. (2005) Quantification of toxic *Microcystis* spp. during the 2003 and 2004 blooms in western Lake Erie using quantitative real-time PCR, *Environmental Science & Technology*, **39** (11), 4198–4205.

Rothrock, M.J. and Garcia-Pichel, F. (2005) Microbial diversity of benthic mats along a tidal desiccation gradient, *Environmental Microbiology*, **7** (4), 593–601.

Saker, M.L., Jungblut, A.D., Neilan, B.A., et al. (2005) Detection of microcystin synthetase genes in health food supplements containing the freshwater cyanobacterium *Aphanizomenon flos-aquae*, *Toxicon*, **46** (5), 555–562.

Saker, M.L., Vale, M., Kramer, D., and Vasconcelos, V.M. (2007b) Molecular techniques for the early warning of toxic cyanobacteria blooms in freshwater lakes and rivers, *Applied Microbiology and Biotechnology*, **75** (2), 441–449.

Saker, M.L., Welker, M., and Vasconcelos, V.M. (2007a) Multiplex PCR for the detection of toxigenic cyanobacteria in dietary supplements produced for human consumption, *Applied Microbiology and Biotechnology*, **73** (5), 1136–1142.

Salmaso, N., Cerasino, L., Boscaini, A., and Capelli, C. (2016) Planktic *Tychonema* (Cyanobacteria) in the large lakes south of the Alps: Phylogenetic assessment and toxigenic potential, *FEMS Microbiology Ecology*, doi: 10.1093/femsec/fiw155.

Sekar, R., Mills, D.K., Remily, E.R., et al. (2006) Microbial communities in the surface mucopolysaccharide layer and the black band microbial mat of black band-diseased *Siderastrea siderea*, *Applied and Environmental Microbiology*, **72** (9), 5963–5973.

Shams, S., Capelli, C., Cerasino, L., et al. (2015) Anatoxin-a producing *Tychonema* (Cyanobacteria) in European waterbodies, *Water Research*, **69**, 68–79.

Shih, P.M., Wu, D., Latifi, A., et al. (2013) Improving the coverage of the cyanobacterial phylum using diversity-driven genome sequencing, *Proceedings of the National Academy of Sciences*, **110** (3), 1053–1058.

Sipari, H., Rantala-Ylinen, A., Jokela, J., et al. (2010) Development of a chip assay and quantitative PCR for detecting microcystin synthetase E gene expression, *Applied and Environmental Microbiology*, **76** (12), 3797–3805.

Sotton, B., Guillard, J., Anneville, O., et al. (2014) Trophic transfer of microcystins through the lake pelagic food web: Evidence for the role of zooplankton as a vector in fish contamination, *Science of the Total Environment*, **466–467**, 152–163.

Steffen, M.M., Li, Z., Effler, T.C., et al. (2012) Comparative metagenomics of toxic freshwater cyanobacteria bloom communities on two continents, *PLOS ONE*, **7** (8), e44002.

Steven, B., McCann, S., and Ward, N.L. (2012) Pyrosequencing of plastid 23S rRNA genes reveals diverse and dynamic cyanobacterial and algal populations in two eutrophic lakes, *FEMS Microbiology Ecology*, **82** (3), 607–615.

Stucken, K., Murillo, A.A., Soto-Liebe, K., et al. (2009) Toxicity phenotype does not correlate with phylogeny of *Cylindrospermopsis raciborskii* strains, *Systematic and Applied Microbiology*, **32** (1), 37–48.

Sukenik, A., Quesada, A., and Salmaso, N. (2015) Global expansion of toxic and non-toxic cyanobacteria: Effect on ecosystem functioning, *Biodiversity and Conservation*, **24**, 889–908.

Takenaka, H. and Yamaguchi, Y. (2014) Commercial-scale culturing of cyanobacteria: An industrial experience, in N. Sharma, A. Rai, and L. Stal (eds), *Cyanobacteria: An economic perspective*, John Wiley & Sons, Ltd, Chichester, UK.

Tanabe, Y., Kasai, F., and Watanabe, M.M. (2007) Multilocus sequence typing (MLST) reveals high genetic diversity and clonal population structure of the toxic cyanobacterium *Microcystis aeruginosa*, *Microbiology*, **153** (11), 3695–3703.

Te, S.H. and Gin, K.Y.H. (2011) The dynamics of cyanobacteria and microcystin production in a tropical reservoir of Singapore, *Harmful Algae*, **10**, 319–329.

Tillett, D., Parker, D.L., and Neilan, B.A. (2001) Detection of toxigenicity by a probe for the microcystin synthetase A gene (*mcy*A) of the cyanobacterial genus *Microcystis*: Comparison of toxicities with 16S rRNA and phycocyanin operon (phycocyanin intergenic spacer) phylogenies, *Applied and Environmental Microbiology*, **67** (6), 2810–2818.

Vaitomaa, J., Rantala, A., Halinen, K., et al. (2003) Quantitative real-time PCR for determination of microcystin synthetase E copy numbers for *Microcystis* and *Anabaena* in lakes, *Applied and Environmental Microbiology*, **69** (12), 7289–7297.

Valério, E., Chambel, L., Paulino, S., et al. (2009) Molecular identification, typing and traceability of cyanobacteria from freshwater reservoirs, *Microbiology*, **155** (2), 642–656.

Valério, E., Chambel, L., Paulino, S., et al. (2010) Multiplex PCR for detection of microcystins-producing cyanobacteria from freshwater samples, *Environmental Toxicology*, **25** (3), 251–260.

Valério, E., Pereira, P., Saker, M.L., et al. (2005) Molecular characterization of *Cylindrospermopsis raciborskii* strains isolated from Portuguese freshwaters, *Harmful Algae*, **4** (6), 1044–1052.

Vareli, K., Jaeger, W., Touka, A., et al. (2013) Hepatotoxic seafood poisoning (HSP) due to microcystins: A threat from the ocean?, *Marine Drugs*, **11** (8), 2751–2768.

Vareli, K., Zarali, E., Zacharioudakis, G.S.A., et al. (2012) Microcystin producing cyanobacterial communities in Amvrakikos Gulf (Mediterranean Sea, NW Greece) and toxin accumulation in mussels (*Mytilus galloprovincialis*), *Harmful Algae*, **15**, 109–118.

Vichi, S., Lavorini, P., Funari, E., et al. (2012) Contamination by *Microcystis* and microcystins of blue–green algae food supplements (BGAS) on the Italian market and possible risk for the exposed population, *Food and Chemical Toxicology*, **50** (12), 4493–4499.

Villeneuve, A., Laurent, D., Chinain, M., et al. (2012) Molecular characterization of the diversity and potential toxicity of cyanobacterial mats in two tropical lagoons in the South Pacific Ocean, *Journal of Phycology*, **48** (2), 275–284.

Wang, X., Sun, M., Xie, M., et al. (2013) Differences in microcystin production and genotype composition among *Microcystis* colonies of different sizes in Lake Taihu, *Water Research*, **47** (15), 5659–5669.

Water Research Foundation (2007a) *Development of molecular reporters for microcystis activity and toxicity, Project #2818*. AwwaRF, Denver, http://www.waterrf.org/PublicReportLibrary/2818.pdf, accessed 22 February 2017.

Water Research Foundation (2007b) *Early Detection of Cyanobacterial Toxins Using Genetic Methods, Project #2881*. AwwaRF, Denver, http://www.waterrf.org/ExecutiveSummaryLibrary/91198_2881_profile.pdf, accessed 22 February 2017.

Wilhelm, S.W., Farnsley, S.E., LeCleir, G.R., et al. (2011) The relationships between nutrients, cyanobacterial toxins and the microbial community in Taihu (Lake Tai), China, *Harmful Algae*, **10**, 207–215.

Wilson, K.M., Schembri, M.A., Baker, P.D., and Saint, C.P. (2000) Molecular characterization of the toxic cyanobacterium *Cylindrospermopsis raciborskii* and design of a species-specific PCR, *Applied and Environmental Microbiology*, **66** (1), 332–338.

Wood, S.A., Jentzsch, K., Rueckert, A., et al. (2009) Hindcasting cyanobacterial communities in Lake Okaro with germination experiments and genetic analyses, *FEMS Microbiology Ecology*, **67** (2), 252–260.

Wood, S.A., Kuhajek, J.M., de Winton, M., and Phillips, N.R. (2012a) Species composition and cyanotoxin production in periphyton mats from three lakes of varying trophic status, *FEMS Microbiology Ecology*, **79** (2), 312–326.

Wood, S.A., Mountfort, D., Selwood, A.I., et al. (2008) Widespread distribution and identification of eight novel microcystins in antarctic cyanobacterial mats, *Applied and Environmental Microbiology*, **74** (23), 7243–7251.

Wood, S.A., Smith, F.M., Heath, M.W., et al. (2012b) Within-mat variability in anatoxin-a and homoanatoxin-a production among benthic *Phormidium* (cyanobacteria) strains, *Toxins*, **4** (10), 900–912.

Wood, S.A., Smith, K.F., Banks, J.C., et al. (2013) Molecular genetic tools for environmental monitoring of New Zealand's aquatic habitats, past, present and the future, *New Zealand Journal of Marine and Freshwater Research*, **47** (1), 90–119.

Yoshida, M., Yoshida, T., Takashima, Y., et al. (2005) Genetic diversity of the toxic cyanobacterium *Microcystis* in Lake Mikata, *Environmental Toxicology*, **20** (3), 229–234.

Yoshida, M., Yoshida, T., Takashima, Y., et al. (2007) Dynamics of microcystin-producing and non-microcystin-producing *Microcystis* populations is correlated with nitrate concentration in a Japanese lake, *FEMS Microbiology Letters*, **266** (1), 49–53.

Zhang, W., Lou, I., Ung, W.K., et al. (2014) Analysis of cylindrospermopsin- and microcystin-producing genotypes and cyanotoxin concentrations in the Macau storage reservoir, *Hydrobiologia*, **741** (1), 51–68.

Appendix

Supplementary Tables

Table S6.1 *Reference strains: positive and negative controls for PCR performed in SOPs 6.1–6.9*

SOP	Culture collection[1]	Strain Number	Genus, Species	Positive reference for	Negative reference for	Reference	Genome sequence access. no.
mcy gene detection							
6.1, 6.2, 6.3, 6.7, 6.8, 6.9	NIVA-CYA	126/8	*Planktothrix agardhii*	*mcyA-Cd, mcyB, mcyE,* single filament analysis,	n/a	Christiansen *et al.* (2014)	ASAK00000000.1
6.1, 6.2, 6.3, 6.7, 6.8	NIVA-CYA	98	*Planktothrix rubescens*	*mcyA-Cd, mcyB, mcyE,* single filament analysis,	n/a	Rounge *et al.* (2009); Chen *et al.* (2016)	AVFZ00000000.1
6.1, 6.2, 6.3	UHCC	90	*Anabaena* sp	*mcyA-Cd, mcyE*	n/a	Wang *et al.* (2012)	CP003284 (Chromosome I), CP003285 (Chromosome II)
6.1, 6.2, 6.3, 6.9	PCC	7806	*Microcystis aeruginosa*	*mcyA-Cd, mcyE*	n/a	Tillett *et al.* (2000), Frangeul *et al.* (2008)	AM778843– AM778958
6.1, 6.2, 6.3, 6.9	PCC	7941	*Microcystis aeruginosa*	*mcyA-Cd, mcyE*	n/a	Hisbergues *et al.* (2003); Humbert *et al.* (2013)	CAIK00000000.1
6.1, 6.2, 6.3, 6.9	PCC	7005	*Microcystis aeruginosa*	n/a	*mcyA-Cd, mcyB, mcyE*	Hisbergues *et al.* (2003); Sandrini *et al.* (2014)	AQPY00000000.1
6.1, 6.2, 6.3, 6.9	NIES-MCC	44	*Microcystis aeruginosa*	n/a	*mcyA-Cd, mcyB, mcyE*	Hisbergues *et al.* (2003), Okano *et al.* (2015)	BBPA00000000.1
6.1, 6.2, 6.3, 6.9	NIES-MCC	89	*Microcystis aeruginosa*	*mcyA-Cd, mcyE*	n/a	Hisbergues *et al.* (2003)	n/a

						Reference	Genome accession
6.1, 6.2, 6.3, 6.9	NIES-MCC	98	*Microcystis aeruginosa*	n/a	*mcyA-Cd, mcyB, mcyE*	Hisbergues *et al.* (2003),	MDZH00000000.1
6.1, 6.2, 6.3, 6.9	PCC	7805	*Planktothrix agardhii*	n/a	*mcyA-Cd, mcyB, mcyE*	Christiansen *et al.* (2008)	n/a
6.1, 6.2, 6.3, 6.9	PCC	7811	*Planktothrix agardhii*	n/a	*mcyA-Cd, mcyB, mcyE*	Christiansen *et al.* (2008)	n/a
6.1, 6.2, 6.3, 6.7, 6.8	PCC	7821	*Planktothrix rubescens*	*mcyA-Cd, mcyE*, single filament analysis	n/a	Christiansen *et al.* (2008)	n/a
6.2, 6.3, 6.9	PCC	9237	*Nostoc* sp.	*mcyE*	n/a	Oksanen *et al.* (2004)	n/a
6.2, 6.3	PCC	7804	*Nodularia* sp.	*ndaF*	n/a	Koskenniemi *et al.* (2007); Moffitt *et al.* (2001)	n/a
6.2, 6.3	PCC	73104	*Nodularia spumigena*	n/a	*ndaF*	Moffitt *et al.* (2001)	n/a
6.2, 6.3	UHCC	0039	*Nodularia spumigena*	*mcyE*	n/a	Lyra *et al.* (2005)	n/a
6.2, 6.3	UHCC	0038	*Nodularia sphaerocarpa*	n/a	*ndaF*	Lyra *et al.* (2005)	n/a
6.2, 6.3	PCC	6307	*Cyanobium* sp.	n/a	*mcyE*	Głowacka *et al.* (2011)	n/a
6.2, 6.3	PCC	6501	*Gloeothece* sp.	n/a	*mcyE*	Głowacka *et al.* (2011)	n/a
6.2, 6.3	PCC	7326	*Dermocarpella*	n/a	*mcyE*	Głowacka *et al.* (2011)	n/a

(Continued)

Table S6.1 (Continued)

SOP	Culture collection[1]	Strain Number	Genus, Species	Positive reference for	Negative reference for	Reference	Genome sequence access. no.
ana gene detection							
6.4, 6.6	PCC	6506	*Oscillatoria* sp.	*anaC, cyrJ*	*mcyE*	Mejean et al. (2009), Mazmouz et al. (2010)	CACA01000001 -CACA01000377
6.4	NIVA-CYA	711	*Cuspidothrix issatschenkoi*	*anaC, anaF*	*mcyE*	Ballot et al. (2010).	n/a
6.4	UHCC	37	*Anabaena* sp.	*anaC*	*mcyE*	Rantala-Ylinen et al. (2011)	n/a
6.4	PCC	10702	*Oscillatoria* sp.	n/a	*anaC*	Rantala-Ylinen et al. (2011)	
cyr gene detection							
6.4, 6.6	PCC	6506	*Oscillatoria* sp.	*anaC, cyrJ*	*mcyE*	Mejean et al. (2009), Mazmouz et al. (2010)	CACA01000001 -CACA01000377
6.6	CSIRO-ANACC	CS-505	*Cylindrospermopsis raciborskii*	*cyrJ*	n/a	Stucken et al. (2010)	ACYA00000000.1
6.6	CSIRO-ANACC	CS-506	*Cylindrospermopsis raciborskii*	*cyrJ*	n/a	Sinha et al. (2014)	n/a
6.6	CSIRO-ANACC	CS-509	*Cylindrospermopsis raciborskii*	n/a	*cyrJ*	Sinha et al. (2014)	n/a
6.6	UHCC	0966	*Anabaena lapponica*	*cyrJ*	n/a	Spoof et al. (2006)	n/a

sxt gene detection						
6.5	NIVA-CYA	666	Aphanizomenon gracile	sxtA, sxtG, sxtH, sxtI, sxtX	n/a	Casero et al. (2014)
6.5	NIVA-CYA	675	Aphanizomenon gracile	sxtA, sxtG, sxtH, sxtI, sxtX	n/a	Casero et al. (2014)
6.5	NIVA-CYA	698	Aphanizomenon flos-aquae	sxtA, sxtG, sxtH, sxtX	sxtI	Casero et al. (2014)
6.5	CSIRO-ANACC	CS-337/01	Anabaena circinalis	sxtA, sxtB, sxtG	n/a	Savela et al. (2015)
6.5	CSIRO-ANACC	CS-537/13	Anabaena circinalis	sxtA, sxtB, sxtG	n/a	Savela et al. (2015)
6.5	CSIRO-ANACC	CS-530/05	Anabaena circinalis	n/a	sxtA, sxtB, sxtG	Savela et al. (2015)
6.5	CSIRO-ANACC	CS-533/12	Anabaena circinalis	n/a	sxtA, sxtB, sxtG	Savela et al. (2015)

n/a = not applicable.

[1] Culture collections providing clonal isolates maintained under the respective strain number.

NIVA-CCA, Norwegian Institute for Water Research, Culture Collection of Algae, https://niva-cca.no/.

PCC, Biological Resource Center of Institute Pasteur (CRBIP)-Pasteur Culture Collection of Cyanobacteria, http://cyanobacteria.web.pasteur.fr/.

NIES-MCC, National Institute of Environmental Studies, Microbial Culture collection, http://mcc.nies.go.jp/.

UHCC, Helsinki University Cyanobacterial culture collection, University of Helsinki, Faculty of Agriculture and Forestry, Division of Microbiology and Biotechnology, http://www.helsinki.fi/hambi/HAMBI_eng/online%20cat_eng/HAMBI_UHCC/index.html.

CSIRO-ANACC, Commonwealth Scientific and Industrial Research Organisation (CSIRO), Australian National Algae Culture Collection (ANACC), http://www.csiro.au/en/Research/Collections/ANACC.

References Table S6.1

Ballot, A., Fastner, J., Lentz, M., and Wiedner, C. (2010) First report of anatoxin-a producing cyanobacterium *Aphanizomenon issatschenkoi* in northeastern Germany, *Toxicon*, **56**, 964–971.

Casero, M.C., Ballot, A., Agha, R., *et al.* (2014) Characterization of saxitoxin production and release and phylogeny of *sxt* genes in paralytic shellfish poisoning toxin-producing *Aphanizomenon gracile, Harmful Algae*, **37**, 28–37.

Chen, Q., Christiansen, G., Deng, L., and Kurmayer, R (2016) Emergence of nontoxic mutants as revealed by single filament analysis in bloom-forming cyanobacteria of the genus *Planktothrix, BMC Microbiology*, **16** (1), 1–12.

Christiansen, G., Goesmann, A., and Kurmayer, R. (2014) Elucidation of insertion elements carried on plasmids and in vitro construction of shuttle vectors from the toxic cyanobacterium *Planktothrix, Applied Environmental Microbiology*, **80** (16), 4887–4897.

Christiansen, G., Molitor, C., Philmus, B., and Kurmayer, R. (2008) Nontoxic strains of cyanobacteria are the result of major gene deletion events induced by a transposable element, *Molecular Biology and Evolution*, **25** (8), 1695–1704.

Frangeul, L., Quillardet, P., Castets, A.M., *et al.* (2008) Highly plastic genome of *Microcystis aeruginosa* PCC 7806: A ubiquitous toxic freshwater cyanobacterium, *BMC Genomics*, **9**, 274.

Głowacka, J., Szefel-Markowska, M., Waleron, M., *et al.* (2011) Detection and identification of potentially toxic cyanobacteria in Polish water bodies, *Acta Biochimica Polonica*, **58**, 321–333.

Hisbergues, M., Christiansen, G., Rouhiainen, L., *et al.* (2003) PCR-based identification of microcystin-producing genotypes of different cyanobacterial genera, *Archives of Microbiology*, **180**, 402–410.

Humbert, J.F., Barbe, V., Latifi, A., *et al.* (2013) A tribute to disorder in the genome of the bloom-forming freshwater cyanobacterium *Microcystis aeruginosa, PLOS One*, **8** (8), e70747.

Koskenniemi, K., Lyra, C., Rajaniemi-Wacklin, P., *et al.* (2007) Quantitative real-time PCR detection of toxic *Nodularia* cyanobacteria in the Baltic Sea, *Applied and Environmental Microbiology*, **73** (7), 2173–2179.

Lyra, C., Laamanen, M., Lehtimaki, J.M., *et al.* (2005) Benthic cyanobacteria of the genus *Nodularia* are non-toxic, without gas vacuoles, able to glide and genetically more diverse than planktonic *Nodularia, International Journal of Systematic and Evolutionary Microbiology*, **55** (2), 555–568.

Mazmouz, R., Chapuis-Hugon, F., Mann, S., *et al.* (2010) Biosynthesis of cylindrospermopsin and 7-Epicylindrospermopsin in *Oscillatoria* sp strain PCC 6506: Identification of the *cyr* gene cluster and toxin analysis, *Applied and Environmental Microbiology*, **76** (15), 4943–4949.

Mejean, A., Mann, S., Maldiney, T., *et al.* (2009) Evidence that biosynthesis of the neurotoxic alkaloids anatoxin-a and homoanatoxin-a in the cyanobacterium *Oscillatoria* PCC 6506 occurs on a modular polyketide synthase initiated by L-proline, *Journal of the American Chemical Society*, **131** (22), 7512–7513.

Moffitt, M.C., Blackburn, S.I., and Neilan, B.A. (2001) rRNA sequences reflect the eco-physiology and define the toxic cyanobacteria of the genus *Nodularia, International Journal of Systematic and Evolutionary Microbiology*, **501**, 505–512.

Okano, K., Miyata, N., and Ozaki, Y. (2015) Whole genome sequence of the non-microcystin-producing *Microcystis aeruginosa* strain NIES-44, *Genome Announcements*, **3** (2), e00135–15.

Oksanen, I., Jokela, J., Fewer, D.P., *et al.* (2004) Discovery of rare and highly toxic micro-cystins from lichen-associated cyanobacterium *Nostoc* sp strain IO-102-I, *Applied and Environmental Microbiology*, **70** (10), 5756–5763.

Rantala-Ylinen, A., Kana, S., Wang, H., *et al.* (2011) Anatoxin-a synthetase gene cluster of the cyanobacterium *Anabaena* sp Strain 37 and molecular methods to detect potential producers, *Applied and Environmental Microbiology*, **77** (20), 7271–7278.

Rounge, T.B., Rohrlack, T., Nederbragt, A.J., *et al.* (2009) A genome-wide analysis of non-ribosomal peptide synthetase gene clusters and their peptides in a *Planktothrix rubescens* strain, *BMC Genomics*, **10**, 396.

Sandrini, G., Matthijs, H.C.P., Verspagen, J.M.H., *et al.* (2014) Genetic diversity of inorganic carbon uptake systems causes variation in CO2 response of the cyanobacterium *Microcystis*, *ISME Journal*, **8** (3), 589–600.

Savela, H., Spoof, L., Perälä, N., *et al.* (2015) Detection of cyanobacterial *sxt* genes and paralytic shellfish toxins in freshwater lakes and brackish waters on Åland Islands, Finland, *Harmful Algae*, **46**, 1–10.

Sinha, R., Pearson, L.A., Davis, T.W., *et al.* (2014) Comparative genomics of *Cylindrospermopsis raciborskii* strains with differential toxicities, *BMC Genomics*, **15** (1), 1–14.

Spoof, L., Berg, K.A., Rapala, J., *et al.* (2006) First observation of cylindrospermopsin in *Anabaena lapponica* isolated from the boreal environment (Finland), *Environmental Toxicology*, **21** (6), 552–560.

Stucken, K., John, U., Cembella, A., *et al.* (2010) The smallest known genomes of multicellular and toxic cyanobacteria: Comparison, minimal gene sets for linked traits and the evolutionary implications, *PLOS One*, **5** (2): e9235.

Tillett, D., Dittmann, E., Erhard, M., *et al.* (2000) Structural organization of microcystin biosynthesis in *Microcystis aeruginosa* PCC7806: An integrated peptide-polyketide synthetase system. *Chemistry & Biology*, **7** (10), 753–764.

Wang, H., Sivonen, K., Rouhiainen, L., *et al.* (2012) Genome-derived insights into the biology of the hepatotoxic bloom-forming cyanobacterium *Anabaena* sp. strain 90, *BMC Genomics*, **13** (1), 1–17.

Table S6.2 *Oligonucleotides to observe the mutations occurring within the mcy gene cluster of Planktothrix sp. Chen et al. (2016), see SOP 6.7*

Gene locus	Forward Primer				Reverse Primer				product length (bp)
	Name	Sequence (5'–3')	bp number[1]	Tm (°C)	Name	Sequence	bp number[1]	Tm (°C)	
Entry PCR									
psaA, psaB	Psafwd[2]	GGGTGGTACTTGCCAAGTCTCT	—	58	Psarev[2]	CGACGTGTTGTCGGGTCTT	—	58	669
Screening of the entire mcy gene cluster in steps of 3.5 kbp									
mcyTD	Fmcy1+	ATTTTCCAAGCATTCTAGG	481–500	60.06	Fmcy1-	TCCAATTTCTAGGAAGATTTG	3674–3694	60.90	3213
mcyD	Fmcy2+	AGCAGTTAATAATGATGGCG	2314–2333	59.72	Fmcy2-	GGCTGGGTTCAATAATATTAA	5922–5942	60.98	3628
mcyD	Fmcy3+	GCCACAGCAGATATTCAAAA	5696–5715	60.77	Fmcy3-	TGAATATTGCGATCCTTGTC	9333–9352	60.90	3656
mcyD	Fmcy4+	CCAAAAGGTGAGGATTCTC	9253–9272	60.59	Fmcy4-	ATCTGTTCAATTCCTGAGAT	12775–12795	60.27	4542
mcyDE	Fmcy5+	TTCCGCCAAATTATTACAGA	12715–12734	60.06	Fmcy5-	ATGGGTAAGGTTTGCTTGAT	16409–16428	60.35	3713
mcyE	Fmcy6+	TCTCAATTACAACCTATTGCA	16313–16333	60.87	Fmcy6-	AATTCGGCTTAAAAAAGCTG	19890–19909	60.35	3576
mcyE	Fmcy7+	CAGCTTTTTTAAGCCCGAATT	19890–19909	60.35	Fmcy7-	TTCAAGTATAATTCTCCTTCCTC	23441–23463	60.60	3554
mcyEG	Fmcy8+	AACTCCATTGACCCAGAGAT	23384–23403	60.50	Fmcy8-	ATTAAACCCACAATACCGGA	27032–27043	60.95	3640
mcyG	Fmcy9+	AATATCGGACATATGCAAATT	27003–27023	60.94	Fmcy9-	AGTAAATCAGCAATATTTTGAGTAA	30481–30505	60.64	3502
mcyGHA	Fmcy10+	GCGCCGATATTACTCAAAAT	30472–30491	60.94	Fmcy10-	GAATCGGCAAAAAATTAAGC	34041–34060	60.54	3588
mcyA	Fmcy11+	GGAGCCTTAAATATAGAAACATC	34013–34035	60.13	Fmcy11-	GATTCCAAACAATAGCGACA	39010–39029	60.63	5016
mcyAB	Fmcy12+	TTTGAAATCCAAGTTGAACG	38977–38997	60.63	Fmcy12-	AATTTGTAAAACTTCCCCATA	42681–42701	60.43	3724
mcyB	Fmcy13+	TATGGGGAAGTTTTACAAATT	42681–42701	60.15	Fmcy13-	TGGCAATATCCAACTCAGAT	46081–46100	60.06	3419
mcyBC	Fmcy14+	ATCTGAGTTGGATATTGCCA	46081–46100	60.06	Fmcy14-	TGGATGATTTCTTCCAGA	49580–49597	60.16	3516
mcyCJ	Fmcy15+	TTCTCAAGCAGCTTCATATAC	49411–49431	60.90	Fmcy15-	AACTCTTCTTGAGCAATTTC	53086–53106	60.31	3695
mcyJ	Fmcy16+	AAGCTGGAGATCAAGTTTTAG	52970–52990	60.34	Fmcy16-[4]	TTTGTGTCTTGGTTAGGG	56573–56593	60.56	3624

Screening of the entire *mcy* gene cluster in shorter intervals (1.75 kbp)

mcyD	Fmcym1+	ATCAGATGTTAGCCTCCGAT	2163–2182	60.61	Fmcym1-	ATCCGGAGGCTAACATCTGAT	2163–2182	60.61
mcyD	Fmcym2+	CCTTACTAGGTGATCGTTTTC	3984–4004	60.45	Fmcym2-	GAAAACGATCACCTAGTAAGG	3984–4004	60.45
mcyD	Fmcym3+	AAATGATGAAACTATTGCTCC	7468–7488	60.78	Fmcym3-	GGAGCAATAGTTTCATCATT	7468–7488	60.78
mcyD	Fmcym4+	TAGAATTATTCCGAGATCCTG	11061–11081	60.87	Fmcym4-	CAGGATCTCGGAATAATTCTA	11061–11081	60.87
mcyE	Fmcym5+	CAACCCCTTCCTAGTCATCT	14576–14595	60.09	Fmcym5-	AGATGACTAGGAACGGCGGTTG	14576–14595	60.09
mcyE	Fmcym6+	ATATTTTCCGGCTCCTATCA	18149–18168	60.40	Fmcym6-	TGATAGGAGCCGCGAAAATAT	18149–18168	60.40
mcyE	Fmcym7+	CCGGTGTCGATTGAATTTATG	21657–21676	60.74	Fmcym7-	CATAAATTCAATCACACCGG	21657–21676	60.74
mcyG	Fmcym8+	AGCAATGGGTAAAAGTCGTT	25417–25436	60.52	Fmcym8-	AACGACTTTACCCATTGCT	25417–25436	60.52
mcyG	Fmcym9+	AAACGGTATCGGCTATTGTT	28735–28754	60.36	Fmcym9-	AACAATACCCGATACCGTTT	28735–28754	60.36
mcyH	Fmcym10+	GTTTGGGGCTGTTGTTAATT	32204–82436	60.75	Fmcym10-	AATTAACAACAGCCCCAAAC	32204–82436	60.75
mcyA	Fmcym11+	TAGGCATGATCTTCCACAGT	37629–37648	60.25	Fmcym11-	ACTGTGGAAGATCATGCCTA	37629–37648	60.25
mcyA	Fmcym12+	ACATAGATTCTATCGGGGATA	40701–40721	60.11	Fmcym12-	TATCCCGATAGAATCTATGT	40701–40721	60.11
mcyB	Fmcym13+	TTCTCTGCCGGACTTTAC	44351–44370	60.30	Fmcym13-	GTAAAAGTCGCAGGAGAGAA	44351–44370	60.30
mcyB	Fmcym14+	AATATTTCCTATGCCGCGAGT	47799–47818	60.59	Fmcym14-	ACTCGCGGCATAGGAAATATT	47799–47818	60.59
mcyC	Fmcym15+	TGCCAATACCCAAATTTATAT	51094–51114	60.88	Fmcym15-	ATATAAATTTGGGTATTGGCA	51094–51114	60.88
IS element[5]	Fmcym16+	GGGATGGAAAATCAACAATT	54637–54656	60.53	Fmcym16-	AATTGTTGATTTTCCATCCC	54637–54656	60.53

[1] according to *P. agardhii* NIVA-CYA126/8 (AJ441056), (Christiansen *et al.* (2003);

[2] Primers designed by (Christiansen *et al.* 2008);

[3] (Taton *et al.* 2003);

[4] F*mcy*16- binds 2,948 bp downstream of *mcyJ* as revealed from NIVA-CYA126/8 genome (Access No. ASAK01000000);

[5] IS element group I flanking the *mcy* gene cluster (Christiansen *et al.* 2014).

References Table S6.2

Chen, Q., Christiansen, G., Deng, L., and Kurmayer, R. (2016) Emergence of nontoxic mutants as revealed by single filament analysis in bloom-forming cyanobacteria of the genus *Planktothrix*, *BMC Microbiology*, **16**, 1–12.

Christiansen, G., Fastner, J., Erhard, M., *et al.* (2003) Microcystin biosynthesis in *Planktothrix*: Genes, evolution, and manipulation. *Journal of Bacteriology*, **185** (2), 564–572.

Christiansen, G., Goesmann, A., and Kurmayer, R. (2014) Elucidation of insertion elements carried on plasmids and *in vitro* construction of shuttle vectors from the toxic cyanobacterium *Planktothrix*, *Applied Environmental Microbiology*, **80** (16), 4887–4897.

Christiansen, G., Molitor, C., Philmus, B., and Kurmayer, R. (2008) Nontoxic strains of cyanobacteria are the result of major gene deletion events induced by a transposable element, *Molecular Biology and Evolution*, **25** (8), 1695–1704.

Taton, A., Grubisic, S., Brambilla, E., *et al.* (2003) Cyanobacterial diversity in natural and artificial microbial mats of Lake Fryxell (McMurdo Dry Valleys, Antarctica): A morphological and molecular approach, *Applied and Environmental Microbiology*, **69** (9), 5157–5169.

## Chapter 7

Table S7.1 Examples of quantitative PCR assays to detect and quantify microcystin-, nodularin-, cylindrospermopsin-, and saxitoxin-producing cyanobacteria

Target gene	Target organism	Method	Primers + probe, amplicon length	Quantification	Application	Reference
Microcystin-producing cyanobacteria						
mcyA	Microcystis	SYBR Green	MSF/MSR, 1369 bp[1]; MSF/MSR-2R, 190 bp	External standard curve made with dilution series of DNA (in cell numbers)	Quantification of toxic Microcystis in lake water samples. Lower detection limit with the MSF/MSR-2R primer pair.	Furukawa et al. (2006)
mcyA	Planktothrix agardhii	Hydrolysis probes: FAM/BHQ-1	MAPF/MAPR + MAP Taq, 140 bp	ΔC_q equation*	Determination of relative mcyA genotype proportions of the total Planktothrix population (PC-IGS genotype) in lake water samples and in competition experiments with a nontoxic strain. mcyA and PC-IGS assays performed by a single multiplex qPCR	Briand et al. (2008)
mcyA; mcyB	Planktothrix: Mdha, Dhb, Hty genotypes	Hydrolysis probes	mdha+/mdha- + Mdha, 77 bp; dhb+/dhb- + Dhb, 83 bp; HtyS/HtyA + Hty, 98 bp	External standard curve made with dilution series of DNA (cell number equivalents)	Quantification of Planktothrix genotypes producing microcystin variants [Asp, Mdhd]-MC-RR, [Asp, Dhb]-MC-RR, and MC-HtyR in lake water samples	Ostermaier and Kurmayer (2010)

(Continued)

Table S7.1 *(Continued)*

Target gene	Target organism	Method	Primers + probe, amplicon length	Quantification	Application	Reference
mcyB	*Microcystis*	Hydrolysis probes: FAM/TAMRA	30F/108R + probe, 78 bp	External standard curve made with dilution series of DNA (cell number equivalents)	Quantification of toxic genotypes (*mcyB*) and determination of *mcyB* genotype proportions of the total *Microcystis* population (PC-IGS genotype) in lake water samples	Kurmayer and Kutzenberger (2003)
mcyB	*Microcystis*	Hydrolysis probes: FAM/TAMRA	30F/108R + probe, 78 bp	ΔC_q equation*	Determination of relative *mcyB* genotype proportions of the total *Microcystis* population (PC-IGS genotype) in reservoir water samples. *mcyB* and PC-IGS assays performed by a single multiplex qPCR	Briand *et al.* (2009)
mcyB	*Planktothrix*	Hydrolysis probes: FAM/TAMRA	mcyBA1 + probe, 76 bp	External standard curve made with dilution series of DNA (in cell biovolume equivalents)	Quantification of toxic genotypes (*mcyB*) and determination of *mcyB* genotype proportions of the total *Planktothrix* population (16S rRNA genotype) in lake water samples. In addition, assays for quantification of three insertions in *mcyA*, and *mcyD* genes, and one deletion in *mcyHA* gene region described. (See SOP 7.1)	Ostermaier and Kurmayer (2009)

Gene	Organism	Chemistry	Primers/Probe, amplicon	Standard	Application	Reference
mcyB	*Microcystis*	Hydrolysis probe (UPL probe**)	#04F/#04R + #04P, 95 bp	External standard curve	Quantification of toxic genotypes (*mcyB*) and determination of *mcyB* genotype proportions of the total *Microcystis* population (16S rRNA genotype) in reservoir water samples. qPCR assays performed in multiplex using a portable qPCR machine on-site.	Michinaka et al. (2012)
mcyB	*Microcystis, Anabaena, Planktothrix*	Hydrolysis probes + quencher probes: Lanthanide chelate/BHQ1	*mcyB*HF03A, *mcyB*HF03M, *mcyB*HF03P/ *mcyB*HR04 + *mcyB*-aP/Q, *mcyB*-mP/Q, *mcyB*-pP/Q 103/105 bp	External standard curve made with dilution series of PCR amplicons	Quantification of *Anabaena*, *Microcystis* and *Planktothrix mcyB* genotypes in lake, river and reservoir samples using both a conventional plate-based qPCR setup and dry chemistry qPCR chips.	Hautala et al. (2013); Savela et al. (2014)
mcyD	*Microcystis*	Hydrolysis probes: FAM/BHQ1	*mcyD*F2/*mcyD*DR2[2] +*mcyD*F2 probe, 297 bp	External standard curve made with dilution series of gDNA (in cell number equivalents)	Quantification of toxic *Microcystis* in lake water and sediment samples. *mcyD* assay used in combination with qPCR assays for quantitation of total cyanobacteria (16S rRNA) and total *Microcystis* (16S rRNA)	Rinta-Kanto et al. (2005, 2009a, 2009b)
mcyD	*Microcystis*	SYBR Green	$mcyD_{KS}R1/mcyD_{KS}F1$, $mcyD_{DH}R1/mcyD_{DH}F1$, 107 bp, 129 bp	External standard curve made with dilution series of target sequence containing plasmids	Quantification of two target sequences within the *mcyD* gene; β-ketoacyl synthase and first dehydratase subunits in freshwater lake samples	Fortin et al. (2010)

(Continued)

Table S7.1 (Continued)

Target gene	Target organism	Method	Primers + probe, amplicon length	Quantification	Application	Reference
mcyD	Microcystis PCC7806	Hydrolysis probes: Taqman MGB	RmcyD-For/RmcyD-Rev + mcyD MGB probe	16S RNA used as reference gene	Relative quantification of mcyD transcripts in iron-replete and iron-deplete conditions; Relative quantification of mcyD transcripts in different nitrate conditions; Relative quantification of mcyD transcripts in different light and oxidative stress conditions	Sevilla et al. (2008, 2010, 2012)
mcyE	Microcystis, Anabaena, Planktothrix	SYBR Green	mcyE-F2/MicmcyE-R8, 247 bp; mcyE-F2/AnamcyE-12R, 247 bp mcyE-F2/mcyE-plaR3, 249 bp	External standard curve made with dilutions series of DNA (gene copy number)	Quantification of toxic Microcystis Anabaena, and Planktothrix in lake water samples; determination of the main microcystin producer	Vaitomaa et al. (2003); Rantala et al. (2008)
mcyE	Microcystis	Hydrolysis probes: FAM/BHQ1	mcyE-F2/MicmcyE-R8 + probe	Internal armored RNA standard	Relative quantification of mcyE transcripts during all growth stages in batch cultures; Relative quantification of mcyE transcripts in lake water samples	Rueckert and Cary (2009); Wood et al. (2011)
mcyE	Anabaena, Microcystis	Hydrolysis probes: FAM/TAMRA	611F/737R + 672P, 126 bp; 127F/247R + 186P, 120 bp	External standard curve made with dilutions series of DNA (gene copy number)	Detection of mcyE in environmental DNA and RNA samples (See SOP 7.1)	Sipari et al. (2010)
mcyE; mcyB	Microcystis	SYBR Green	mcyB F/R, 126 bp; mcyE F/R, 200 bp	External standard curve made with dilutions series of DNA	Quantification of Microcystis mcyB and mcyE genotypes and total Microcystis population density (16S rRNA qPCR)	Conradie and Barnard (2012)

Gene	Organism	Chemistry	Primers/probes	Standard	Application	Reference
mcyE	Microcystis, Anabaena, Planktothrix	Hydrolysis probes: FAM/ZEN/IBFQ, HEX/IBFQ, Cy5/BHQ2	AnamcyE-424F/ AnamcyE-583R + AnamcyE-FAM, 160 bp; MicmcyE-415 F/MicmcyE-581R + MicmcyE-Hex , 167 bp; PlamcyE-427 F/PlamcyE-610R + PlamcyE-Cy5, 184 bp	External standard curve made with dilution series of target sequence containing plasmid	Multiplex quantification of Microcystis, Anabaena and Planktothrix mcyE genotypes in lake water samples using genus-specific primers and probes	Ngwa et al. (2014a)
mcyG; mcyE; mcyA	Microcystis	SYBR Green	mcyG-67F/mcyG-310R, 244 bp	External standard curve made with dilutions series of DNA (gene copy numbers)	Quantification of mcyG, mcyE and mcyA copy numbers in lake water samples[4]	Ngwa et al. (2014b)
mcyJ	Microcystis	Hydrolysis probe: FAM/BHQ	mcyJ MF/mcyJ MR + probe	External standard curve made with dilution series of DNA (in cell number equivalents)	Quantification of mcyJ genotypes in reservoir water samples	Joung et al. (2011)
Nodularin-producing cyanobacteria						
ndaF	Nodularia	SYBR Green	ndaF8452/ndaF8640, 189 bp	External standard curve made with dilution series of DNA (in gene copy numbers)	Quantification of toxic (ndaF) genotypes in Baltic Sea water samples	Koskenniemi et al. (2007)
ndaA, -B, -C, -D, -E, -F, -G, -H, -I	Nodularia	SYBR Green	FP/RP for each gene	External standard curve made with dilution series of DNA; normalization with 16S rRNA gene	Relative quantification of nda transcripts in N. spumigena AV1 during phosphate depletion and ammonia supplementation; relative quantification of ndaF transcripts in Baltic Sea water samples	Jonasson et al. (2008)

(Continued)

Table S7.1 (Continued)

Target gene	Target organism	Method	Primers + probe, amplicon length	Quantification	Application	Reference
Cylindrospermopsin-producing cyanobacteria						
cyrA	Cylindrospermopsis raciborskii, Aphanizomenon spp.	Hydrolysis probe + quencher probe: Lanthanide chelate/BHQ1	cyrAdF1, cyrAdR1 + cyrAPC/cyrAPQ, 103 bp	External standard curve made with dilution series of PCR amplicons	Quantification of cyrA genotypes in freshwater lakes	Kokocinski et al. (2013)
aoaC(pks)	Cylindrospermopsis raciborskii, Anabaena bergii, Aphanizomenon ovalisprorum	Hydrolysis probes: FAM/BHQ1	$k18^3/m4^3$ + PKS, 422 bp	External standard curve made with dilution series of PCR amplicons	Quantification of pks genotype in lake, reservoir, and river water samples. Assay performed in multiplex with a C. raciborskii specific assay (rpoC1)	Rasmussen et al. (2008)
cyrC	C. raciborskii	Hydrolysis probes: FAM/BHQ1	$k18^3/m4^3$ + PKS, 422 bp	External standard curve made with dilution series of PCR amplicons	Quantification of cyrC genotype in reservoir water samples. Assay performed in multiplex with a C. raciborskii specific assay (rpoC1).	Orr et al. (2010)
Saxitoxin-producing cyanobacteria						
sxtA	Anabaena circinalis	SYBR Green	sxtA–F/sxtA–R, 125 bp	External standard curve made with dilution series of DNA (in gene copy numbers)	Quantification of sxtA genotype in bloom samples. In addition, a cyanobacteria-specific qPCR assay for 16S rRNA gene was designed	Al-Tebrineh et al. (2010)
sxtB	Anabaena circinalis	Hydrolysis probe: Lanthanide chelate/BHQ1	sxtB–F2/sxtB–R2, 125 bp	External standard curve made with dilution series of PCR amplicons	Quantification of sxtB genotype in freshwater lake and coastal brackish water samples. (See SOP 7.7)	Savela et al. (2015)

Multiplex assays detecting producers of several toxins

Gene	Species	Chemistry	Primers (product)	Standard curve	Application	Reference
mcyE; ndaF	Microcystis, Anabaena, Planktothrix, Nodularia	SYBR Green	DQmcyEf/DQmcyER, 128 bp	External standard curve made with dilutions series of DNA (a mixture of Microcystis, Anabaena, Planktothrix, and Nodularia DNAs)	Detection of mcyE/ndaF in freeze-dried bloom samples	Al-Tebrineh et al. (2011)
mcyE/ndaF; cyrA; sxtA	Anabaena, Microcystis, Planktothrix/Nodularia; C. raciborskii, Aph. ovalisporum, Umezakia natans; A. circinalis, A. flos-aquae, C. raciborskii, Lyngbya wollei	Hydrolysis probes: CY5/BHQ1; CY3/BHQ2; Texas Red/BHQ2	mcyF/mcyR + mcyP, 128 bp; cyrF/cyrR + cyrP, 71 bp, sxtF/sxtR + sxtP, 148 bp	External standard curve made with dilution series of DNA	Quantification of mcyE/ndaF, cyrA, and sxtA genotypes in pooled DNA extracts from cyanobacterial cultures of A. circinalis, C. raciborskii and in river water samples. qPCR assays performed in multiplex with qPCR assay for 16S rRNA gene	Al-Tebrineh et al. (2012a, 2012b)

*ΔC_q equation is obtained by relating the calculated ΔC_q (change in Cq-values of PC-IGS and mcyA) to percentage of cells with the mcyA genotype.

**Universal ProbeLibrary, Roche includes 165 short hydrolysis probes substituted with Locked Nucleic Acids. Probes are labeled at the 5' end with fluorescein (FAM) and at the 3' end with a dark quencher dye.

[1] Primers first published by Neilan et al. (1997).
[2] Primers first published by Kaebernick et al. (2000) [3] Primers first published by Fergusson and Saint (2003).
[2] Primers first published by Schembri et al. (2001).
[4] mcyA and mcyE primers first published by Furukawa et al. (2006) and Vaitomaa et al. (2003), respectively.

References, Table S7.1

Al-Tebrineh, J., Gehringer, M.M., Akcaalan, R., and Neilan, B.A. (2011) A new quantitative PCR assay for the detection of hepatotoxigenic cyanobacteria, *Toxicon*, **57**, 546–554.

Al-Tebrineh, J., Merrick, C., Ryan, D., *et al.* (2012b) Community composition, toxigenicity, and environmental conditions during a cyanobacterial bloom occurring along 1,100 kilometers of the Murray River, *Applied and Environmental Microbiology*, **78**, 263–272.

Al-Tebrineh, J., Mihali, T.K., Pomati, F., and Neilan, B.A. (2010) Detection of saxitoxin-producing cyanobacteria and *Anabaena circinalis* in environmental water blooms by quantitative PCR, *Applied and Environmental Microbiology*, **76**, 7836–7842.

Al-Tebrineh, J., Pearson, L.A., Yasar, S.A., and Neilan, B.A. (2012a) A multiplex qPCR targeting hepato- and neurotoxigenic cyanobacteria of global significance, *Harmful Algae*, **15**, 19–25.

Briand, E., Escoffier, N., Straub, C., *et al.* (2009) Spatiotemporal changes in the genetic diversity of a bloom-forming *Microcystis aeruginosa* (cyanobacterium) population, *The ISME Journal*, **3**, 419–429.

Briand, E., Gugger, M., François, J.C., *et al.* (2008) Temporal variations in the dynamics of potentially microcystin-producing strains in a bloom-forming *Planktothrix agardhii* (cyanobacterium) population, *Applied and Environmental Microbiology*, **74**, 3839–3848.

Conradie, K.R. and Barnard, S. (2012) The dynamics of toxic *Microcystis* strains and microcystin production in two hypertrophic South African reservoirs, *Harmful Algae*, **20**, 1–10.

Fergusson, K.M., and Saint, C.P. (2003) Multiplex PCR assay for *Cylindrospermopsis raciborskii* and cylindrospermopsin-producing cyanobacteria, *Environmental Toxicology*, **18**, 120–125.

Fortin, N., Aranda-Rodriguez, R., Jing, H., *et al.* (2010) Detection of microcystin-producing cyanobacteria in Missisquoi Bay, Quebec, Canada, using quantitative PCR, *Applied and Environmental Microbiology*, **76**, 5105–5112.

Furukawa, K., Noda, N., Tsuneda, S., *et al.* (2006) Highly sensitive real-time PCR assay for quantification of toxic cyanobacteria based on microcystin synthetase A gene, *Journal of Bioscience and Bioengineering*, **102**, 90–96.

Hautala, H., Lamminmäki, U., Spoof, L., *et al.* (2013) Quantitative PCR detection and improved sample preparation of microcystin-producing *Anabaena, Microcystis, and Planktothrix, Ecotoxicology and Environmental Safety*, **87**, 49–56.

Jonasson, S., Vintila, S., Sivonen, K., and El-Shehawy, R. (2008) Expression of the nodularin synthetase genes in the Baltic Sea bloom-former cyanobacterium *Nodularia spumigena* strain AV1, *FEMS Micbiology Ecology*, **65**, 31–39.

Joung, S.H., Oh, H.M., Ko, S.R., and Ahn, C.Y. (2011) Correlations between environmental factors and toxic and non-toxic *Microcystis* dynamics during bloom in Daechung Reservoir, *Korea, Harmful Algae*, **10**, 188–193.

Kaebernick, M., Neilan, B.A., Börner, T., and Dittmann, E. (2000) Light and the transcriptional response of the microcystin biosynthesis gene cluster, *Applied and Environmental Microbiology*, **66**, 3387–3392.

Kokocinski, M., Mankiewicz-Boczek, J., Jurczak, T., *et al.* (2013) *Aphanizomenon gracile* (Nostocales): A cylindrospermopsin-producing cyanobacterium in Polish lakes, *Environmental Science and Pollution Research International*, **20** (8), 5243–5264.

Koskenniemi, K., Lyra, C., Rajaniemi-Wacklin, P., *et al.* (2007) Quantitative real-time PCR detection of toxic *Nodularia* cyanobacteria in the Baltic Sea, *Applied and Environmental Microbiology*, **73**, 2173–2179.

Kurmayer, R. and Kutzenberger, T. (2003) Application of real-time PCR for quantification of microcystin genotypes in a population of the toxic cyanobacterium *Microcystis* sp., *Applied and Environmental Microbiology*, **69**, 6723–6830.

Michinaka, A., Yen, H.K., Chiu, Y.T., *et al.* (2012) Rapid on-site multiplex assays for total and toxigenic *Microcystis* using real-time PCR with microwave cell disruption, *Water Science & Technology*, **66**, 1247–1252.

Neilan, B.A., Jacobs, D., del Dot, T. *et al.* (1997) rRNA sequences and evolutionary relationships among toxic and nontoxic cyanobacteria of the genus Microcystis, *International Journal of Systematic Bacteriology*, **47**, 693–697.

Ngwa, F., Madramootoo, C., and Jabaji, S. (2014b) Monitoring toxigenic *Microcystis* strains in the Missisquoi Bay, Quebec, by PCR targeting multiple toxic gene loci, *Environmental toxicology*, **29** (4), 440–451.

Ngwa, F.F., Madramootoo, C.A., and Jabaji, S. (2014a) Development and application of a multiplex qPCR technique to detect multiple microcystin-producing cyanobacterial genera in a Canadian freshwater lake, *Journal of Applied Phycology*, **26** (4), 1675–1687.

Orr, P.T., Rasmussen, J.P., Burford, M.A., *et al.* (2010) Evaluation of quantitative real-time PCR to characterise spatial and temporal variations in cyanobacteria, *Cylindrospermopsis raciborskii* (Woloszynska) Seenaya *et* Subba Raju and cylindrospermopsin concentrations in three subtropical Australian reservoirs, *Harmful Algae*, **9**, 243–254.

Ostermaier, V. and Kurmayer, R. (2009) Distribution and abundance of nontoxic mutants of cyanobacteria in lakes of the Alps, *Microbial Ecology*, **58**, 323–333.

Ostermaier, V. and Kurmayer, R. (2010) Application of real-time PCR to estimate toxin production by the cyanobacterium *Planktothrix* sp., *Applied and Environmental Microbiology*, **76**, pp. 3495–3502.

Rantala, A., Rizzi, E., Castiglioni, B., *et al.* (2008) Identification of hepatotoxin-producing cyanobacteria by DNA-chip, *Environmental Microbiology*, **10**, 653–664.

Rasmussen, J.P., Giglio, S., Monis, P.T., *et al.* (2008) Development and field testing of a real-time PCR assay for cylindrospermopsin-producing cyanobacteria, *Journal of Applied Microbiology*, **104**, 1503–1515.

Rinta-Kanto, J.M., Konopko, E.A., DeBruyn, J.M., *et al.* (2009b) Lake Erie *Microcystis*: Relationships between microcystin production, dynamics of genotypes and environmental parameters in a large lake, *Harmful Algae*, **8**, 665–673.

Rinta-Kanto, J.M., Ouellette, A.J.A., Boyer, G.L., *et al.* (2005) Quantification of toxic *Microcystis* during the 2003 and 2004 blooms in western Lake Erie using quantitative real-time PCR, *Environmental Science and Technology*, **39**, pp. 4198–4205.

Rinta-Kanto, J.M., Saxton, M.A., DeBruyn, J.M., *et al.* (2009a) The diversity and distribution of toxigenic *Microcystis* sin present day and archived pelagic and sediment samples from Lake Erie, *Harmful Algae*, **8**, 385–394.

Rueckert, A. and Cary, S.G. (2009) Use of an armored RNA standard to measure micro-cystin synthetase E gene expression in toxic *Microcystis* sp. by reverse-transcription QPCR, *Limnology and Oceanography Methods*, **7**, 509–520.

Savela, H., Spoof, L., Perälä, N., *et al.* (2015) Detection of cyanobacterial *sxt* genes and par-alytic shellfish toxins in freshwater lakes and brackish waters on Åland Islands, Finland, *Harmful Algae*, **46**, 1–10.

Savela, H., Vehniäinen, M., Spoof, L., *et al.* (2014) Rapid quantification of *mcy*B copy numbers on dry chemistry PCR chips and predictability of microcystin concentrations in freshwater environments, *Harmful Algae*, **39**, 280–286.

Schembri, M.A., Neilan, B.A., and Saint, C.P. (2001) Identification of genes implicated in toxin production in the cyanobacterium *Cylindrospermopsis raciborskii*, *Environmental Toxicology*, **16**, 413–421.

Sevilla, E., Martin-Luna, B., Vela, L., *et al.* (2008) Iron availability affects *mcy*D expres-sion and microcystin-LR synthesis in *Microcystis aeruginosa* PCC7806, *Environmental Microbiology*, **10**, 2476–2483.

Sevilla, E., Martin-Luna, B., Vela, L., *et al.* (2010) Microcystin-LR synthesis as response to nitrogen: transcriptional analysis of the *mcy*D gene in *Microcystis aeruginosa* PCC7806, *Ecotoxicology*, **19**, 1167–1173.

Sevilla, E., Martin-Luna, B., Vela, L., *et al.* (2012) An active photosynthetic electron transfer chain required for *mcy*D transcription and microcystin synthesis in *Microcystis aeruginosa* PCC7806, *Ecotoxicology*, **21**, 811–819.

Sipari, H., Rantala-Ylinen, A., Jokela, J., *et al.* (2010) Development of a chip assay and quantitative PCR for detecting microcystin synthetase E gene expression, *Applied and Environmental Microbiology*, **76**, 3797–3805.

Vaitomaa, J., Rantala, A., Halinen, K., *et al.* (2003) Quantitative real-time PCR for deter-mination of microcystin synthetase E copy numbers for *Microcystis* and *Anabaena* in lakes, *Applied and Environmental Microbiology*, **69**, 7289–7297.

Wood, S.A., Rueckert, A., Hamilton, D.P., *et al.* (2011) Switching toxin production on and off: Intermittent microcystin synthesis in a *Microcystis* bloom, *Environmental Microbi-ology Reports*, **3**, 118–124.

Chapter 11

Table S11.1 List of studies showing the main different applications (and some methodological features) in which the detection/identification of tox(igen)ic cyanobacteria have been pursued, by molecular-based method used. Note: this is the full form of Table 11.1

Method[%]	Application	Type of sample	Targeted molecular marker(s)[&]	Cyanotoxin biosynthesis genes[#]	Complementary evaluation/validation[$]	(Putative) Toxigenic cyanobacteria	Location	Habitat	References
Conventional PCR (chapter 6)	surface water	isolate(s)	16S rRNA gene	n.d.	ELISA (+, for MC), LC-MS/MS (+, for MC)	Synechococcus sp.	North America: USA	brackish to saline water	Carmichael and Li (2006)
	surface water	water	mcyA; mcyB; mcyC; mcyD; mcyE; mcyG; ndaF; cyrB; cyrC; sxtI	MC (+), NOD (+), CYL (+), STX (+)	MALDI-TOF/MS (−)	Cyanobium sp., Synechococcus cf. nidulans, Synechocystis salina, Leptolyngbya spp., Microcoleus sp.	Europe: Portugal	brackish water	Lopes et al. (2012)
	surface water	isolate(s)	aoaA; aoaB; aoaC	CYL (−)	ELISA (−), HPLC (−)	n.d.	Africa: Tunisia	freshwater	Fathalli et al. (2010)
	surface water	water	mcyA; mcyB; mcyC; mcyD; mcyE	MC (+)	ELISA (+)	Microcystis sp.	Asia: Singapore	freshwater	Te and Gin (2011)
	surface water	isolate(s)	mcyB	MC (+)	MALDI-TOF (+)	Microcystis sp.	Europe: Germany	freshwater	Kurmayer et al. (2002)
	surface water	isolate(s)	sxtA; sxtG; sxtH; sxtI; sxtX	STX (+)	ELISA (+) and reported in previous studies	Aphanizomenon gracile	Europe: Germany and Spain	freshwater	Casero et al. (2014)
	surface water	water	mcyA; mcyB; mcyD; mcyE	MC (+)	HPLC (+), PPIA (+)	Planktothrix agardhii, Microcystis aeruginosa	Europe: Poland	freshwater	Mankiewicz-Boczek et al. (2006)
	surface water	water	cyrJ	CYL (+)	HPLC (+)	Aphanizomenon sp.	Europe: Poland	freshwater	Mankiewicz-Boczek et al. (2012)

(Continued)

Table S11.1 (Continued)

Method%	Application	Type of sample	Targeted molecular marker(s)&	Cyanotoxin biosynthesis genes#	Complementary evaluation/ validation$	(Putative) Toxigenic cyanobacteria	Location	Habitat	References
	surface water	isolate(s) and water	cyrI, pks	CYL (+)	ELISA (+), HPLC (+), HPLC MS/MS(+)	Aphanizomenon gracile	Europe: Poland	freshwater	Kokociński et al. (2013)
	surface water	isolate(s)	aoaB; aoaC	CYL (−)	HPLC (−)	n.d.	Europe: Portugal	freshwater	Valério et al. (2005)
	surface water	isolate(s) and water	mcyA; mcyB; mcyE; aoaA; aoaB; aoaC	MC (+)	ELISA (+)	n.d.	Europe: Portugal	freshwater	Martins et al. (2010)
	surface water	water	mcyA	MC (+)	ELISA (+)	n.d.	Europe: Portugal	freshwater	Saker et al. (2007a)
	surface water	water	mcyA; mcyB	MC (+)	PPIA (+), LC-MS (+)	Planktothrix spp., Microcystis spp.	North America: USA.	freshwater	Allender et al. 2009
	surface water	water (from food fish farming)	mcyB; 16S rRNA gene	MC (+)	HPLC (+)	Microcystis aeruginosa	North America: USA.	freshwater	Nonneman and Zimba (2002)
	surface water	bloom	mcyE; ndaF; sxtA; cyrA; 16S rRNA gene	MC and/or NOD (+), STX (+), CYL (+)	ELISA (−;+;+)	Anabaena circinalis, Microcystis flos-aquae, Cylindrospermopsis raciborskii	Oceania: Australia	freshwater	Al-Tebrineh et al. (2012a)
	surface water	bloom	mcyA; mcyC	MC (+)	Mouse bioassay (+), PPIA (+)	Microcystis flos-aquae, Anabaena sp.	Oceania: Australia	freshwater	Baker et al. (2001)
	surface water	bloom	mcyA; 16S rRNA gene	MC (+), STX (+)	HPLC (+;+)	Microcystis sp., Anabaena sp.	Oceania: Australia	freshwater	Baker et al. (2002)

Sample	Type	Genes	Toxins	Method	Species	Location	Environment	Reference
surface water	water	cyrC; mcyA; mcyC; sxtI	CYL (+), MC (−), STX (−)	HPLC-MS/MS (+)	*Raphidiopsis mediterranea*	Oceania: Australia	freshwater	McGregor et al. (2011)
surface water	water	mcyE	MC (+)	HPLC (+)	*Anabaena sp., Microcystis sp.*	Asia: Russia	freshwater	Belykh et al. (2011)
surface water	bloom	mcyB	MC (+)	ELISA (+)	*Microcystis sp.*	South America: Brazil	freshwater	Bittencourt-Oliveira (2003)
surface water	isolate(s)	cyrA; cyrB; cyrC; cyrI; sxtA; sxtB; sxtI	CYL (−), STX (+)	LC-MS (−;+)	*Cylindrospermopsis raciborskii*	South America: Brazil	freshwater	Hoff-Risseti et al. (2013)
surface water	mat	mcyE	MC (+)	LC-MS (+)	*Nostoc sp.*	Antarctica	freshwater and terrestrial	Wood et al. (2008)
surface water	isolate(s)	mcyA; mcyB; mcyC; ndaF; CYL biosynthesis-related genes; 16S rRNA gene	MC (+), CYL (−)	Artemia salina bioassay (+), MALDI-TOF/MS (−)	*Leptolyngbya sp., Oscillatoria sp.*	Europe: Portugal	marine	Frazão et al. (2010)
surface water	isolate(s)	16S rRNA gene	n.d.	PPIA (+), ELISA (+)	*Synechococcus, Geitlerinema, Leptolyngbya, Phormidium, Pseudanabaena, Spirulina*	North America: diverse locations/Asia: Philippines	marine	Gantar et al. (2009)
surface water	mat (coral tissue)	mcyA, mcyE; ndaF; cyrJ; sxtI	MC (+), NOD (−), CYL (−), STX (−)	LC-MS (−;−;−;−)	n.d.	Oceania: Australia	marine	Glas et al. (2010)
surface water	herbarium specimens	mcyD	MC (+)	HPLC (+), ELISA (+), MALDI-TOF (+)	*Nostoc commune, Oscillatoria amphibia, Anabaena flos-aquae, Anabaena oscillarioides*	several origins	—	Metcalf et al. (2012a)

(Continued)

Table S11.1 (*Continued*)

Method[%]	Application	Type of sample	Targeted molecular marker(s)[&]	Cyanotoxin biosynthesis genes[#]	Complementary evaluation/validation[$]	(Putative) Toxigenic cyanobacteria	Location	Habitat	References
	food (additive)	dietary supplements	*mcyE*; *ndaF*; *cpcBA*-IGS	MC and/or NOD (+)	PPIA, ELISA and LC-MS (+, for MC), LC-MS (−, for CYLN, ATX and STX/PSPs)	n.d.	n/a, different producers	—	Heussner *et al.* (2012)
	food (additive)	dietary supplements	*mcyA*; *mcyE*; *ndaF*; *cpcBA*-IGS; 16S rRNA gene	MC (+), MC and/or NOD (+)	ELISA (+, for MC)	*Microcystis aeruginosa*	n/a, different producers	—	Saker *et al.* (2005)
	food web	bloom and grazers (gut content)	*mcyA*; *mcyB*	MC (+)	PPIA (+), ELISA (+)	*Microcystis aeruginosa*	North America: USA.	freshwater	Oberholster *et al.* (2006)
	Terrestrial	cryptogamic (soil) crust	*mcyD*; *aoaB*; *aoaC*	MC (+), CYL (−)	HPLC (+; −), ELISA (+, for MC)	n.d.	Asia: Qatar	terrestrial (desert)	Metcalf *et al.* (2012b)
	Terrestrial	cyanobiont isolate(s) and *in planta*	*mcyE*; *ndaA*; *ndaB*; *ndaF*; 16S rRNA gene	NOD (+), MC and/or NOD (+)	HPLC HESI-MS/MS (+, for NOD) (−, for MC)	*Nostoc* sp. (endosymbiont of cycad)	Oceania: Australia	terrestrial	Gehringer *et al.* (2012)
	Terrestrial	cyanobiont isolate(s) and lichen tissue	*mcyE*	MC (+)	LC-MS (+, for MC and NOD)	different *Nostoc* phylotypes	several origins	terrestrial	Kaasalainen *et al.* (2012)
	Terrestrial	cyanobiont isolate(s)	*mcyA*; *mcyD*; *mcyE*; *rpoC1*; 16S rRNA gene	MC (+)	PPIA (+), ELISA (+), HPLC-MS/MS (+)	*Nostoc* sp.	Europe: Finland	terrestrial	Oksanen *et al.* (2004)

Multiplex-PCR (chapter 6)	surface water	water	mcyA; mcyB; mcyC; mcyD; mcyE; mcyG; mcyI; cyrC	MC (+), CYL (+)	HPLC-MS/MS (+;+)	*Microcystis* sp., *Cylindrospermopsis raciborskii*	Asia: Macau (China)	freshwater	Zhang et al. (2014)
	surface water	bloom	mcyA; mcyB; mcyC; mcyD; mcyE; mcyG	MC (+)	MALDI-TOF (+), HPLC (+)	*Microcystis* sp.	Europe: Spain	freshwater	Ouahid and Fernández del Campo (2009)
	surface water	water	mcyB; mcyD	MC (+)	n.d.	*Microcystis* sp.	North America	freshwater	Ouellette et al. (2006)
	food (additive)	dietary supplements	mcyA; cpcBA-IGS	MC (+)	MALDI-TOF/MS (+)	*Microcystis* spp., *Planktothrix* sp.	n/a, different producers	—	Saker et al. (2007b)
Multiplexed-tandem-PCR (chapter 6)	surface water	bloom	mcyE; ndaF; cyrC; sxtI	MC (+), NOD (+), CYL (+), STX (+)	ELISA (+;+;+;+)	n.d.	Oceania: Australia	fresh and brackish waters	Baker et al. (2013)
Quantitative PCR (chapter 7)	surface water	water	mcyA	MC (+)	n.d.	*Microcystis* sp.	Asia: China	freshwater	Li et al. (2012)
	surface water	water	mcyA	MC (+)	HPLC (+), ELISA (+)	*Microcystis* sp.	Asia: Japan	freshwater	Yoshida et al. (2005)
	surface water	water	mcyA	MC (+)	n.d.	*Microcystis aeruginosa*	Asia: Japan	freshwater	Yoshida et al. (2007)
	surface water	water	mcyA	MC (+)	ELISA (+)	*Microcystis* sp.	Asia: Japan	freshwater	Ha et al. (2011)
	surface water	water	mcyA; mcyB; mcyC; mcyD; mcyE; mcyG; mcyI; cyrC	MC (+), CYL (+)	HPLC-MS/MS (+;+)	*Microcystis* sp., *Cylindrospermopsis raciborskii*	Asia: Macau (China)	freshwater	Zhang et al. (2014)
	surface water	water	mcyA; 16S rRNA gene	MC (+)	HPLC (+), PPIA (+), ELISA (+), HPLC (+), HPLC-MS/MS (+)	*Microcystis aeruginosa*	Europe: Poland	freshwater	Gągała et al. (2014)
	surface water	water	cyrA	CYL (+)		*Aphanizomenon gracile*	Europe: Poland	freshwater	Kokociński et al. (2013)

(Continued)

Method%	Application	Type of sample	Targeted molecular marker(s)&	Cyanotoxin biosynthesis genes#	Complementary evaluation/validation$	(Putative) Toxigenic cyanobacteria	Location	Habitat	References
	surface water	water	mcyE	MC (+)	ELISA (+)	Microcystis sp.	Asia: Singapore	freshwater	Te and Gin (2011)
	surface water	water	mcyJ	MC (+)	HPLC (+)	Microcystis sp.	Asia: South Korea	freshwater	Joung et al. (2011)
	surface water	bloom	mcyE; ndaF	MC (+), NOD (+)	HPLC (+;+)	Planktothrix agardhii, Planktothrix rubescens, Microcystis aeruginosa, Anabaenopsis elenkinii, Nodularia spumigena	Europe and Asia: Turkey	freshwater	Al-Tebrineh et al. (2011)
	surface water	water	mcyE	MC (+)	n.d.	Microcystis sp., Anabaena sp.	Europe: Finland	freshwater	Vaitomaa et al. (2003)
	surface water	sediment	mcyB	MC (+)	PPIA (+)	n.d.	Europe: France	freshwater	Misson et al. (2012)
	surface water	water	mcyA; cpc operon	MC (+)	PPIA (+)	Planktothrix agardhii	Europe: France	freshwater	Briand et al. (2008)
	surface water	water	mcyB; cpc operon	MC (+)	n.d.	Microcystis sp.	Europe: Germany	freshwater	Kurmayer and Kutzenberger (2003)
	surface water	water	mcyA	MC (+)	HPLC (+)	Planktothrix sp. genotypes	Europe: several countries	freshwater	Ostermaier and Kurmayer (2010)
	surface water	water	aoaA; aoaB; aoaC	CYL (+)	HPLC LC-MS (+)	n.d.	Oceania: Australia	freshwater	Rasmussen et al. (2008)

Method	Matrix	Sample	Target gene(s)	Toxin	Confirmation	Species	Origin	Water	Reference
	food (additive)	dietary supplements	mcyB; cpcBA-IGS; 16S rRNA gene	MC (+)	ELISA (+), LC-MS/MS (+)	Microcystis sp.	n/a, different producers	—	Vichi et al. (2012)
	food web	gut content of grazers and edible fish	16S rRNA gene	n.d.	ELISA (+, for MC), HPLC/PDA (+, for MC)	Planktothrix rubescens	Europe: Switzerland	freshwater	Sotton et al. (2014)
	food web	bloom	mcyA; mcyB	MC (+)	PPIA (+), ELISA (+)	M. aeruginosa	North America: USA	freshwater	Oberholster et al. (2006)
Multiplex real-time PCR (chapter 7)	surface water	isolate(s)	mcyE; ndaF; sxtA; cyrA	MC and/or NOD (+), STX (+), CYL (+)	toxigenicity of strains (STX-producer) reported in previous studies	Anabaena, Aphanizomenon, Cylindrospermopsis, Lyngbya, Microcystis, Planktothrix, Nodularia	several origins	—	Al-Tebrineh et al. (2012b)
	surface water	bloom	mcyB; cpc operon	MC (+)	n.d.	Microcystis aeruginosa	Europe: France	freshwater	Briand et al. (2009)
	surface water	bloom	mcyE; ndaF; sxtA; cyrA	MC and/or NOD (+), STX (+), CYL (+)	ELISA (−;+;+)	Anabaena circinalis, Microcystis flos-aquae, Cylindrospermopsis raciborskii	Oceania: Australia	freshwater	Al-Tebrineh et al. (2012a)
Microarrays (chapter 8)	surface water	isolate(s) and water	mcyE; 16S rRNA gene HIP1; putative Na^+-dependent transporter gene	MC (+)	HPLC-MS (+), LC-MS (+)	Microcystis sp., Anabaena sp.	Europe: Finland	freshwater	Sipari et al. (2010)
	surface water	isolate(s)		n.d.	toxigenicity of strains (STX-producer) reported in previous studies	Anabaena circinalis	Oceania: Australia	freshwater	Pomati et al. (2004)

(Continued)

Table S11.1 (Continued)

Method%	Application	Type of sample	Targeted molecular marker(s)&	Cyanotoxin biosynthesis genes#	Complementary evaluation/ validation$	(Putative) Toxigenic cyanobacteria	Location	Habitat	References
	surface water	isolate(s) and water	mcyE; ndaF	MC and/or NOD (+)	n.d.	Anabaena sp., Microcystis sp., Planktothrix spp., Nostoc sp., Nodularia sp.	several origins	fresh and brackish waters	Rantala et al. (2008)
DGGE/ clone libraries/ DNA typing methods (chapter 9)	surface water	isolate(s)	cpcBA-IGS; rRNA ITS	n.d.	toxigenicity of strains reported in previous studies	Anabaena spp., Microcystis spp., others?	several origins	—	Neilan (2002)
	surface water	water	mcyE; ndaF; 16S rRNA gene	MC and/or NOD (+)	ELISA (–;–)	Nodularia spumigena, Microcystis aeruginosa	Africa: Egypt	freshwater	Amer et al. (2009);
	surface water	isolate(s)	RAPD markers	n.d.	toxigenicity of strains (MC-producer) reported in previous studies	Microcystis spp.	Asia: Japan	freshwater	Nishihara et al. (1997)
	surface water	sediment	mcyB; 16S rRNA gene; rRNA ITS; rpoC1	MC (–)?	n.d.	Microcystis sp.	Asia: Japan	freshwater	Innok et al. (2005)
	surface water	isolate(s)	STRR sequences; 16S rRNA gene	n.d.	HPLC-MS (+, for CYL)	Cylindrospermopsis raciborskii	Asia: Thailand and Japan	freshwater	Chonudomkul et al. (2004)
	surface water	sediment	mcyB; rRNA ITS	MC (+)	PPIA (+)	Microcystis sp.	Europe: France	freshwater	Misson et al. (2012)

Sample	Material	Gene target	PCR result	Analytical method	Organism	Origin	Water type	Reference
surface water	isolate(s) and water	mcyA	MC (+)	ELISA (+), MALDI-TOF (+)	Anabaena sp., Microcystis sp., Planktothrix sp.	Europe: Germany	freshwater	Hisbergues et al. (2003)
surface water	water	rRNA ITS	n.d.	HPLC (+, for MC)	Microcystis sp., toxic and non-toxic genotypes	Europe: The Netherlands	freshwater	Kardinaal et al. (2007)
surface water	isolate(s) and water	mcyB; mcyD; 16S rRNA gene	MC (+)	toxigenicity of strains (MC-producer) reported in previous studies	Microcystis spp.	North America: USA	freshwater	Ouellette et al. (2006)
surface water	isolate(s)	uncharacterized PS and PKS genes, 16S rRNA gene	PS/PKS (+)	HPLC (+, for CYL) (–, for STX (i.e. PSP))	Cylindrospermopsis raciborskii, Aphanizomenon ovalisporum	Oceania: Australia/ Asia: Israel	freshwater	Valério et al. (2005)
surface water	mat	mcyE; ndaF; rRNA ITS; 16S rRNA gene	MC and/or NOD (+)	LC-MS (+, NOD), LC-MS (–, for MC and ATX)	n.d.	Oceania: New Zealand	freshwater	Wood et al. (2012)
surface water	sediment	rRNA ITS	n.d.	LC-MS (+, for ATX)	Aphanizomenon issatschenkoi	Oceania: New Zealand	freshwater	Wood et al. (2009)
surface water	isolate(s)	mcyE; ndaF; STRR sequences; 16S rRNA gene	MC and/or NOD (+)	PPIA (+); ELISA (+)	Nodularia spp.	several origins	freshwater	Lyra et al. (2005)
surface water	isolate(s)	HIP1; 16S rRNA gene	n.d.	Mouse bioassay (+), HPLC (+, for CYL)	Cylindrospermopsis raciborskii	several origins	freshwater	Neilan et al. (2003)

(Continued)

Table S11.1 (Continued)

Method%	Application	Type of sample	Targeted molecular marker(s)&	Cyanotoxin biosynthesis genes#	Complementary evaluation/validation$	(Putative) Toxigenic cyanobacteria	Location	Habitat	References
	surface water	naturally occurring colonies and water	rRNA ITS	n.d.	MALDI-TOF/MS (+, for MC)	*Microcystis* spp.	several origins	freshwater	Janse et al. (2004)
	surface water	mat	*mcyE*; rRNA ITS; 16S rRNA gene	MC (+)	PPIA (+), ELISA (+), LC-MS (+)	probably *Nostoc* sp.	Antarctica	freshwater and terrestrial	Wood et al. (2008)
	surface water	isolate(s) and mat (coral tissue)	*mcyA*; 16S rRNA gene	MC (+)	PPIA (+), ELISA (+), HPLC-MS (+), HPLC-MS (–, for ATX and STX)	*Geitlerinema* sp., *Leptolyngbya* sp.	North America: USA. and Bahamas	marine	Richardson et al. (2007)
	surface water	mat	*mcyE*; *ndaF*; ATX biosynthesis-related gene; *cyrJ*; *sxtG*; rRNA ITS; 16S rRNA gene	MC and/or NOD (–), ATX-related (–), CYL (–), STX (+)	n.d.	n.d, mat mainly composed by 4 cyanobacterial species	Oceania: French Polynesia	marine	Villeneuve et al. (2012)
	food web	seawater and edible mussels	*mcyA*; *mcyE*; uncharacterized PKS; rRNA ITS	MC (+) and PKS (+)	ELISA (+, for MC), HPLC-MS: (+, for MC)	n.d, pico-cyanobacteria?	Europe: Greece	marine	Vareli et al. (2012)
NGS and/or metagenomics techniques (chapter 10)	surface water	bloom	*mcyA*; 16S rRNA gene	MC (+)	PPIA (+), LC-MS (+), LC-MS (–, for CYL, STX and BMAA)	*Microcystis* sp. and Cyanobacteria (at the phylum level)	Asia: China	freshwater	Steffen et al. (2012)

Method	Sample	Target gene / library	Toxigenicity of strains (MC-producer) reported in previous studies	Toxin detection	Organism	Region	Reference
	surface water	bloom	*Microcystis* metagenome	n.d.		*Microcystis* sp.	North America: Canada and USA/Asia: China	Wilhelm *et al.* (2011)
	surface water	bloom	16S rRNA gene; BAC library	n.d.		bloom dominated by *Aphanizomenon* and *Cylindrospermopsis* species	Oceania: Australia	Pope and Patel (2008)
FISH	surface water	isolate(s) and water	*mcyA*	MC (+)	ELISA (+), HPLC (+)	*Microcystis* sp.	Asia: China	Gan *et al.* (2010)
	surface water	isolate(s) and water	*mcyA*	MC (+)	ELISA (+)	*Microcystis aeruginosa*	Europe: U.K.	Metcalf *et al.* (2009)

% For more information on the method, see the respective Chapter; FISH – fluorescence *in situ* hybridization.

References, Table S11.1

Allender, C.J., LeCleir, G.R., Rinta-Kanto, J.M., *et al.* (2009) Identifying the source of unknown microcystin genes and predicting microcystin variants by comparing genes within uncultured cyanobacterial cells, *Applied and Environmental Microbiology*, **75** (11), 3598–3604.

Al-Tebrineh, J., Gehringer, M.M., Akcaalan, R., and Neilan, B.A. (2011) A new quantitative PCR assay for the detection of hepatotoxigenic cyanobacteria, *Toxicon*, **57** (4), 546–554.

Al-Tebrineh, J., Merrick, C., Ryan, D., *et al.* (2012a) Community composition, toxigenicity, and environmental conditions during a cyanobacterial bloom occurring along 1,100 kilometers of the Murray River, *Applied and Environmental Microbiology*, **78** (1), 263–272.

Al-Tebrineh, J., Pearson, L.A., Yasar, S.A., and Neilan, B.A. (2012b) A multiplex qPCR targeting hepato- and neurotoxigenic cyanobacteria of global significance, *Harmful Algae*, **15**, 19–25.

Amer, R., Díez, B., and El-Shehawy, R. (2009) Diversity of hepatotoxic cyanobacteria in the Nile Delta, Egypt, *Journal of Environmental Monitoring*, **11** (1), 126–133.

Baker, J.A., Entsch, B., Neilan, B.A., and McKay, D.B. (2002) Monitoring changing toxigenicity of a cyanobacterial bloom by molecular methods, *Applied and Environmental Microbiology*, **68** (12), 6070–6076.

Baker, J.A., Neilan, B.A., Entsch, B., and McKay, D.B. (2001) Identification of cyanobacteria and their toxigenicity in environmental samples by rapid molecular analysis, *Environmental Toxicology*, **16** (6), 472–482.

Baker, L., Sendall, B.C., Gasser, R.B., *et al.* (2013) Rapid, multiplex-tandem PCR assay for automated detection and differentiation of toxigenic cyanobacterial blooms, *Molecular and Cellular Probes*, **27**, 208–214.

Belykh, O.I., Sorokovikova, E.G., Fedorova, G.A., *et al.* (2011) Presence and genetic diversity of microcystin-producing cyanobacteria (*Anabaena* and *Microcystis*) in Lake Kotokel (Russia, Lake Baikal Region), *Hydrobiologia*, **671** (1), 241–252.

Bittencourt-Oliveira, M.C. (2003) Detection of potential microcystin-producing cyanobacteria in Brazilian reservoirs with a *mcy*B molecular marker, *Harmful Algae*, **2** (1), 51–60.

Briand, E., Escoffier, N., Straub, C., *et al.* (2009) Spatiotemporal changes in the genetic diversity of a bloom-forming *Microcystis aeruginosa* (Cyanobacteria) population, *The ISME Journal*, **3**, 419–429.

Briand, E., Gugger, M., François, J.C., *et al.* (2008) Temporal variations in the dynamics of potentially microcystin-producing strains in a bloom-forming *Planktothrix agardhii* (cyanobacterium) population, *Applied and Environmental Microbiology*, **74** (12), 3839–3848.

Carmichael, W.W. and Li, R. (2006) Cyanobacteria toxins in the Salton Sea, *Saline Systems*, **2**, 5.

Casero, M.C., Ballot, A., Agha, R., *et al.* (2014) Characterization of saxitoxin production and release and phylogeny of *sxt* genes in paralytic shellfish poisoning toxin-producing *Aphanizomenon gracile*, *Harmful Algae*, **37**, 28–37.

Chonudomkul, D., Yongmanitchai, W., Theeragool, G., *et al.* (2004) Morphology, genetic diversity, temperature tolerance and toxicity of *Cylindrospermopsis raciborskii* (Nostocales, Cyanobacteria) strains from Thailand and Japan, *FEMS Microbiology Ecology*, **48** (3), 345–355.

Fathalli, A., Jenhani, A.B., Moreira, C., *et al.* (2011) Genetic variability of the invasive cyanobacteria *Cylindrospermopsis raciborskii* from Bir M'cherga reservoir (Tunisia), *Archives of Microbiology*, **193**, 595–604.

Frazão, B., Martins, R., and Vasconcelos, V. (2010) Are known cyanotoxins involved in the toxicity of picoplanktonic and filamentous North Atlantic marine cyanobacteria?, *Marine Drugs*, **8** (6), 1908–1919.

Gągała, I., Izydroczyk, K., Jurczak, T., *et al.* (2014) Role of environmental factors and toxic genotypes in the regulation of microcystins-producing cyanobacterial blooms, *Microbial Ecology*, **67** (2), 465–479.

Gan, N.Q., Huang, Q., Zheng, L.L., and Song, L.R. (2010) Quantitative assessment of toxic and nontoxic *Microcystis* colonies in natural environments using fluorescence *in situ* hybridization and flow cytometry, *Science China Life Sciences*, **53** (8), 973–980.

Gantar, M., Sekar, R., and Richardson, L.L. (2009) Cyanotoxins from black band disease of corals and from other coral reef environments, *Microbial Ecology*, **58** (4), 856–864.

Gehringer, M.M., Adler, L., Roberts, A.A., *et al.* (2012) Nodularin, a cyanobacterial toxin, is synthesized *in planta* by symbiotic *Nostoc* sp., *The ISME Journal*, **6** (10), 1834–1847.

Glas, M.S., Motti, C.A., Negri, A.P., *et al.* (2010) Cyanotoxins are not implicated in the etiology of coral black band disease outbreaks on Pelorus Island, Great Barrier Reef, *FEMS Microbiology Ecology*, **73** (1), 43–54.

Ha, J.H., Hidaka, T., and Tsuno, H. (2011) Analysis of factors affecting the ratio of microcystin to chlorophyll-a in cyanobacterial blooms using real-time polymerase chain reaction, *Environmental Toxicology*, **26** (1), 21–28.

Heussner, A.H., Mazija, L., Fastner, J., and Dietrich, D.R. (2012) Toxin content and cytotoxicity of algal dietary supplements, *Toxicology and Applied Pharmacology*, **265** (2), 263–271.

Hisbergues, M., Christiansen, G., Rouhiainen, L., *et al.* (2003) PCR-based identification of microcystin-producing genotypes of different cyanobacterial genera, *Archives of Microbiology*, **180** (6), 402–410.

Hoff-Risseti, C., Dörr, F.A., Schaker, P.D.C., *et al.* (2013) Cylindrospermopsin and saxitoxin synthetase genes in *Cylindrospermopsis raciborskii* strains from Brazilian freshwater, *PLoS ONE*, **8** (8), e74238.

Innok, S., Matsumura, M., Boonkerd, N., and Teaumroong, N. (2005) Detection of *Microcystis* in lake sediment using molecular genetic techniques, *World Journal of Microbiology and Biotechnology*, **21** (8–9), 1559–1568.

Janse, I., Kardinaal, W.E.A., Meima, M., *et al.* (2004) Toxic and nontoxic *Microcystis* colonies in natural populations can be differentiated on the basis of rRNA gene internal transcribed spacer diversity, *Applied and Environmental Microbiology*, **70** (7), 3979–3987.

Joung, S.H., Oh, H.M., Ko, S.R., and Ahn, C.Y. (2011) Correlations between environmental factors and toxic and non-toxic *Microcystis* dynamics during bloom in Daechung Reservoir, Korea, *Harmful Algae*, **10**, 188–193.

Kaasalainen, U., Fewer, D.P., Jokela, J., *et al.* (2012) Cyanobacteria produce a high variety of hepatotoxic peptides in lichen symbiosis, *Proceedings of the National Academy of Sciences of the USA*, **109** (15), 5886–5891.

Kardinaal, W.E.A., Janse, I., Kamst-van Agterveld, M., *et al.* (2007) *Microcystis* genotype succession in relation to microcystin concentrations in freshwater lakes, *Aquatic Microbial Ecology*, **48** (1), 1–12.

Kokociński, M., Mankiewicz-Boczek, J., Jurczak, T., *et al.* (2013) *Aphanizomenon gracile* (Nostocales): A cylindrospermopsin-producing cyanobacterium in Polish lakes, *Environmental Science and Pollution Research International*, **20**, 5243–5264.

Kurmayer, R. and Kutzenberger, T. (2003) Application of real-time PCR for quantification of microcystin genotypes in a population of the toxic cyanobacterium *Microcystis* sp., *Applied and Environmental Microbiology*, **69** (11), 6723–6730.

Kurmayer, R., Dittmann, E., Fastner, J., and Chorus, I. (2002) Diversity of microcystin genes within a population of the toxic cyanobacterium *Microcystis* sin Lake Wannsee (Berlin, Germany), *Microbial Ecology*, **43** (1), 107–118.

Li, D., Kong, F., Shi, X., *et al.* (2012) Quantification of microcystin-producing and non-microcystin producing Microcystis populations during the 2009 and(2010) blooms in Lake Taihu using quantitative real-time PCR, *Journal of Environmental Sciences*, **24** (2), 284–290.

Lopes, V.R., Ramos, V., Martins, A., *et al.* (2012) Phylogenetic, chemical and morphological diversity of cyanobacteria from Portuguese temperate estuaries, *Marine Environmental Research*, **73**, 7–16.

Lyra, C., Laamanen, M., Lehtimäki, J.M., *et al.* (2005) Benthic cyanobacteria of the genus *Nodularia* are non-toxic, without gas vacuoles, able to glide and genetically more diverse than planktonic *Nodularia*, *International Journal of Systematic and Evolutionary Microbiology*, **55** (2), 555–568.

Mankiewicz-Boczek, J., Izydorczyk, K., Romanowska-Duda, Z., *et al.* (2006) Detection and monitoring toxigenicity of cyanobacteria by application of molecular methods, *Environmental Toxicology*, **21**, 380–387.

Mankiewicz-Boczek, J., Kokociński, M., Gągała, I., *et al.* (2012) Preliminary molecular identification of cylindrospermopsin- producing Cyanobacteria in two Polish lakes (Central Europe), *FEMS Microbiology Letters*, **326** (2), 173–179.

Martins, J., Peixe, L., and Vasconcelos, V. (2010) Cyanobacteria and bacteria co-occurrence in a wastewater treatment plant: Absence of allelopathic effects, *Water Science and Technology*, **62** (8), 1954–1962.

McGregor, G.B., Sendall, B.C., Hunt, LT., and Eaglesham, G.K. (2011) Report of the cyanotoxins cylindrospermopsin and deoxy-cylindrospermopsin from *Raphidiopsis mediterranea* Skuja (Cyanobacteria/Nostocales), *Harmful Algae*, **10**, 402–410.

Metcalf, J.S., Beattie, K.A., Purdie, E.L., *et al.* (2012a) Analysis of microcystins and microcystin genes in 60–170-year-old dried herbarium specimens of cyanobacteria, *Harmful Algae*, **15**, 47–52.

Metcalf, J.S., Reilly, M., Young, F.M., and Codd, G.A. (2009) Localization of microcystin synthetase genes in colonies of the cyanobacterium *Microcystis* using fluorescence *in situ* hybridization, *Journal of Phycology*, **45** (6), 1400–1404.

Metcalf, J.S., Richer, R., Cox, P.A., and Codd, G.A. (2012b) Cyanotoxins in desert environments may present a risk to human health, *Science of the Total Environment*, **421–422**, 118–123.

Misson, B., Donnadieu-Bernard, F., Godon, J.J., *et al.* (2012) Short- and long-term dynamics of the toxic potential and genotypic structure in benthic populations of *Microcystis, Water Research,* **46** (5), 1438–1446.

Neilan, B.A. (2002) The molecular evolution and DNA profiling of toxic cyanobacteria, *Current Issues in Molecular Biology,* **4** (1), 1–11.

Neilan, B.A., Saker, M.L., Fastner, J., *et al.* (2003) Phylogeography of the invasive cyanobacterium *Cylindrospermopsis raciborskii, Molecular Ecology,* **12** (1), 133–140.

Nishihara, H., Miwa, H., Watanabe, M., *et al.* (1997) Random amplified polymorphic DNA (RAPD) analyses for discriminating genotypes of *Microcystis* cyanobacteria, *Bioscience, Biotechnology and Biochemistry,* **61** (7), 1067–1072.

Nonneman, D. and Zimba, P.V. (2002) A PCR-based test to assess the potential for microcystin occurrence in channel catfish production ponds, *Journal of Phycology,* **38** (1), 230–233.

Oberholster, P.J., Botha, A.M., and Cloete, T.E. (2006) Use of molecular markers as indicators for winter zooplankton grazing on toxic benthic cyanobacteria colonies in an urban Colorado lake, *Harmful Algae,* **5** (6), 705–716.

Oksanen, I., Jokela, J., Fewer, D.P., *et al.* (2004) Discovery of rare and highly toxic microcystins from lichen-associated cyanobacterium *Nostoc* sp. strain IO-102-I, *Applied and Environmental Microbiology,* **70** (10), 5756–5763.

Ostermaier, V. and Kurmayer, R. (2010) Application of real-time PCR to estimate toxin production by the cyanobacterium *Planktothrix sp., Applied and Environmental Microbiology,* **76** (11), 3495–3502.

Ouahid, Y. and Fernández del Campo, F. (2009) Typing of toxinogenic *Microcystis* from environmental samples by multiplex PCR, *Applied Microbiology and Biotechnology,* **85**, 405–412.

Ouellette, A.J., Handy, S.M., and Wilhelm, S.W. (2006) Toxic *Microcystis* is widespread in Lake Erie: PCR detection of toxin genes and molecular characterization of associated cyanobacterial communities, *Microbial Ecology,* **51** (2), 154–165.

Pomati, F., Burns, B.P., and Neilan, B.A. (2004) Identification of an Na^+-dependent transporter associated with saxitoxin-producing strains of the cyanobacterium *Anabaena circinalis, Applied and Environmental Microbiology,* **70** (8), 4711–4719.

Pope, P.B. and Patel, B.K.C. (2008) Metagenomic analysis of a freshwater toxic cyanobacteria bloom, *FEMS Microbiology Ecology,* **64** (1), 9–27.

Rantala, A., Rizzi, E., Castiglioni, B., *et al.* (2008) Identification of hepatotoxin-producing cyanobacteria by DNA-chip, *Environmental Microbiology,* **10** (3), 653–664.

Rasmussen, J.P., Giglio, S., Monis, P.T., *et al.* (2008) Development and field testing of a real-time PCR assay for cylindrospermopsin-producing cyanobacteria, *Journal of Applied Microbiology,* **104** (5), 1503–1515.

Richardson, L.L., Sekar, R., Myers, J., *et al.* (2007) The presence of the cyanobacterial toxin microcystin in black band disease of corals, *FEMS Microbiology Letters,* **272**, 182–187.

Saker, M.L., Jungblut, A.D., Neilan, B.A., *et al.* (2005) Detection of microcystin synthetase genes in health food supplements containing the freshwater cyanobacterium *Aphanizomenon flos-aquae, Toxicon,* **46** (5), 555–562.

Saker, M.L., Vale, M., Kramer, D., and Vasconcelos, V.M. (2007a) Molecular techniques for the early warning of toxic cyanobacteria blooms in freshwater lakes and rivers, *Applied Microbiology and Biotechnology,* **75** (2), 441–449.

Saker, M.L., Welker, M., and Vasconcelos, V.M. (2007b) Multiplex PCR for the detection of toxigenic cyanobacteria in dietary supplements produced for human consumption, *Applied Microbiology and Biotechnology*, **73** (5), 1136–1142.

Sipari, H., Rantala-Ylinen, A., Jokela, J., *et al.* (2010) Development of a chip assay and quantitative PCR for detecting microcystin synthetase E gene expression, *Applied and Environmental Microbiology*, **76** (12), 3797–3805.

Sotton, B., Guillard, J., Anneville, O., *et al.* (2014) Trophic transfer of microcystins through the lake pelagic food web: Evidence for the role of zooplankton as a vector in fish contamination, *Science of the Total Environment*, **466–467**, 152–163.

Steffen, M.M., Li, Z., Effler, T.C., *et al.* (2012) Comparative metagenomics of toxic freshwater cyanobacteria bloom communities on two continents, *PLoS ONE*, **7** (8), e44002.

Te, S.H., and Gin, K.Y.H. (2011) The dynamics of cyanobacteria and microcystin production in a tropical reservoir of Singapore, *Harmful Algae*, **10**, 319–329.

Vaitomaa, J., Rantala, A., Halinen, K., *et al.* (2003) Quantitative Real-Time PCR for determination of microcystin synthetase E copy numbers for *Microcystis* and *Anabaena* in lakes, *Applied and Environmental Microbiology*, **69** (12), 7289–7297.

Valério, E., Pereira, P., Saker, M.L., *et al.* (2005) Molecular characterization of *Cylindrospermopsis raciborskii* strains isolated from Portuguese freshwaters, *Harmful Algae*, **4** (6), 1044–1052.

Vareli, K., Zarali, E., Zacharioudakis, G.S.A., *et al.* (2012) Microcystin producing cyanobacterial communities in Amvrakikos Gulf (Mediterranean Sea, NW Greece) and toxin accumulation in mussels (*Mytilus galloprovincialis*), *Harmful Algae*, **15**, 109–118.

Vichi, S., Lavorini, P., Funari, E., *et al.* (2012) Contamination by *Microcystis* and microcystins of blue–green algae food supplements (BGAS) on the Italian market and possible risk for the exposed population, *Food and Chemical Toxicology*, **50** (12), 4493–4499.

Villeneuve, A., Laurent, D., Chinain, M., *et al.* (2012) Molecular characterization of the diversity and potential toxicity of cyanobacterial mats in two tropical lagoons in the South Pacific Ocean, *Journal of Phycology*, **48** (2), 275–284.

Wilhelm, S.W., Farnsley, S.E., LeCleir, G.R., *et al.* (2011) The relationships between nutrients, cyanobacterial toxins and the microbial community in Taihu (Lake Tai), China, *Harmful Algae*, **10** (207–215).

Wood, S.A., Jentzsch, K., Rueckert, A., *et al.* (2009) Hindcasting cyanobacterial communities in Lake Okaro with germination experiments and genetic analyses, *FEMS Microbiology Ecology*, **67** (2), 252–260.

Wood, S.A., Kuhajek, J.M., de Winton, M., and Phillips, N.R. (2012) Species composition and cyanotoxin production in periphyton mats from three lakes of varying trophic status, *FEMS Microbiology Ecology*, **79** (2), 312–326.

Wood, S.A., Mountfort, D., Selwood, A.I., *et al.* (2008) Widespread distribution and identification of eight novel microcystins in antarctic cyanobacterial mats, *Applied and Environmental Microbiology*, **74** (23), 7243–7251.

Yoshida, M., Yoshida, T., Takashima, Y., *et al.* (2005) Genetic diversity of the toxic cyanobacterium *Microcystis* in Lake Mikata, *Environmental Toxicology*, **20** (3), 229–234.

Yoshida, M., Yoshida, T., Takashima, Y., *et al.* (2007) Dynamics of microcystin-producing and non-microcystin-producing *Microcystis* populations is correlated with nitrate concentration in a Japanese lake, *FEMS Microbiology Letters*, **266** (1), 49–53.

Zhang, W., Lou, I., Ung, W.K., *et al.* (2014) Analysis of cylindrospermopsin- and microcystin-producing genotypes and cyanotoxin concentrations in the Macau storage reservoir, *Hydrobiologia*, **741** (1), 51–68.

Table S11.2 Examples of other applications/studies where molecular methods are being applied for the detection/identification of cyanobacteria (including potentially toxic or harmful taxa) and of microbial communities related to them. Note: this is the full form of Table 11.2

Application/ type of study	Method%	Type of sample	Targeted molecular marker(s)	Presence of cyanotoxin biosynthesis genes	Results from biochemical and analytical methods	Studied organism(s)	Location	Habitat	References
ecosystem health and conservation (coral reefs monitoring)	DGGE/clone libraries/DNA typing methods (chapter 9)	coral tissue	16S rRNA gene	n.d.	n.d.	*Lyngbya*-like species and unidentified cyanobacterium	Nort America: Bahamas	marine (coral reef)	Sekar et al. (2006)
food web bioaccumulation	Conventional PCR (chapter 6)	water and grazers	16S rRNA gene	n.d.	n.d.	*Nodularia spumigena*	Europe: Gulf of Finland	brackish water	Gorokhova (2009)
genetics of secondary metabolites	Microarrays (chapter 8)	isolates	several, genes encoding unknown/ putative proteins	n.d.	toxigenicity of strains (SXT-producer) reported in previous studies	*Anabaena circinalis*	Oceania: Australia	freshwater	Pomati and Neilan (2004)
microbial community structure	DGGE/clone libraries/ DNA typing methods (chapter 9)	tree leaves	16S rRNA gene	n.d.	n.d.	several epiphytic cyanobacteria	South America: Brazil	terrestrial (subaerial)	Rigonato et al. (2012)

microbial community structure	NGS and/or metagenomics techniques (chapter 10)	water	16S rRNA gene	n.d.	n.d.	Cyanobacteria (at the phylum level)	South America: Brazil	freshwater	Ghai et al. (2011)
microbial community structure	NGS and/or metagenomics techniques (chapter 10)	water	16S rRNA gene	n.d.	n.d.	Cyanobium, Merismopedia, Synechococcus, Pseudanabaena	Europe: Spain	fresh and hypersaline waters	Ghai et al. (2012)
microbial community structure	NGS and/or metagenomics techniques (chapter 10)	bloom	16S rRNA gene	n.d.	n.d.	Microcystis sp.	Asia: China	freshwater	Chen et al. (2011)
microbial community structure	NGS and/or metagenomics techniques (chapter 10)	bloom	several, including putative genes of MC-LR cleavage pathway (mlr) and xenobiotic metabolisms (GST genes)	n.d.	ELISA: MC (+)	transforming and detoxifying microcystins microbiota	North America: U.S.A.	freshwater	Mou et al. (2013)
microbial community structure	NGS and/or metagenomics techniques (chapter 10)	bloom	23S rRNA gene	n.d.	n.d.	Cyanobacteria (at the order level) and eukaryotic plastids (at the phylum level)	North America: U.S.A.	freshwater	Steven et al. (2012)

(Continued)

Table S11.2 (Continued)

Application/ type of study	Method[%]	Type of sample	Targeted molecular marker(s)	Presence of cyanotoxin biosynthesis genes	Results from biochemical and analytical methods	Studied organism(s)	Location	Habitat	References
taxonomy	Conventional PCR (chapter 6)	herbarium specimens	16S rRNA gene	n.d.	n.d.	several cyanobacteria	several origins	different environments	Palinska et al. (2006)
taxonomy (DNA barcoding of toxigenic cyanobacteria)	DGGE/clone libraries/DNA typing methods (chapter 9)	water	16S rRNA gene	n.d.	n.d.	several potential toxin-producing cyanobacteria	North America: U.S.A.	freshwater	Kurobe et al. (2013)
water quality (cyanobacterial ecotypes as bioindicators)	DGGE/clone libraries/DNA typing methods (chapter 9)	epilithic biofilms	16S rRNA gene	n.d.	n.d.	several epilithic cyanobacteria	Europe: Spain	freshwater	Loza et al. (2013)
water quality (for hemodialysis)	DGGE/clone libraries/DNA typing methods (chapter 9)	hemodialysis water	16S rRNA gene	n.d.	n.d.	several unidentified cyanobacteria	Europe: Spain?	freshwater	Gomila et al. (2006)

[%] For more information on the method, see the respective chapter.
* MC-LR – microcystin-LR; GST – glutathione S-transferase.
n.d. – not determined.

References, Table S11.2

Chen, C., Zhang, Z., Ding, A., *et al.* (2011) Bar-coded pyrosequencing reveals the bacterial community during microcystis water bloom in Guanting Reservoir, Beijing, *Procedia Engineering*, **18**, 341–346.

Ghai, R., Hernandez, C.M., Picazo, A., *et al.* (2012) Metagenomes of Mediterranean coastal lagoons, *Scientific Reports*, **2**, 490.

Ghai, R., Rodŕíguez-Valera, F., McMahon, K.D., *et al.* (2011) Metagenomics of the water column in the pristine upper course of the Amazon River, *PLOS ONE*, **6** (8), e23785.

Gomila, M., Gascó, J., Gil, J., *et al.* (2006) A molecular microbial ecology approach to studying hemodialysis water and fluid, *Kidney International*, **70** (9), 1567–1576.

Gorokhova, E. (2009) Toxic cyanobacteria *Nodularia spumigena* in the diet of Baltic mysids: Evidence from molecular diet analysis, *Harmful Algae*, **8** (2), 264–272.

Kurobe, T., Baxa, D.V., Mioni, C.E., *et al.* (2013) Identification of harmful cyanobacteria in the Sacramento-San Joaquin Delta and Clear Lake, California by DNA barcoding, *SpringerPlus*, **2** (1), 1–12.

Loza, V., Perona, E., and Mateo, P. (2013) Molecular fingerprinting of cyanobacteria from river biofilms as a water quality monitoring tool, *Applied and Environmental Microbiology*, **79** (5), 1459–1472.

Mou, X., Lu, X., Jacob, J., *et al.* (2013) Metagenomic identification of bacterioplankton taxa and pathways involved in microcystin degradation in Lake Erie, *PLOS ONE*, **8** (4), e61890.

Palinska, K.A. (2006) Phylogenetic evaluation of cyanobacteria preserved as historic herbarium exsiccata, *International Journal of Systematic and Evolutionary Microbiology*, **56** (10), 2253–2263.

Pomati, F. and Neilan, B.A. (2004) PCR-based positive hybridization to detect genomic diversity associated with bacterial secondary metabolism, *Nucleic Acids Research*, **32** (1), e7.

Rigonato, J., Alvarenga, D.O., Andreote, F.D., *et al.* (2012) Cyanobacterial diversity in the phyllosphere of a mangrove forest, *FEMS Microbiology Ecology*, **80** (2), 312–322.

Sekar, R., Mills, D.K., Remily, E.R., *et al.* (2006) Microbial communities in the surface mucopolysaccharide layer and the black band microbial mat of black band-diseased *Siderastrea siderea*, *Applied and Environmental Microbiology*, **72** (9), 5963–5973.

Steven, B., McCann, S., and Ward, N.L. (2012) Pyrosequencing of plastid 23S rRNA genes reveals diverse and dynamic cyanobacterial and algal populations in two eutrophic lakes, *FEMS Microbiology Ecology*, **82** (3), 607–615.

Cyanobacterial Species Cited in the Book

Taxonomic orders are those of Komárek *et al.* (2014). Current names of the species and their most widely known synonym(s) are from Guiry and Guiry (2017) and Salmaso *et al.* (2017). The classification, especially at the order level, is in a dynamic state of change and should be considered only indicative.

Order/Species	Synonyms
Synechococcales	
Leptolyngbya boryana (Gomont) K. Anagnostidis and J. Komárek 1988	*Plectonema boryanum* Gomont 1899
Pseudanabaena redekei (Goor) B.A.Whitton 2011	*Oscillatoria redekei* Goor 1918
Pseudanabaena rutilus-viridis H.J.Kling, H.D.Laughinghouse and J.Komárek in Kling *et al.* 2012	
Snowella litoralis (Häyrén) Komárek and Hindák 1988	*Gomphosphaeria litoralis* Häyrén 1921
Synechococcus nidulans (Pringsheim) Komárek in Bourrelly 1970	*Lauterbornia nidulans* Pringsheim 1968
Synechocystis salina Wislouch 1924	
Woronichinia compacta (Lemmermann) Komárek and Hindák 1988	*Gomphosphaeria lacustris* var. *compacta* Lemmermann 1899; *Gomphosphaeria compacta* (Lemmermann) Ström 1923; *Gomphosphaeria lacustris* f. *compacta* (Lemmermann) Elenkin 1938
Chroococcales	
Microcystis aeruginosa (Kützing) Kützing 1846	
Oscillatoriales	
Arthrospira platensis Gomont 1892	*Spirulina platensis* (Gomont) Geitler 1925; *Oscillatoria platensis* (Gomont) Bourrelly 1970
Geitlerinema amphibium (C. Agardh ex Gomont) Anagnostidis 1989	*Oscillatoria amphibia* C. Agardh ex Gomont 1892
Lyngbya wollei (Farlow ex Gomont) B.J. Speziale and L.A. Dyck 1992	*Plectonema wollei* Farlow ex Gomont 1892
Phormidium uncinatum Gomont ex Gomont 1892	*Lyngbya uncinata* (Gomont ex Gomont) Compère 1980
Planktothrix agardhii (Gomont) K. Anagnostidis and J. Komárek 1988	*Oscillatoria agardhii* Gomont 1892
Planktothrix pseudagardhii Suda and Watanabe in Suda *et al.* 2002	

Order/Species	Synonyms
Planktothrix rubescens (de Candolle ex Gomont) K. Anagnostidis and J. Komárek 1988	*Oscillatoria rubescens* de Candolle ex Gomont 1892
Tychonema bourrellyi (J.W.G. Lund) K. Anagnostidis and J. Komárek 1988	*Oscillatoria bourrellyi* J.W.G. Lund 1955
Nostocales	
Anabaena lapponica Borge 1913	
Anabaena oscillarioides Bory ex Bornet and Flahault 1888	
Anabaenopsis elenkinii V.V. Miller 1923	
Aphanizomenon flos-aquae Ralfs ex Bornet and Flahault 1886	
Aphanizomenon gracile (Lemmermann) Lemmermann 1907	
Chrysosporum bergii (Ostenfeld) E. Zapomelová, O. Skácelová, P. Pumann, R. Kopp and E. Janecek 2012	*Aphanizomenon bergii* (Ostenfeld) Komárek 1983; *Anabaena bergii* Ostenfeld 1908
Chrysosporum ovalisporum (Forti) E. Zapomelová, O. Skácelová, P. Pumann, R. Kopp and E. Janecek 2012	*Aphanizomenon ovalisporum* Forti 1911
Cuspidothrix issatschenkoi (Usachev) P. Rajaniemi, Komárek, R. Willame, P. Hrouzek, K. Kastovská, L. Hoffmann and K. Sivonen 2005	*Aphanizomenon issatschenkoi* (Usačev) Proshkina-Lavrenko 1968
Cylindrospermopsis raciborskii (Woloszynska) Seenayya and Subba Raju 1972	*Anabaena raciborskii* Woloszynska 1912*
Cylindrospermum stagnale Bornet and Flahault 1888	
Dolichospermum circinale (Rabenhorst ex Bornet and Flahault) P. Wacklin, L. Hoffmann and J. Komárek 2009	*Anabaena circinalis* Rabenhorst ex Bornet and Flahault 1888
Dolichospermum crassum (Lemmermann) P. Wacklin, L. Hoffmann and J. Komárek 2009	*Anabaena crassa* (Lemmermann) J. Komárková-Legnerová and G. Cronberg 1992
Dolichospermum flos-aquae (Brébisson ex Bornet and Flahault) P. Wacklin, L. Hoffmann and J. Komárek 2009	*Anabaena flos-aquae* Brébisson ex Bornet and Flauhault 1888
Dolichospermum lemmermannii (P.G. Richter) P. Wacklin, L. Hoffmann and J. Komárek 2009	*Anabaena lemmermannii* P.G. Richter 1903; *Anabaena flos-aquae* f. *lemmermannii* (P.G. Richter) Canabaeus 1929; *Anabaena utermoehlii* Geitler 1925

Order/Species	Synonyms
Dolichospermum mendotae (W. Trelease) P. Wacklin, L. Hoffmann and J. Komárek 2009	*Anabaena mendotae* W. Trelease 1889
Dolichospermum planctonicum (Brunnthaler) P. Wacklin, L. Hoffmann and J. Komárek 2009	*Anabaena planctonica* Brunnthaler 1903; *Anabaena limnetica* G.M. Smith 1916
Nodularia spumigena Mertens ex Bornet and Flahault 1888	
Nostoc commune Vaucher ex Bornet and Flahault 1888	
Nostoc flagelliforme Harvey ex Molinari-Novoa, Calvo-Pérez and Guiry in Calvo-Pérez, Molinari-Novoa and Guiry 2016	*Nostoc commune* var. *flagelliforme* Bornet and Flahault 1886
Petalonema alatum (Borzì ex Bornet and Flahault) Correns 1889	*Scytonema alatum* Borzì ex Bornet and Flahault 1886
Raphidiopsis brookii P.J. Hill 1972	
Raphidiopsis curvata F.E. Fritsch and M.F. Rich 1929	*Raphidiopsis spiralis* F.E.Fritsch and M.F. Rich in Skuja 1949
Raphidiopsis mediterranea Skuja 1937	*Raphidiopsis subrecta* Frémy ex Skuja 1949
Sphaerospermopsis torques-reginae (Komárek) Werner, Laughinghouse IV, Fiore and Sant'Anna in Werner *et al.* 2012	*Anabaena torques-reginae* Komárek 1984; *Dolichospermum torques-reginae* (Komárek) Wacklin, Hoffmann and Komárek 2009
Umezakia natans M. Watanabe 1987	

Also cited as *Anabaenopsis raciborskii* (see Fig. 10 in Woloszynska, 1912, in Guiry and Guiry, 2017).

References

Guiry, M.D. and Guiry, G.M. (2017) *AlgaeBase: World-wide electronic publication*. National University of Ireland, Galway, http://www.algaebase.org.

Komárek, J., Kaštovský, J., Mareš, J., and Johansen, J.R. (2014) Taxonomic classification of cyanoprokaryotes (cyanobacterial genera) 2014, using a polyphasic approach, *Preslia*, **86**, 295–335.

Salmaso, N., Akçaalan, R., Bernard, C., *et al.* (2017) Appendix 1: Cyanobacterial species and recent synonyms, in J. Meriluoto, L. Spoof, and G.A. Codd (eds), *Handbook on Cyanobacterial Monitoring and Cyanotoxin Analysis*, John Wiley and Sons, Ltd, Chichester, UK, 491–502.

Glossary

List of common terms and abbreviations used in the book. The citation of specific products or manufacturers does not imply that they are recommended in preference to others.

16S rRNA The small subunit of prokaryotic ribosomes. The gene coding the small subunit is referred to as the 16S rRNA gene. Owing to its slow rate of evolution and absence of (or very limited) HGT, this gene is used in the study of phylogenies.

18S rRNA The small subunit of eukaryotic ribosomes. The coding gene (18S rRNA gene) is used in the study of eukaryotic phylogenies.

A Spectrophotometric absorbance.

AFLP Amplified fragment length polymorphism. Requires axenic strains. Rarely used in cyanobacterial studies.

Agar Jelly-like substance, derived from the polysaccharide agarose and smaller molecules of agaropectin. When mixed with culture media, it is used as a solid substrate for culturing microorganisms.

Agarose Polysaccharide polymer that contains repeated agarobiose units. In the preparation of agarose gel, the polymers form a network with pores of different sizes, which are dependent on the concentration of the gel (larger pores are obtained by decreasing the gel concentration, and *vice versa*). This characteristic is used in gel electrophoresis to separate DNA of various sizes.

Agarose LE Agarose low electroendosmosis. Agarose contains negatively charged groups that retard the movement of DNA (electroendosmosis). Agarose LE can minimize this process.

AlgaeBase Database of information on algae that includes terrestrial, marine, and freshwater organisms (http://www.algaebase.org/).

Allophycocyanin Blue phycobiliprotein. Absorption maximum, 650 nm.

Amplification efficiency (E) Expresses the quality of a qPCR process. The highest amplification efficiency is reached when the number of target molecules doubles with every qPCR cycle (i.e. with an efficiency of 2, or 100%). E (in %) can be calculated from the

Molecular Tools for the Detection and Quantification of Toxigenic Cyanobacteria, First Edition.
Edited by Rainer Kurmayer, Kaarina Sivonen, Annick Wilmotte and Nico Salmaso.
© 2017 John Wiley & Sons Ltd. Published 2017 by John Wiley & Sons Ltd.

slope of the standard curve relating C_T and the log (DNA concentration) with the formula: $E = (10^{-1/slope} - 1) \times 100$.

ana Anatoxin biosynthesis gene cluster.

ANA Anatoxin.

Annealing temperature (T_a) Temperature that allows annealing of the primers to the single-stranded DNA template. If the temperature is too high, hybridization is prevented. Conversely, if T_a is too low, primers can anneal to unwanted sequences or bind imperfectly. T_a values are primer specific.

APS Ammonium persulfate solution. Along with TEMED, APS begins the polymerization of the acrylamide gel (used in DGGE). When added, the gel solution should be poured into the vessel within a few minutes.

ASN-III Liquid medium commonly used for the culturing of marine cyanobacteria.

Axenic culture Culture with only a single strain present, without any other contaminating organisms (e.g. bacteria).

Background DNA Diverse DNA that can interfere with the measurement of the target genotype. In qPCR optimizations, its effects can be evaluated by adding environmental DNA free of the target genotype, or DNA of most closely related strains that do not have the target genotype.

Barcodes Sample-specific oligonucleotide sequence introduced into the end of the primers.

BG11 Liquid medium commonly used for the culturing of freshwater cyanobacteria.

Bidest Bidistilled water. Also abbreviated or DDW.

Bioconductor Bioconductor uses the R statistical programming language. It provides several tools for the analysis of high-throughput genomic data.

BLAST Basic Local Alignment Search Tool. Finds regions of similarity between biological sequences.

Bovine serum albumin (BSA) BSA increases PCR yields from low-purity templates that can contain PCR inhibitors (e.g. genomic DNA isolated from environmental samples).

bp Base pair; measurement unit of the physical length of double-stranded nucleic acid sequences.

BSCs Biological soil crusts.

cDNA Complementary DNA. Double-stranded DNA synthesized in a reaction catalyzed by the enzyme reverse transcriptase from a single-stranded RNA (e.g. mRNA).

Chimera False sequences. Artifacts originated from the combination of two phylogenetically distinct parent sequences.

Classification The arrangement of biological organisms in hierarchical categories (see AlgaeBase). Modern biological classifications are based on the evolutionary relationships between organisms.

Clonal strain Group of genetically identical organisms grown from a single cell through isolation and purification techniques.

Clone library A gene bank (database) containing clones (DNA sequences) and associated information. Allows identification of unknown DNA sequences. Collection of DNA sequences obtained from cloning of PCR products in *E. coli*.

***cpc*BA-IGS** Intergenic sequence (IGS) between the terminal end of the *cpc*B gene and the proximal end of the *cpc*A gene. *cpc* genes encode phycocyanin.

Cryopreservation Process where organisms are conserved by cooling to very low temperatures.

Cryovials Tubes used for the cryopreservation of organisms and able to withstand temperatures down to $-196°C$.

C_T/C_q Threshold cycle/quantification cycle in qPCR analysis. Also known as C_p (crossing point). Since the target sequence is doubled at each qPCR cycle, the start of amplification reflects the amount of target sequence present (i.e. the more target sequence, the earlier the amplification exceeds a set threshold). C_T/C_q expresses the number of cycles when the threshold is reached.

CTAB Cetyltrimethylammonium bromide (cetrimonium bromide); used to precipitate polysaccharides and purify DNA during DNA isolation.

Culture chamber/culture cabinet Chambers of different size for culturing of living organisms in controlled laboratory conditions (e.g. temperature, light, photoperiod).

Cyanobacteria Term introduced by R. Stanier in 1974.

Cyanophage Viruses infecting cyanobacteria.

Cyanophyceae Algal class name introduced by J. Sachs in 1874.

Cyanophyta Algal division name introduced by F. Steinecke in 1931.

CYL Cylindrospermopsin.

cyr Cylindrospermopsin biosynthesis gene cluster.

DDBJ DNA Data Bank of Japan.

Deep amplicon sequencing High-throughput sequencing of PCR amplified gene loci using (next-generation) parallel sequencing tools. Relevant in NGS and metabarcoding studies.

Demultiplexing In NGS analyses. Grouping reads to samples according to a barcode sequence.

DEPC Diethyl pyrocarbonate; DEPC can inactivate ribonucleases, providing water suitable for the preparation and analysis of RNA.

DGGE Denaturing gradient gel electrophoresis. Methods used to separate amplicons of different lengths obtained from DNA isolated from different strains, multispecies assemblages, and environmental samples. In cyanobacterial ecology and toxicology, PCR fragments (e.g. 16S rRNA) are typically separated in a polyacrylamide gel containing a gradient of denaturants. DGGE allows separation of single base pair differences.

Diacritical character Character capable of distinguishing.

DMSO (dimethyl sulfoxide) DMSO decreases the melting point of the primers. It improves strand separation in GC-rich regions. Betaine can be also used as an alternative for GC-rich regions.

DNA chip A solid surface with attached microscopic DNA "spots." DNA chips can be used to detect DNA or RNA. The functioning is based on the hybridization between two DNA strands.

DNA ladder Molecular-weight-sized markers used to identify the sizes of DNA molecules run on an agarose gel during electrophoresis. Lower molecular weight DNA (low bp) will have faster migration rates, and *vice versa*.

DNA microarray DNA chip.

DNA/DNA hybridizations Annealing of single-stranded DNA or RNA molecules to complementary DNA or RNA. This technique is still required by the ICNP. Relative binding ratios of at least 70% (or more), and a melting temperature increment of less than 5°C, have been set as the limits for strains related at the species level.

DNase Deoxyribonuclease. Nucleases that catalyze the degradation of DNA.

dNTPs Deoxynucleotide triphosphates.

DTT Dithiothreitol (Cleland's reagent). Used to stabilize enzymes and other proteins possessing free sulfhydryl groups.

eDNA Environmental DNA.

EDTA Ethylenediaminetetraacetic acid.

ELISA Enzyme-linked immunosorbent assay. With ELISA, a substance is identified by using antibodies and color change/intensity.

EMBL-EBI The European Bioinformatics Institute.

Epifluorescence microscopy The specimen is irradiated with a specific band of wavelengths (using selective excitation filters), and the weaker emitted fluorescence is separated from the excitation light and detected. In epifluorescence-equipped microscopes, only the emission light reaches the eye or detector.

Ethidium bromide (EtBr) Nucleic acid staining solution. Binds to the DNA, allowing its visualization under ultraviolet light (e.g. under a transilluminator). EtBr is toxic.

EtOH Ethanol; it is frequently used to precipitate DNA following enzymatic reactions.

EzTaxon Database with curated 16S rRNA sequences of bacteria and archaea.

FACS Fluorescence Activated Cell Sorting.

FACSAria™ (BD Biosciences) Flow cytometer for measuring and sorting fluorescently labelled cells.

FACSCalibur™ (BD Biosciences) Flow cytometer performing both cell analysis and cell sorting.

FASTA Standard format used to represent nucleotide and peptide sequences. Text-based format in which the nucleotides or amino acids are represented by letter codes. The name stands for FAST-All, indicating a format suitable for fast protein or nucleotide comparisons.

FASTQ Format used to store nucleotide sequences and the corresponding quality scores (Phred). Present standard for storing high-throughput data obtained in NGS analyses.

Flow cytometry (FCM) High-throughput technology used in cell counting and sorting. Cells are suspended in a stream of fluid, and counted by an electronic detection apparatus at a speed of thousands of particles every second.

Fluorochrome/Fluorophore Fluorescent chemical compound that emits light upon light excitation.

Formaldehyde Used to preserve phytoplankton samples. Toxic.

GC-content Indicates the number/percentage of either guanine or cytosine in a sequence of DNA, RNA, or in the whole genome. While the GC pair is linked by three hydrogen bonds, the AT pair is linked by two hydrogen bonds. As a consequence, DNA with higher GC-content is more stable than DNA with lower GC-content. This difference has important effects on the thermodynamic properties of DNA (e.g. T_m, primer design, annealing). GC-content is also used in the classification of bacteria.

gDNA Genomic DNA.

GelRed™ and GelGreen™ (Biotium, Canada) Fluorescent DNA stains used to replace ethidium bromide.

GF/C Whatman® glass microfiber filters, borosilicate glass, grade GF/C, approx. pore size 1.2 μm

GF/F Whatman® glass microfiber filters, borosilicate glass, grade GF/F, approx. pore size 0.7 μm

GOLD Genomes OnLine Database.

Greengenes Database with curated 16S rRNA sequences of bacteria, archaea, and fungi.

Hairpin loop Secondary structure in a single-stranded nucleic acid sequence due to the presence of complementary sequences within its length. The presence of hairpin loops can reduce the efficiency of the reaction by limiting the ability of a primer to bind to a specific target site.

***het*R** Gene involved in the control of heterocyte differentiation. *het*R sequences are used for identification and classification purposes.

HGT Horizontal (or lateral) gene transfer. Movement/transmission of DNA between different genomes. It is distinguished from the vertical gene transfer that occurs during cell division. HGT has been widely documented in a majority of cyanobacterial proteins, causing potential problems in the reconstruction of major evolutionary transitions (using e.g. housekeeping genes).

High-throughput methods   Techniques that allow rapid parallel analyses, increasing the acquisition of data by orders of magnitude compared to conventional approaches (e.g. NGS techniques).

Housekeeping genes   Genes required for the vital functioning of organisms.

I_A Index of association. Assesses linkage disequilibrium between loci. The standardized version (r_D) corrects for the number of observed loci. Implemented in the poppr R package.

ICBN   International Code of Botanical Nomenclature (Botanical Code). See ICN.

ICN   International Code of Nomenclature for algae, fungi, and plants. Formerly called the International Code of Botanical Nomenclature (ICBN).

ICNB   International Code of Nomenclature of Bacteria (Bacteriological Code). See ICNP.

ICNP   International Code of Nomenclature of Prokaryotes. Formerly called the International Code of Nomenclature of Bacteria (ICNB).

Identification   Assigning taxon names to individual biological organisms. It can be based on phenotypic traits or molecular markers. In the polyphasic approach, the identification process can also include ecological features.

Illumina MiSeq   NGS platform used for amplicon sequencing. Data output, 0.3–15 Gb.

Indel   Insertion or deletion of bases in a DNA molecule.

INSDC   International Nucleotide Sequence Database Collaboration between DDBJ, EMBL-EBI, and NCBI.

ITS   Internal transcribed spacer. Spacer DNA that separates the small and the large subunits in rRNA genes.

LC-MS (HPLC-MS)   Liquid chromatography–mass spectrometry. Analytical chemistry technique that combines liquid chromatography (HPLC) and mass spectrometry (MS).

LDR   Ligation detection reaction.

leBIBI^{QBPP}   Ribosomal rRNA databases and other selected target genes (e.g. *rpo*B, *gyr*B).

LGT   Lateral (or horizontal) gene transfer. See HGT.

Linkage disequilibrium   Non-random association of alleles at different loci. It happens when the frequency of association of the different alleles in specific loci is different from the frequency expected if the loci were independent and associated randomly.

Logarithmic growth phase   Under favorable conditions, bacterial populations grow exponentially, following a geometric progression.

LSU   Large subunit of the ribosomal RNA.

Lugol's solution Solution of iodine and potassium iodide in water used to preserve (1:100 v/v) phytoplankton samples. It is named after the French physician J.G.A. Lugol.

Lysozymes Glycoside hydrolases able to damage bacterial cell walls.

Macronutrients Nutrients that are needed in large amounts for survival and growth (e.g. N, P and (mostly diatoms) Si).

MC Microcystin.

mcy Microcystin biosynthesis gene cluster.

Melbourne Code International Code of Nomenclature for algae, fungi, and plants (ICN).

Melting curve analysis In qPCR, evaluates the dissociation-characteristics of double-stranded DNA during heating. The relation between fluorescence and temperature provides indications about, for example, the presence of SNPs or unwanted by-products (primer dimers) in qPCR.

Melting temperature (T_m) The temperature at which one-half of a double-stranded DNA dissociates into single strands.

MeOH Methanol; it can be used as a cryopreservation agent.

Metagenomics Culture-independent genetic analysis of the genomes contained in an environmental sample using NGS techniques. "Full shotgun metagenomics" refers to the sequence-based analysis of the microbial genomes present in an environmental sample. "Marker gene amplification metagenomics" refers to NGS based on PCR amplification of specific genes (e.g. 16S rRNA).

MICCA Microbial Community Analysis. Bioinformatic pipeline used to process high-throughput sequencing data.

Midori Green (Nippon Genetics Co. Ltd.) Nucleic acid staining solution. Binds to the DNA, allowing its visualization under UV light or blue LED gel illuminators. Midori Green can replace EtBr. Midori in Japanese means "green."

Milli-Q® (Merck Millipore) Either ultrapure water of "Type 1" (e.g. ISO 3696), or the apparatus produced by Merck Millipore used to produce it; conductivity of Milli-Q® water is close to that of pure water (i.e. 0.055 µS cm^{-1} (18.2 MΩ·cm) at 25°C).

MLSA Multilocus sequence analysis. A range of housekeeping gene sequences determined for each strain concatenated and used in phylogenetic analyses.

MLST Multilocus sequence typing. Procedure for characterizing isolates of bacterial species based on a range of housekeeping gene sequences ("alleles") that differ from each other.

Morphotype Taxon defined by morphological features.

MOTHUR Bioinformatic pipeline used to process high-throughput sequencing data.

Motuporin Nodularin structural analogue.

mRNA Messenger RNA. Transfer the genetic information from DNA to the ribosome, specifying the amino acid sequence.

multiplex PCR PCR is used to amplify simultaneously different DNA sequences (amplicons) in one reaction using different sets of primers.

NanoDrop™ (Thermo Fisher Scientific) Measure nucleic acid concentrations using tiny amount (e.g. 1 μL) of sample. Among others, typical uses include estimation of nucleic acid purity by examination of 260/230 and 260/280 absorbance ratios; enhanced quantification of double-stranded DNA using dyes (PicoGreen®, Thermo Fisher Scientific Inc.) for downstream qPCR analyses.

NCBI National Center for Biotechnology Information; GenBank.

nda Nodularin biosynthesis gene cluster.

NGS Next-generation sequencing, also known as second-generation sequencing (SGS) or massively parallel sequencing (MPS). High-throughput sequencing, performed by a number of different sequencing technologies, which include Illumina (Solexa) sequencing, Roche 454 pyrosequencing, Ion torrent: Proton/PGM sequencing, and SOLiD sequencing. DNA and RNA are sequenced much more quickly and cheaply than the previous technologies (Sanger sequencing), but with shorter length (a few hundred base pairs long). Third-generation DNA sequencing technologies can produce longer reads, between 5000 bp and > 15000 bp.

Nicotinic acetylcholine receptor Neuron receptors that respond to acetylcholine, a neurotransmitter. Anatoxin-a is an agonist of acetylcholine, with the difference that the affinity for the receptors is around 20 times higher, and that the anatoxin-a binding is irreversible, because it cannot be interrupted by acetylcholinesterase.

*nif*H Gene encoding the dinitrogenase reductase (component of the nitrogenase enzyme complex). *nif*H sequences are used for identification and classification purposes.

Niskin sampler Plastic tube equipped at both ends with opening caps controlled by an elastic rope. The bottle can be lowered vertically and closed by a "messenger" at specific desired depths along the water column, allowing for the collection of deep-water samples.

NOD Nodularin.

Nomenclature Rules governing the naming of organisms and the use of names.

NRPS Nonribosomal peptide synthesis, means peptide synthesis by thio-template mechanism through modular multifunctional enzymes consisting minimally of adenylation, thiolation and condensation domains.

nt Units of nucleotides in single-stranded DNA/RNA. See also bp.

NTC Non-template control; gDNA without the target gene.

Nucleases Enzymes that hydrolyze phosphodiester bonds linking nucleotides.

OD Optical density (absorbance). At 750 nm (OD_{750}) this measure is roughly related to the density (cell numbers) of an algal culture.

OTU Operational taxonomic units.

PAR Photosynthetically active radiation. Radiation between 400 and 700 nm that can be used by photosynthetic organisms in the process of photosynthesis.

PAR quantum meter Sensors that measure PAR (photosynthetically active radiation) (μmol photons m^{-2} s^{-1}) both outside and inside the water.

PC Phycocyanin. Blue accessory pigment present in cyanobacteria. The intergenic sequence (IGS) between the terminal end of the *cpc*B gene and the proximal end of the *cpc*A gene (cpcBA-IGS or PC-IGS) has been used as a reference marker to test the presence of cyanobacteria.

PC-IGS See *cpc*BA-IGS.

PCR Polymerase chain reaction.

PCR master mix Mixed solution containing sterile Milli-Q$^{\circledR}$ water, PCR buffer, MgCl$_2$, dNTPs, primers, and DNA polymerase at concentrations optimized for amplification of specific DNA templates.

PCR products cleanup/DNA purification Purification of the DNA obtained during a PCR reaction by removal of unincorporated dNTPs, short primers, and other unwanted products/salts. If not removed, these components can interfere with DNA sequencing and other downstream analyses.

PEG Polyethylene glycol. Used to precipitate DNA during isolation.

Pfu polymerase High-fidelity DNA polymerase isolated and adapted from *Pyrococcus furiosus*, a thermophilic archaebacterium able to withstand temperatures greater than 100°C. *P. furiosus* was isolated for the first time on the island of Vulcano (Sicily) in 1986. Besides 5′ to 3′ DNA polymerase activity, Pfu has 5′ to 3′ exonuclease (proof-reading) activity.

Phototaxis Movement of organisms toward or away from light sources.

Phred quality score Quality scores of nucleotide sequences. This information is part of the FASTQ format.

Phycobiliproteins Pigmented water-soluble proteins assembled into large supramolecular complexes called "phycobilisomes." Phycobiliproteins capture light energy that is transferred to chlorophylls during the photosynthetic process.

Phycobilisomes Large supramolecular complexes of pigmented water-soluble proteins fixed to the thylakoid membranes.

Phycocyanin Blue phycobiliprotein. Absorption maximum 620 nm. The chromophore of phycocyanin is the blue phycocyanobilin.

Phycoerythrin Red phycobiliprotein. R-phycoerythrin has absorbance maxima at 565 nm, and secondary peaks at 495 nm and 545 nm; B-phycoerythrin has absorbance maxima at 545 nm, with a secondary peak at 563 nm. The red chromophores of phycoerythrin are phycourobilin and phycoerythrobilin.

Phylotypes OTUs sharing more than a determined (and arbitrary) level of similarity for a particular gene.

PicoGreen Fluorescent nucleic acid stain used for the quantification of double-stranded DNA.

Picoplankton The smallest (0.2–2 µm) fraction of phytoplankton composed of both eukaryotic and cyanobacterial organisms.

PKS Polyketide synthase. PKS are multifunctional enzymes synthesizing polyketides from acetate units by thio-template mechanism which are part both of primary and secondary metabolism.

Planktic Referring to organisms living in pelagic environments, with no or limited capability of movement.

Plasmid In bacteria, small, circular, double-stranded DNA molecules separated from the chromosomal DNA. Plasmids are able to replicate independently.

Polyacrylamide gel Polyacrylamide is a water-soluble polymer formed from acrylamide subunits. Polyacrylamide gel electrophoresis is used to separate proteins or nucleic acids, as in DGGE.

PPIA Protein phosphatase inhibition assay.

Primer Short sequence (min. 20 bp) used as a starting point for DNA synthesis in PCR reactions.

Prochlorophyta Group of cyanobacteria that contain both chlorophyll-a and -b, without phycobilin pigments. Genera: *Prochlorococcus* (0.6 µm responsible for a large fraction of the global primary production) and *Prochloron* (commonly found as an extracellular symbiont) are marine genera. *Prochlorothrix* is a filamentous freshwater strain.

Proteinase K Proteinase K derives its name from its ability to digest keratin (e.g. hairs).

PSP Paralytic shellfish poisoning. Illness caused by consumption of shellfish which have accumulated PST produced by toxic microalgae.

PST Paralytic shellfish toxins. Saxitoxins and analogues (neurotoxic alkaloids).

Python High-level interpreted programming language.

QIIME Quantitative Insights Into Microbial Ecology. Bioinformatic pipeline used to process high-throughput sequencing data.

qPCR Quantitative real-time polymerase chain reaction. The amplification of a targeted DNA molecule is monitored in real time (continuously) during the PCR. qPCR allows determination of the number of target genes (copy number) in the template.

R Software for statistical computing and graphics. It contains many packages explicitly created to perform a wide range of bioinformatic analyses, including the analysis of single DNA sequences (e.g. ape) and high-throughput data (e.g. phyloseq).

RAPD Random amplified polymorphic DNA. Genomic DNA is amplified by PCR with random short primers that have a high probability to anneal frequently in the entire genome. Banding patterns are not easily reproducible. Method no longer recommended. Requires axenic strains.

*rbc***LX** Part of the operon encoding the small (*rbc*S) and large subunits (*rbc*L) of the D-ribulose 1,5-bisphosphate carboxylase-oxygenase. *rbc*LX sequences are used for identification and classification purposes.

r_D Standardized version of I_A. The r_D values provide an estimate of linkage disequilibrium, and range from 0 (panmixia) to 1 (clonality). Implemented in the poppr R package.

RDP Ribosomal Database Project. Database with curated 16S rRNA sequences of bacteria, archaea, and fungi.

Restriction enzymes Cut the DNA at or near short palindromic sequences resulting in nucleotide-specific fragmentation which can be visualized using gel electrophoresis.

Reverse transcriptase Enzyme involved in the process of reverse transcription to generate complementary DNA (cDNA) from an RNA template.

RFLP Restriction fragment length polymorphism. PCR products of specific genes/loci are digested by restriction enzyme(s) prior to electrophoresis.

RNAlater® Aqueous tissue storage reagent that rapidly permeates tissues to stabilize and protect cellular RNA. Minimize the need to analyze immediately or to freeze samples in liquid nitrogen.

RNases Ribonucleases. Nucleases that catalyze the degradation of RNA.

*rpo***B** Gene encoding the beta subunit of the RNA polymerase. *rpo*B sequences are used for identification and classification purposes.

*rpo***C1** Gene encoding the gamma subunit of the cyanobacterial (or chloroplast) RNA polymerase. *rpo*C1 sequences are used for identification and classification purposes.

RT-qPCR Reverse transcription-qPCR.

Ruttner sampler Similar to the Niskin sampler, but with a different closing mechanism. Lowered vertically.

SDS Sodium dodecyl sulfate.

Secchi disk White or black and white disk of 30 cm diameter used to estimate water transparency by determining the depth at which the disk is no longer visible when lowered with a rope into water. Devised by the priest Pietro Angelo Secchi in 1865.

Sedgewick Rafter chamber Counting chambers made of a rectangular frame mounted on a thick clear base. After filling, a cover slip is placed onto the top.

SEM Scanning electron microscope.

Sequencing platform Platform able to read the nucleotides present in DNA or RNA sequences (RNA-seq library preparation usually includes reverse transcription). The two types of sequencing technologies are Sanger sequencing and next-generation sequencing.

SFF format In 454 pyrosequencing platform, raw data can be formatted in the SFF format, which describes the nucleotides and the associated light intensity peaks of the sequencing chromatogram.

Sheath fluid Solution that runs in a flow cytometer.

SILVA Database with curated 16S rRNA sequences of bacteria, archaea, and eukaryotes. From Latin *silva*, meaning forest.

SNP Single nucleotide polymorphism. Variation of a single nucleotide in a specific position in the genome of an organism.

Solid medium Liquid medium made solid after agar addition.

SSC Saline-sodium citrate buffer. Used as a hybridization buffer.

SSU Small subunit of the ribosomal RNA.

Stationary growth phase In a closed system, the growth is limited by consumption of nutrients or other limiting/stressing factors. Growth rates and death rates are counterbalanced.

sxt Saxitoxin biosynthesis gene cluster.

SXT Saxitoxin.

SYBR Green (SG) Nucleic acid stain. The DNA-SG-complex absorbs blue light (497 nm) and emits green light (520 nm). In molecular biology, SG is used in the quantification of double-stranded DNA in some qPCR methods.

Systematics Comparative study of organisms and their relationships through time. Relationships may be visualized as phylogenetic trees (cladograms). Diacritical characters of organisms are used in taxonomy.

TAE buffer Mixture of Tris base, acetic acid, and EDTA. The TAE buffer is used in gel electrophoresis. TBE can be used instead of TAE. There is little difference between the two.

Taq/Taq polymerase DNA polymerase isolated from *Thermus aquaticus*, a thermophilic bacterium living in hot springs and hydrothermal vents. Taq is frequently used in PCR, owing to its property to resist high temperatures and protein-denaturing conditions. In contrast to Pfu polymerase, the Taq polymerase has no proofreading ability.

Taxon Plural: taxa. A taxonomic group at any rank.

Taxonomy Definition and naming of groups of biological organisms based on the identification of common characteristics. Includes identification, classification, and nomenclature. Taxonomy is part of systematics.

TBE buffer Mixture of Tris base, boric acid, and EDTA. The TBE buffer is used in gel electrophoresis. TAE can be used instead of TBE. There is little difference between the two.

TE buffer Tris-EDTA buffer solution (i.e. Tris(hydroxymethyl)aminomethane and Ethylenediaminetetraacetic acid).

TEMED N,N,N′,N′-Tetramethylethylenediamine. Along with APS, catalyzes the polymerization of the acrylamide gel. See APS.

TES buffer Tris-HCl, EDTA, Sucrose buffer (see SOP 5.1).

TGGE Technique similar to DGGE; a temperature gradient rather than a chemical gradient is used to denature the DNA.

Trace element Micronutrient. Element required by organisms in very low quantities (e.g. Cu, Zn, Mn).

Transcriptome Set of all gene transcripts of one organism.

Transilluminator UV-transilluminators are used to visualize (and record) DNA (or RNA) separated by agarose gel electrophoresis. Usually, they are equipped with digital cameras that enable documentation of gels.

Tris Tris(hydroxymethyl)aminomethane, component of buffer solutions for nucleic acids.

TRIzol® Reagent Isolates total RNA from tissues or cells.

Ultraplankton Term indicating the smaller fraction of phytoplankton. Differently defined (size < 3–5 μm). Disused term.

UNG Uracil N-glycosylase. Enzyme utilized to eliminate carryover PCR products in qPCR.

Universal PCR primer Universal primers are designed from a highly conserved motif in determined gene domains. They are used to amplify sequences of entire groups of species (e.g. prokaryotes, or cyanobacteria).

UPARSE OTU clustering. Bioinformatic technique for generating clusters (OTUs) from next-generation sequencing reads of marker genes (e.g. 16S rRNA). The method is implemented in the USEARCH sequence analysis tool.

USEARCH Sequence analysis software providing search and clustering algorithms suitable for the analysis of high-throughput data.

Van Dorn sampler Similar to a Niskin sampler, but mounted horizontally, allowing sampling water near the bottom of lakes and streams.

WDCM World Data Center for Microorganisms.

WFCC World Federation for Culture Collections.

WGS Whole Genome Sequences.

Z8 Liquid medium commonly used for the culturing of freshwater cyanobacteria.

Index